Periodic Table of the Elements with the Gmelin System Numbers

1	2	3	4	5	6	7	8	9	10	11	12	13	14	15	16	17	18
1 H 2																1 H 2	2 He 1
3 Li 20	4 Be 26											5 B 13	6 C 14	7 N 4	8 O 3	9 F 5	10 Ne 1
11 Na 21	12 Mg 27											13 Al 35	14 Si 15	15 P 16	16 S 9	17 Cl 6	18 Ar 1
19 * K 22	20 Ca 28	21 Sc 39	22 Ti 41	23 V 48	24 Cr 52	25 Mn 56	26 Fe 59	27 Co 58	28 Ni 57	29 Cu 60	30 Zn 32	31 Ga 36	32 Ge 45	33 As 17	34 Se 10	35 Br 7	36 Kr 1
37 Rb 24	38 Sr 29	39 Y 39	40 Zr 42	41 Nb 49	42 Mo 53	43 Tc 69	44 Ru 63	45 Rh 64	46 Pd 65	47 Ag 61	48 Cd 33	49 In 37	50 Sn 46	51 Sb 18	52 Te 11	53 I 8	54 Xe 1
55 Cs 25	56 Ba 30	57** La 39	72 Hf 43	73 Ta 50	74 W 54	75 Re 70	76 Os 66	77 Ir 67	78 Pt 68	79 Au 62	80 Hg 34	81 Tl 38	82 Pb 47	83 Bi 19	84 Po 12	85 At 8a	86 Rn 1
87 Fr 25a	88 Ra 31	89*** Ac 40	104 71	105 71													

NH_4 23 *

Lanthanides 39

58 Ce 39	59 Pr	60 Nd	61 Pm	62 Sm	63 Eu	64 Gd	65 Tb	66 Dy	67 Ho	68 Er	69 Tm	70 Yb	71 Lu

***Actinides**

90 Th 44	91 Pa 51	92 U 55	93 Np 71	94 Pu 71	95 Am 71	96 Cm 71	97 Bk 71	98 Cf 71	99 Es 71	100 Fm 71	101 Md 71	102 No 71	103 Lr 71

A Key to the Gmelin System is given on the Inside Back Cover

Gmelin Handbook of Inorganic and Organometallic Chemistry

8th Edition

Gmelin Handbook of Inorganic and Organometallic Chemistry

8th Edition

Gmelin Handbuch der Anorganischen Chemie

Achte, völlig neu bearbeitete Auflage

PREPARED
AND ISSUED BY

Gmelin-Institut für Anorganische Chemie
der Max-Planck-Gesellschaft
zur Förderung der Wissenschaften

Director: Ekkehard Fluck

FOUNDED BY

Leopold Gmelin

8TH EDITION

8th Edition begun under the auspices of the
Deutsche Chemische Gesellschaft by R. J. Meyer

CONTINUED BY

E. H. E. Pietsch and A. Kotowski, and by
Margot Becke-Goehring

Springer-Verlag Berlin Heidelberg GmbH 1993

Volumes published on "Tungsten" (Syst. No. 54)

Gmelin Handbook of Inorganic and Organometallic Chemistry

8th Edition

W

Tungsten

Supplement Volume A 5 b

Metal, Chemical Reactions with Nonmetals
Nitrogen to Arsenic

With 129 illustrations

AUTHORS

Hermann Jehn
Forschungsinstitut für Edelmetalle und Metallchemie,
Schwäbisch Gmünd (Chapter 1)
Gudrun Bär, Erich Best, Ernst Koch

EDITORS

Jörn von Jouanne, Elisabeth Koch, Ernst Koch

CHIEF EDITORS

Jörn von Jouanne, Ernst Koch

System Number 54

Springer-Verlag Berlin Heidelberg GmbH 1993

LITERATURE CLOSING DATE: MID 1991
IN SOME CASES MORE RECENT DATA HAVE BEEN CONSIDERED

Library of Congress Catalog Card Number: Agr 25-1383

ISBN 978-3-662-08686-5   ISBN 978-3-662-08684-1 (eBook)
DOI 10.1007/978-3-662-08684-1

© by Springer-Verlag Berlin Heidelberg 1993
Originally published by Springer-Verlag, Berlin in 1993.
Softcover reprint of the hardcover 8th edition 1993

Preface

The present volume describes the chemical behavior of elemental tungsten toward the non-metallic elements nitrogen to arsenic (i.e., N, F, Cl, Br, I, S, Se, Te, Po, B, C, Si, P, and As).

The description of the chemical behavior starts with information on the phase diagrams which allow an overview of the existing stable compounds. The major part of the information in this volume is about the kinds of products, the experimental conditions, as well as the kinetics and thermodynamics of their formation. Short descriptions of bulk diffusion in binary systems complement the kinetic data on the mostly diffusion-controlled reactions. Pure surface phenomena on tungsten are not considered.

A large chapter is devoted to the comprehensively studied chemical processes and transport processes in tungsten-halogen lamps. Product formation in the tungsten-carbon and tungsten-silicon systems is also of special interest in view of possible applications.

The present volume concludes the series of volumes devoted to the chemical behavior of elemental tungsten. Interactions and reactions with noble gases, hydrogen, and oxygen are covered in "Wolfram" Erg.-Bd. B 1. Interactions and reactions with metallic elements are described in "Tungsten" Suppl. Vol. A 6 a (elements antimony to barium) and in "Tungsten" Suppl. Vol. A 6 b (elements zinc to actinides).

Frankfurt am Main
October 1993

Ernst Koch
Jörn von Jouanne

Table of Contents

XIV

1 Reactions with Nitrogen

1.1 General

Reactions of W with N_2 gas at any temperature generally involve chemisorption and physisorption processes. N atoms and N_2 molecules are sorbed in different energetic states, depending also on the W surface structure (film, poly- or single crystal). At higher temperatures N is dissolved in W to a rather small extent. No nitrides are formed in the direct reaction of N_2 with W metal, at lower temperatures due to kinetic limitations and at higher temperatures due to the extremely high dissociation pressures of the nitrides, which prevent their formation even at the highest pressures experimentally attainable. Nitrides are only formed with NH_3 or by complex chemical reactions.

The present review does not treat surface reactions and properties of adsorbate layers, rather is limited to a description of solution and diffusion/permeation processes of N in solid W (for liquid W no information is available). Additional information is given on the degassing kinetics, ion implantation of N, and the deposition of N-containing W or W nitride films by chemical or physical vapor deposition techniques.

General References:

Gmelin Handbuch "Wolfram" 1933, pp. 153/4.

Jehn, H.; in: Fromm, E.; Gebhard, E.; Gase und Kohlenstoff in Metallen (Gases and Carbon in Metals), Springer, Berlin 1976, pp. 552/63.

Fromm, E.; Jehn, H.; Nitrogen Solubility at High Temperatures [in German], Z. Metallkd. **62** [1977] 378/81.

Jehn, H.; Ettmayer, P.; High-Pressure Studies on the Solubility of Nitrogen in Tungsten [in German], Monatsh. Chem. **111** [1980] 1437/40.

Rieck, G. D.; Tungsten and Its Compounds, Pergamon, Oxford 1967, 154 pp.

Fromm, E.; Jehn, H.; Equilibrium and Degassing Kinetics in Molybdenum-Nitrogen, Tungsten-Nitrogen and Rhenium-Nitrogen Systems, High Temp.-High Pressures **3** [1971] 553/64.

Toth, L. E.; Transition Metal Carbides and Nitrides; in: Margrave, J. L.; Refractory Materials, Vol. 7, Academic, New York 1971, pp. 85/101, 97/8.

Kieffer, R.; Benesovsky, F.; Hartstoffe, Springer, Wien 1963, pp. 337/40.

Schwarzkopf, P.; Kieffer, R.; Refractory Hard Metals, New York 1953, pp. 252/3.

Yih, S. W. H.; Wang, C. T.; Tungsten, Sources, Metallurgy, Properties, and Applications, Plenum Press, New York – London 1979, pp. 309/10.

1.2 Phase Diagram

The solubility of N in solid W is rather low, even at high temperatures and high N_2 pressures. The solidus and solvus lines are not reached by simple W–N_2 reactions. The nitrides W_2N (fcc) and δ-WN (hcp) are only formed with NH_3 or in a plasma jet (W_2N). Other nitrides (W_3N_4, W_2N_3, WN_2, etc.) reported in the literature cannot be regarded as stable compounds (except perhaps WN_2). A detailed description of the W nitrides can be found in [9]. Quantitative data on the solubility and thermodynamic data are given below. A worked-out W–N phase diagram cannot be given because the necessary information is not available.

Solubility of N in W

N is interstitially dissolved in the bcc W lattice. The solution of N_2 according to the equation $1/2\,N_2 \rightarrow N$ (in W) is an endothermal reaction with the equilibrium condition $c(N) = \sqrt{p(N_2)} \cdot k \cdot \exp(-\Delta H_s^\circ/RT)$, where $c(N) = $ N concentration, $p(N_2) = N_2$ pressure, $k = $ preexponential factor, $\Delta H_s^\circ = $ enthalpy of solution, $R = $ gas constant, $T = $ absolute temperature. Table 1 summarizes the quantitative results by various authors [1 to 5]. The equation above reflects the validity of Sieverts's law $c(N) = \alpha\sqrt{p(N_2)}$ which was experimentally verified for N_2 pressures between 0.05 and 300 bar (5×10^3 to 3×10^7 Pa) [5].

Table 1

Equilibrium Data of the Solubility of N in W.

k in ppm·Torr$^{-1/2}$	k in at%·Pa$^{-1/2}$	ΔH_s° in kcal/mol	t in °C; pressure in Pa	experimental procedure	Ref.
7×10^2	7.97×10^{-2}	46.7	2200 to 3000; 2×10^3 to 5×10^4	degassing, manometry	[1, 2]
1.5×10^2	1.71×10^{-2}	37.35	1200 to 2400; 1.01×10^5	degassing, volumetry	[3]
14.3	1.63×10^{-3}	17.6	1200 to 2000; 133 to 3300	degassing, mass spectroscopy	[4]
3.7×10^2	4.15×10^{-2}	43.5	1200 to 1800; 1×10^6 to 3×10^7	heat extraction analysis	[5]

The techniques employed for the direct experimental determination of the N solubility are either at low pressures (atmospheric and reduced pressure) or high pressures. In the first case, W wire samples are heated in an N_2 atmosphere until the equilibrium is reached; subsequently, the N_2 evolved during a degassing heat treatment is determined by volumetric [3] or manometric [1, 2] and mass spectrometric methods [4]. The high-pressure experiments are performed with thin-walled W tubes (made from foils) which afterwards are subjected to a heat extraction analysis [5]. In **Fig. 1** the solubility of N in solid W extrapolated to atmospheric pressure (1.01×10^5 Pa = 760 Torr) is plotted vs. the reciprocal temperature [9]. The results obtained at lower temperatures have a much larger error limit. N concentrations of <1 ppm (<0.001 at%) are very difficult to determine. Hence, the relatively high values of [4, 6] probably are in error. This is suggested by the high-pressure results when extrapolated to comparable temperatures [5]. The values of [6] were calculated from the diffusion coefficient D and the permeation constant P according to the relation $S = P/D$ where $S = \alpha = c(N)/\sqrt{p(N_2)} = $ solubility constant.

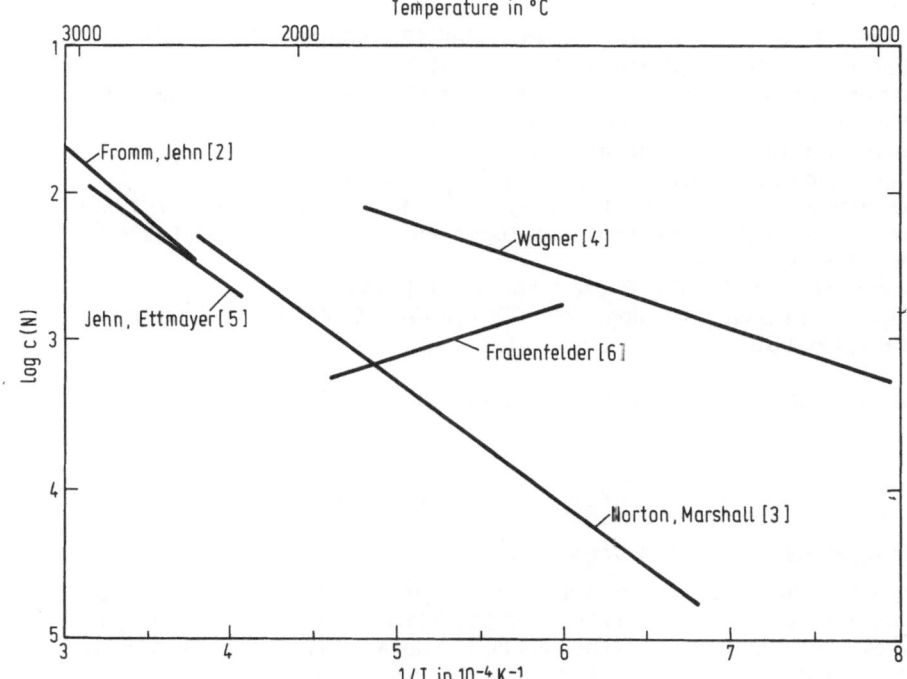

Fig. 1.   Solubility of N in W at $p(N_2) = 10^5$ Pa (~1 atm);
c(N) = N content in at% [9].

Solubility values derived from diffusion measurements on ion-implanted diffusion couples deviated from those obtained by conventional methods. Between 600 and 800°C c(N) increased from $(1.0 \pm 0.2) \times 10^{-2}$ to $(4.0 \pm 0.4) \times 10^{-2}$ at%. Causes for the anomalously high values are discussed in the paper [11].

Nitrides

As mentioned above, W nitrides are inaccessible by direct reaction of W metal with an N_2 atmosphere even at pressures of up to 300 bar (3×10^7 Pa) [5]. However, if an N_2 plasma jet (formed by passing N_2 gas through an electric arc discharge between a Cu nozzle anode and a W cathode) is used, a W_2N layer is obtained. The plasma jet presumably consists mainly of atomic N; the plasma temperature is ~9000 K [7, 8]; for more details see [8]. For the plasma-nitriding of a W layer on a silicon film with NH_3 or N_2, see [10]. Thin W nitride films can also be prepared by physical and chemical vapor deposition; see Chapter 1.6.

The nitride formation by the reaction of W metal with NH_3 or by other reactions not involving elemental nitrogen is not within the scope of the present chapter. Likewise the properties of the W nitrides shall not be treated here.

References:

[1] Fromm, E.; Jehn, H. (J. Less-Common Met. **17** [1969] 124/6).
[2] Fromm, E.; Jehn, H. (Z. Metallkd. **62** [1971] 378/81).
[3] Norton, F. S.; Marshall, A. L. (Trans. Am. Inst. Min. Metall. Pet. Eng. **156** [1944] 351/71).
[4] Wagner, R. L. (Metall. Trans. **1** [1970] 3365/70).
[5] Jehn, H.; Ettmayer, P. (Monatsh. Chem. **111** [1980] 1437/40).
[6] Frauenfelder, R. (J. Chem. Phys. **48** [1968] 3966/71).
[7] Matsumoto, O.; Shirato, Y.; Hayakawa, Y. (J. Electrochem. Soc. Jpn. **37** [1969] 151/64).
[8] Matsumoto, O. (Denki Kagaku Oyobi Kogyo Butsuri Kagaku **35** [1967] 488/92 from C. A. **68** [1968] No. 56039).
[9] Wriedt, H. A. (Bull. Alloy Phase Diagrams **10** [1989] 358/64).
[10] Suzuki, H.; Fujitsu, Ltd. (Jpn. Kokai Tokkyo Koho 2177427 [1988/90] from C. A. **114** [1991] No. 93368).

[11] Keinonen, J.; Räisänen, J.; Anttila, A. (Appl. Phys. A **35** [1984] 227/32).

1.3 Nitrogen Absorption and Degassing

Because of the very low N solubility even at high temperatures and pressures, no direct measurement of the N absorption kinetics is possible. Information can only be gained from a detailed evaluation of degassing experiments. The degassing of N-containing W in general is controlled by diffusion of N atoms dissolved in the bulk towards the W metal surface. Only in the initial stage is the transition of the N atoms from the dissolved into the adsorbed state rate-controlling. The absorption of N in W, in turn, is initially determined by the transition of N atoms from the adsorbed into the dissolved state, i.e., the penetration of the metal surface. Later on and especially in thicker samples, the diffusion of N from the near-surface region into the interior of the sample is rate-determining. The initial absorption rate, v(abs), was calculated to be $v(abs) = \sqrt{p(N_2)} \cdot 2.64 \times 10^4 \cdot exp(-68700/RT)$ with v(abs) in $\mu g \cdot cm^{-2} \cdot min^{-1}$, $p(N_2)$ in Torr, R = gas constant = 1.987 $cal \cdot mol^{-1} \cdot K^{-1}$, T = Kelvin temperature >1673 K. After the equilibrium N_2 (gas) $\rightleftharpoons$ 2N (solid solution) has been established in the near-surface region, the absorption rate follows the well-known diffusion laws [1].

A detailed evaluation of the degassing isotherms (**Fig. 2**) showed, as mentioned above, that the degassing of N-containing W proceeds via the reaction steps (a) diffusion in the bulk towards the surface and (b) transition from the dissolved into the adsorbed state at the metal surface with subsequent desorption as N_2 molecules [2, 3] (for a discussion of the theoretical evaluation, see [4]). The transition from the dissolved to the adsorbed state ist rate-determining only in the very initial degassing stage. During that stage the degassing rate, v(deg), can be represented by the equation $v(deg) = c(N) \cdot 2.8 \times 10^5 \cdot exp(-22000/RT)$ with v(deg) in $\mu g \cdot cm^2 \cdot min^{-1}$, c(N) in at%, R = gas constant, T = Kelvin temperature and the activation energy in cal/mol. The experiments were performed with doped W wires 1 mm in diameter at 1300 to 2200°C [2, 3]. Degassing experiments were also performed by [5] using W wires and sheets heated by electrical current flow and measuring the desorbed N_2 mass spectrometrically. The data were evaluated only with respect to the diffusion coefficient, i.e., for longer degassing periods. Degassing processes, in addition to the formation of a more uniform structure, were assumed to be responsible for the increased plasticity of W rolled in vacuum [6].

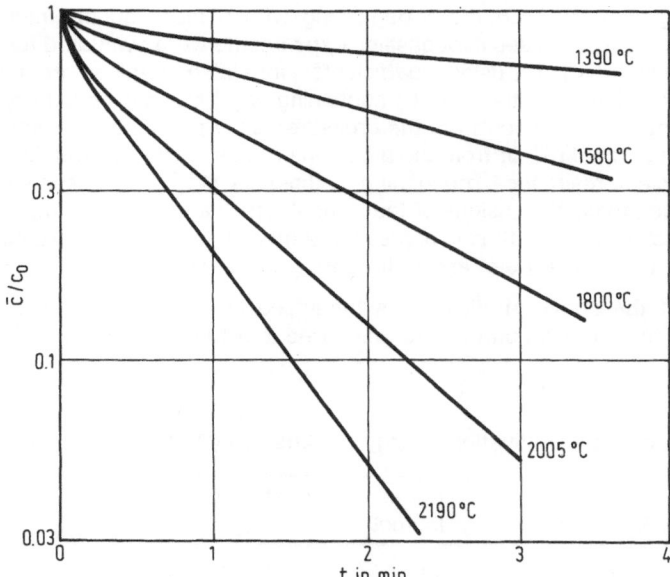

Fig. 2. Degassing isotherms of N-doped W wires; t = time, $\bar{c}$ and c_0 = mean actual and initial N concentration [3].

References:

[1] Jehn, H. (in: Fromm, E.; Gebhard, E.; Gase und Kohlenstoff in Metallen (Gases and Carbon in Metals), Springer, Berlin 1976, pp. 554/5).

[2] Fromm, E.; Jehn, H. (High Temp. – High Pressures **3** [1972] 553/62).

[3] Jehn, H. (Diss. Univ. Stuttgart, FRG, 1970, pp. 1/113).

[4] Hörz, G. (Z. Metallkd. **57** [1966] 703/8).

[5] Ryabchikov, L. M. (Ukr. Fiz. Zh. [Russ. Ed.] **9** [1964] 293/302; C.A. **61** [1964] 2786).

[6] Aleksandrov, A. A.; Ryabchikov, L. M.; Tron, A. S. (Tr. Leningr. Politekh. Inst. im. M. I. Kalinina No. 299 [1968] 16/25 from C.A. **71** [1969] No. 5982).

1.4 Diffusion, Permeation

Diffusion

The diffusion coefficient of the interstitially dissolved N atoms in the W lattice can be determined in degassing and permeation experiments. A direct experimental determination, e.g., by measuring the internal friction, as in other refractory metal–nitrogen systems, is not possible because of the extremely low N solubility in W.

The diffusion theory postulates a linear dependency between log $\bar{c}/c_0$ and t, when $\bar{c}$ is the mean N concentration at t, c_0 the initial N concentration, and t the degassing time in vacuum. For cylindrical samples with radius r_0, the diffusion coefficient D is related to the concentration by the equation $\bar{c}/c_0 = 0.69 \cdot \exp(-5.78 \, Dt/r_0^2)$ [1]. Thus, diffusion data were derived by evaluating the degassing isotherms of N-doped W wires as shown in Fig. 2 [2, 3]. The same technique was used by [4] who followed the N-degassing with a mass spectrometer. Volumetric determinations of the N_2 released in degassing experiments were evaluated for the activation energy of diffusion in [6]. Diffusion coefficients were also obtained from the diffusion-controlled kinetics of internal nitriding of Hf-containing W [5], using the solubility values of [6]. Other diffusion data were derived from the pressure rise in permeation experiments [7, 12], from the permeation rate [13], or from the diffusion profiles in couples with ion-implanted W, measured by nuclear resonance broadening techniques [11]. The activation energy was estimated from the atomic dimensions of the W host lattice and the diffusing N atom [8]. For theoretical discussions of the diffusion process assuming different effective dimensions of the diffusing atom at the saddle point and in its steady state, see [9].

The results of the above studies are summarized in Table 2. In **Fig. 3** the diffusion coefficients from the various sources are compared in a log D vs. 1/T plot [11].

Table 2

Preexponential Factor and Activation Energy of Diffusion ($D = D_0 \cdot \exp[-E_{diff}/RT]$).

D_0 in cm²/s	E_{diff} in kcal/mol	E_{diff} in kJ/mol	t in °C	Ref.
2.4×10^{-3}	28.4	118.8	1400 to 2000	[2, 3]
2.37×10^{-3}	35.8	149.8	1000 to 1800	[4, 10]; see also [14]
1.2×10^{-2}	32.2	134.7	1500 to 2000	[5]
5.4	62	259.4	~1400 to 2230	[7, 12]
–	50.2	210.0	1400 to 2600	[6]
4.3 ± 8.3	53.49 ± 3.69	223.8	600 to 800	[11]
–	35.8	149.8	–	[8]
7.0×10^{-3}	32.6	136.4	1800 to 2600	[3][1], [13]
1.1×10^{-3}	23.3	97.4	1800 to 2600	[3][2], [13]

[1] Values calculated in [3] with the permeation rates from [13] and the solubilities from [6]. – [2] As for [1], but with the solubilities from [15].

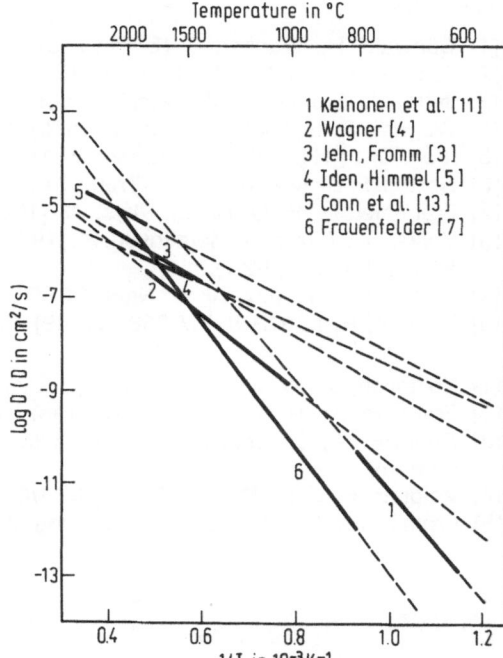

Fig. 3.   Temperature dependence of the diffusion coefficient D of N in W [11]. Curve 5 represents D values calculated by [3] from permeation data of [13] and solubility data of [15].

Permeation

Permeation is defined as gas diffusion through a membrane under the action of a pressure differential. It can be measured directly by determining the amount of gas diffusing through the membrane, but it can also be obtained indirectly from the solubility constant S and the diffusion constant D according to the relation $P = S \cdot D$ (see p. 2). Permeation rates of nitrogen through arc-cast W (membranes 0.26 and 0.53 mm thick) were measured between 1800 and 2600°C at N_2 pressure differentials of 1.0 and 0.1 atm. The quantities of permeating N were determined gas-chromatographically to give the permeation coefficient

$$P = 277 \cdot \exp[(-70\,000 \pm 800)/RT] \quad \text{or} \quad P = 1.27 \times 10^{-2} \cdot \exp[(-70\,000 \pm 800)/RT]$$

with P in $sccm \cdot mm \cdot cm^{-2} \cdot min^{-1} \cdot atm^{-1/2}$ and $Torr \cdot L \cdot cm^{-1} \cdot s^{-1} \cdot Torr^{-1/2}$, respectively, R = gas constant = 1.987 $cal \cdot mol^{-1} \cdot K^{-1}$, T in K, and activation energy in cal/mol; sccm = cm^3 at 0°C and 1 atm [13]. Using ultrahigh vacuum and mass-spectrometric techniques, permeation rates were determined in the temperature range ~1700 to 2500 K resulting in

$$P = 1.6 \times 10^{-4} \cdot \exp[-52\,400/RT]$$

with P in $Torr \cdot L \cdot cm^{-1} \cdot s^{-1} \cdot Torr^{-1/2}$ [7, 12]. Other values for the preexponential coefficient P_0 (in $Torr \cdot L \cdot cm^{-1} \cdot s^{-1} \cdot Torr^{-1/2}$) and the activation energy E_p (in cal/mol) are: $P_0 = 5.0 \times 10^{-4}$, $E_p = 53\,400$ (from $P = S \cdot D$) [4], $P_0 = 1.13 \times 10^{-2}$, $E_p = 69\,600$ (from the diffusion-controlled kinetics of internal nitriding of W–Hf alloys) [5], $E_p = 87\,900$ (from $P = S \cdot D$) [6]. The latter data are taken from [4].

References:

[1] Dünwald, H.; Wagner, C. (Z. Phys. Chem. B **24** [1935] 53/78).
[2] Jehn, H. (Diss. Univ. Stuttgart, FRG, 1970, pp. 1/113, 79).
[3] Jehn, H.; Fromm, E. (J. Less-Common Met. **21** [1970] 333/6).
[4] Wagner, R. L. (Metall. Trans. **1** [1970] 3365/70).
[5] Iden, D. I.; Himmel, L. (Acta Metall. **17** [1969] 1483/99).
[6] Norton, F. S.; Marshall, A. L. (Trans. Am. Inst. Min. Metall. Pet. Eng. **156** [1944] 351/71).
[7] Frauenfelder, R. (J. Chem. Phys. **48** [1968] 3966/71).
[8] Spivak, I. I. (Fiz. Met. Metalloved. **22** [1966] 859/64; Phys. Met. Metallogr. [Engl. Transl.] **22** No. 6 [1966] 52/7).
[9] Kisilishin, V. A. (Izv. Akad. Nauk SSSR Met. **1973** No. 3, pp. 197/202).
[10] Wagner, R. L. (ORNL-TM-2584 [1969] 1/45; N.S.A. **23** [1969] No. 38958).

[11] Keinonen, J.; Raisanen, J.; Anttila, A. (Appl. Phys. A **35** [1984] 227/32).
[12] Frauenfelder, R. (WERL-2823-28 [1967] 1/27; N.S.A. **21** [1967] No. 41580).
[13] Conn, P. K.; Duderstadt, E. C.; Fryxell, R. E. (Transl. Metall. Soc. AIME **242** [1968] 626/30).
[14] Wagner, R. L. (J. Met. **21** [1969] 66A/67A).
[15] Fromm, E.; Jehn, H. (J. Less-Common Met. **17** [1969] 124/6).

1.5 Ion Implantation

Bombarding a W surface with N ions at very low energies (300 and 450 eV) produced an altered surface layer with a saturation concentration of 8×10^{15} and 9×10^{15} N atoms/cm^2, respectively. The mechanisms involved in the formation and decomposition of these layers are discussed [5].

For theoretical calculations of the sputtering and backscattering processes caused by N ions incident on a W sample containing N ions on the surface and within the sample volume and also of the angular and spectral distributions of the sputtered and back-scattered particles as functions of source energy and obliquity, see [6].

Bombarding W under high-vacuum conditions with high-energy nitrogen ions results in implantation of the nitrogen atoms in different depths depending on the primary ion energy. W metal targets were bombarded with N_2^+ ions with energies up to 5000 eV. Subsequent heating (up to 700 to 800 K) recovered the gas from the target. The maximum amount of N taken up increased with increasing ion energy. The high desorption temperature indicates that the binding energy of the trapped atoms is much higher than that typical for physical adsorption [1]. The concentration distribution of $^{15}N^+$ ions implanted with energies of 20 to 100 keV in W substrates was determined by measuring the broadening of the $E_p = 429$ keV resonance yield curve from the $^{15}N(p, \alpha)^{12}C$ reaction. The concentration of implanted ions was $< 0.1\%$. The depth distribution showed a consistency of about 20% with predictions of the Lindhard-Scharff-Schiøtt (LSS) theory. The mean range in the polycrystalline target was larger than that predicted being valid for amorphous targets [2]. Bombardment with N_2^+ ions of higher energy (300 keV) resulted in a nominal maximum saturation of 50 at%. In addition to the nuclear resonance broadening, Rutherford backscattering was applied to determine the concentration profile of the implanted ions. Surface hardening was also observed [3]. Concentration distributions of N_2^+ ions implanted in W were also measured over the wider energy range 200 to 400 keV. The modal (peak position) and mean-range values were compared with the predictions of both a Monte Carlo calculation using the LSS cross sections and taking account of particle reflection at the surface and a simple analytical calculation using Biersack's PRAL-algorithm [4].

References:

[1] Leck, J. H. (Chemisorption Proc. Symp., Keele, Engl., 1956 [1957], pp. 162/8).

[2] Luomajarvi, M.; Keinonen, J.; Bister, M.; Anttila, A. (Phys. Rev. B Condens. Matter **18** [1978] 4657/62).

[3] Anttila, A.; Keinonen, J.; Uhrmacher, M.; Vahvaselka, S. (J. Appl. Phys. **57** [1985] 1423/5).

[4] Anttila, A.; Paltemaa, R.; Varjoranta, T.; Hentela, R. (Radiat. Eff. Lett. Sec. **86** [1984] 179/84).

[5] Winters, H. F. (J. Appl. Phys. **43** [1972] 4909/11).

[6] Goktepe, O. F.; Roush, M. L. (Proc. Summer Comput. Simul. Conf. **1977** 373/5; C.A. **87** [1977] No. 176211).

1.6   Vapor Deposition of W–N Thin Films

1.6.1   Physical Vapor Deposition

During reactive sputtering of W in an Ar–N_2 atmosphere, W atoms are sputtered from a target and deposited on a substrate, forming an N-containing W film or a thin nitride film. These films contain N at concentrations unattainable by direct thermal reaction of N_2 and W. This was used to prepare thin, hard layers containing 2 wt% N on a W substrate, using magnetron cathode sputtering in a reactive N_2 atmosphere. The low substrate temperature of 300°C prevented harmful nitride precipitation at grain boundaries during cooling. The layers showed a (110) texture [1]. Thin films containing a mixture of W and W_2N were formed by reactive sputtering of W in Ar containing 0.3 to 3% N_2. These films exhibited high electrical resistivities and low temperature coefficients of resistance, and were suggested to be used as thin-film resistors [2]. The composition of magnetron-sputtered W–N films deposited on silicon, quartz, and polyimide (Kapton) depends on the $N_2/(Ar + N_2)$ ratio and the total pressure as well as the power input; see **Fig. 4**. X-ray diffractograms showed that at higher N contents a W_2N film is formed, while at lower N contents the bcc W structure is preserved. At N_2 pressures less than 10% of the total pressure and p(Ar) < 40 mTorr, no nitrides are formed; at N_2 pressures above 10% and total pressures above 40 mTorr (resulting in N contents of the films of > 30 at%) W_2N films are obtained. The properties of the films (up to 22 μm thick) likewise depend on the parameters above. The electrical resistivity increases with the N_2 pressure, but decreases with decreasing total pressure and increasing power input. The Vickers hardness ranges between 600 and 3000 kg/mm² and increases with the N_2 pressure. The deposition rate decreases with increasing N_2 pressure, but increases with increasing total pressure and power input due to a higher sputtering rate; see **Fig. 5** [3].

Reactively sputtered WN_x films were deposited on GaAs wafers to make high-temperature-stable Schottky contacts. rf-sputtering of the W target in an Ar–N_2 plasma at 1.33 Pa total pressure and 5.4 W/cm² yielded films with 3.7, 11, and 21 at% N for N_2 contents in the plasma gas of 2, 5, and 50%, respectively. Oxygen contaminations of 0.8 to 1.5 at% were observed in the films at a base pressure of 6.7×10^{-5} Pa. For the dependence of electrical resistivity and barrier height of the as-deposited and annealed films on the N content, see the paper [4].

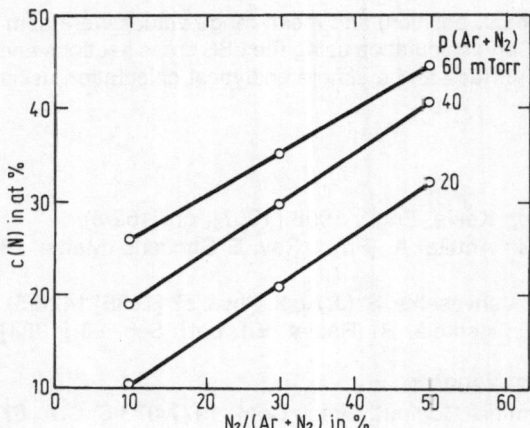

Fig. 4. Composition of magnetron-sputtered W–N films as a function of the N_2 content of the gas atmosphere; c(N) = N content of the films. Parameter at the curves is the total gas pressure p(Ar + N_2). The power input is 0.6 kW [3].

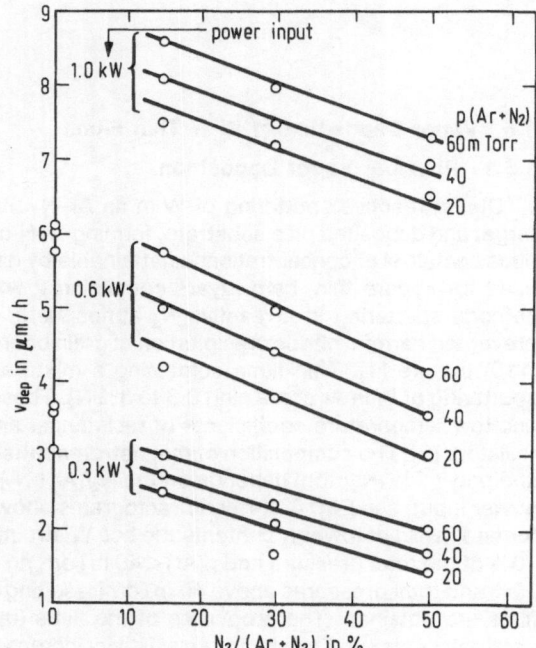

Fig. 5. Deposition rate v_{dep} of W–N films sputtered at different pressures and power inputs vs. the N_2 content in the sputtering gas mixture [3].

WN_x films form the gate material for self alignment GaAs metal-semiconductor field effect transistors. They were deposited by means of different reactive sputtering techniques (rf and d.c. magnetron, S-gun, rf diode). As for all sputter-deposited films, the N content depends on and is easily controlled by the N_2 content of the Ar–N_2 gas atmosphere. X-ray studies showed a varying W_2N proportion in the films. For the electrical properties, see the paper. The WN_x–GaAs system is metallurgically and electrically stable below 800°C [5]. W–N films are also used as diffusion barriers in GaAs metallization with Au, Ag, or Al. The films are prepared by sputtering a W target in an Ar–N_2 gas mixture using a planar rf magnetron. They have N contents of 23, 33, 46, and 55 at% at N_2 pressures of 1×10^{-3}, 2×10^{-3}, 4×10^{-3}, and 6×10^{-3} Torr, respectively (total pressure 1×10^{-2} Torr). W-rich films (23 and 33 at% N) are amorphous with crystallization temperatures above 620°C, while films with 46 and 55 at% are polycrystalline. All these films are useful as diffusion barriers. For Rutherford backscattering spectra (RBS) and electrical properties, see the paper [6].

W/WN$_x$ multilayers on GaAs substrates are prepared by a diode rf-sputtering technique, working alternately in nonreactive and reactive modes. According to RBS measurements, the N/W ratio in the WN$_x$ layers reaches up to 1.5, depending on the $p(N_2)/p(tot)$ pressure ratio, i.e., highly superstoichiometric WN films can be formed. The W layers also contain N up to N/W~0.6. For kinetic ellipsometric trajectories during the deposition of thick WN$_x$ films and W/WN$_x$ stacks, see the paper [7].

Superconductivity could be achieved in reactively sputter-deposited W–N films containing 5 to 20 at% N. The maximum transition temperature T_c of 4.85 K occurred in a film of a composition close to the phase boundary of β-W and W$_2$N (~10 at%) [8]. For studies of other electric properties of radio-frequency-sputtered WN$_x$ films and an investigation of their structure by transmission electron microscopy, see [9].

References:

[1] Zega, B. (PCT Int. Appl. 84-04110 [1984] 1/15 from C.A. **102** [1985] No. 118182).

[2] Rairden, J. R., III. (U.S. 3655544 [1972] 1/6 from C.A. **77** [1972] No. 26332).

[3] Shih, K. K.; Dove, D. B. (J. Vac. Sci. Technol. A **8** [1990] 1359/63).

[4] Geissberger, A. E.; Sadler, R. A.; Leyenaar, F. A.; Balzan, M. L. (J. Vac. Sci. Technol. A **4** [1986] 3091/4).

[5] Uchitomi, N.; Nagaoka, M.; Shimada, K.; Mizoguchi, T.; Toyoda, N. (J. Vac. Sci. Technol. B **4** [1986] 1392/7).

[6] Kolowa, E.; So, F. C. T.; Tandon, J. L.; Nicolet, M.-A. (J. Electrochem. Soc. **134** [1987] 1759/63).

[7] Boher, P.; Houdy, P.; Kaïkati, P.; Van Ijsendoorn, L. J. (J. Vac. Sci. Technol. A **8** [1990] 846/50).

[8] Kilbane, F. M.; Habig, P. S. (J. Vac. Sci. Technol. **12** [1975] 107/9).

[9] Reichelt, K.; Bergmann, G. (J. Appl. Phys. **46** [1975] 2747/51).

1.6.2 Chemical Vapor Deposition

In experiments to prepare fine-grained, non-columnar W metal films by chemical-vapor-deposition (CVD), the deposition of W$_2$N with subsequent decomposition was also investigated. The process involves partial nitriding of W to W$_2$N during deposition. The strategy is to interrupt the columnar growth of W crystals by a W$_2$N phase and then bring about a second phase change or crystal reorientation when the nitride is decomposed to W and gaseous N$_2$. The nitride is formed by CVD of WCl$_6$ with H$_2$ and NH$_3$ at 765 to 885°C (H$_2$/WCl$_6$ = 8; N$_2$/W = 0.4, 60% in excess of the amount needed to form W$_2$N; reactor-chamber pressure 35 Torr). The N concentrations and phases observed in the films are given in the table on the next page.

W$_2$N films can also be prepared using a WF$_6$–NH$_3$–H$_2$ gas mixture. The reduction of WF$_6$ starts at lower temperatures (300°C) compared to those required for WCl$_6$. The films deposited on a carbon plate at 450 to 600°C and WF$_6$: NH$_3$: H$_2$ = 1 : 4 : (1 to 2) show a metallic lustre, while at 700°C the films are black. The crystallite size of W$_2$N increases with temperature. The X-ray diffraction pattern strongly depends on the reaction temperature; the intensity ratio of the diffraction lines for films prepared at 600 to 650°C is almost the same as that for a W_2N powder sample [2]. Amorphous WN films are reported to be formed by CVD using a WF_6-NH_3 gas mixture [3].

deposition temp. in °C	N content in wt%	X-ray diffraction results	calculated amount W_2N in wt%
765	2.123	W strong W_2N strong	58
800	2.167	W strong W_2N strong	59
885	0.421	W strong W_2N weak	11

For details on the heat treatment for the decomposition, see the paper [1].

References:

[1] Landingham, R. L.; Austin, J. H. (J. Less-Common Met. **18** [1969] 229/43).
[2] Nakajima, T.; Watanabe, K.; Watanabe, N. (J. Electrochem. Soc. **134** [1987] 3175/8).
[3] Ikeda, Y. (Jpn. 63002319 [1988] 1/4 from C.A. **109** [1988] No. 46547).

2 Reactions with Halogens

For reactions occurring in a temperature gradient and resulting in the transport of metallic tungsten, see Chapter 2.5, pp. 34/110.

2.1 Fluorine

The behavior of a variety of materials (including tungsten) with F_2 and HF is reviewed [1]. A list of W fluorides and oxide fluorides reported in the literature has been published [2].

Elemental F_2 attacks tungsten severely, even at room temperature [3, p. 60], [4], [5, pp. 220/3], see also [26]. The attack of W by F_2 is more intense than that by other halogens [5]. According to gravimetric studies in the range 200 to 500°C, the reaction of solid W with fluorine gas starts at 200°C with W powder and at 300°C with solid tungsten rods [6]. A spontaneous reaction at room temperature did not occur when tungsten powder was exposed to F_2 of 800 Torr. Even when a W fuse wire was used for ignition, the powder sample was not completely fluorinated even though the fuse wire was [7].

The reaction products of W with F_2 are volatile. Since there is no protective layer formed, the reaction at normal pressure is highly exothermic and violent [8, p. 24]. Quantitative measurements of the tungsten loss rate in flowing F_2 or F were performed [9 to 12, 24]. The rate of W loss from the surface of hot tungsten filaments in an F_2 stream at low pressure was independent of temperature and pressure in the ranges 2000 to 2400 K and $(1 \text{ to } 6) \times 10^{-3}$ Torr. It had a rather high value of $\sim 10^{16}$ atoms $\cdot$ cm$^{-2} \cdot$ s^{-1} [9]. Even when taking into account the differences in experimental techniques and conditions, the magnitudes and trends of the fluorination rate data of [9] could not be reconciled with the results of later studies performed in the temperature range from about 700 K to the W sublimation threshold with both molecular and atomic fluorine in an Ar carrier gas at ~1 Torr total pressure and a flow velocity of ~10⁴ cm/s. The atomic F was generated in a microwave discharge. **Fig. 6** shows the W atom removal probability ε for $p(F_2) = 3 \times 10^{-2}$ and $p(F) = 1.6 \times 10^{-2}$ Torr; ε is defined as the ratio of the specific rate of W loss (in atoms·cm^{-2}·s^{-1}, regardless of the chemical state of aggregation to the F_2 or F arrival rate at the W surface. As can be seen, there is a conspicuous reaction rate "trough" preceding W sublimation. Under these particular ("trough") conditions, the W specimens will suffer

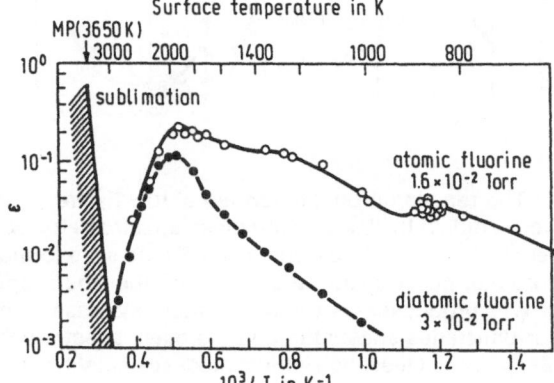

Fig. 6. Temperature dependence of reaction probabilities ε for the attack of tungsten by fluorine atoms and molecules; MP=melting point [10].

a comparatively low loss of material and act as a high-temperature F atom source. The ε vs. T relation for F atoms shows two maxima, which are attributed to the formation – desorption of distinct metal fluorides of widely disparate stoichiometry and thermodynamic stability (WF$_6$ and WF(?)). Near its high-temperature maximum (1980 K), ε is independent of the fluorine pressure in the range studied from 6×10^{-4} to 8×10^{-2} Torr. At lower temperatures, ε remains insensitive to p(F) but decreases with increasing $p(F_2)$. This indicates that under most of the conditions investigated, the F atom reaction on W is approximately first order. In contrast, the corresponding F_2 reaction is first order only near the high-temperature rate maximum, being noticeably of fractional order at lower temperatures. This latter trend, together with the Arrhenius plot of Fig. 6, implies that an increasing fraction of F_2 molecules reflect off the surface as the steady-state coverage of adsorbed fluorine increases. The high reactivity of F atoms under these same coverage conditions reveals the ability of a much larger fraction of incident atoms to successfully form W–F bonds, ultimately leading to the desorption of metal fluoride vapors [10, 11] (preliminary data [12, pp. 6/8]). **Fig. 7** shows the W removal rate from a tungsten filament in 1 bar Ar+10^{-2} or 10^{-3} bar F_2 flows of 40 cm/s. The removal rate is measured as the relative change in the electrical conductivity, dκ/dt·1/κ. The decrease in the removal rates at temperatures above 2000 to 2500 K is ascribed to the low stability of lower valent W fluorides. Similar removal rates, at least at higher temperatures, were observed in gas flows containing 1.33×10^{-2} to 1.33×10^{-3} bar of easily dissociatiable NF$_3$. For experiments in gas flows of 1 bar Ar with additions of O$_2$, SF$_6$, BF$_3$, SiF$_4$, PF$_3$, or WF$_6$, see [24], see also [25].

The main reaction product at lower temperatures is WF$_6$ [7, 13], see the review [14, p. 16]. This was verified by IR spectroscopy [6] and direct synthesis of WF$_6$ from pure W and F_2 at 350 to 400°C [15]. According to [16], the synthesis of WF$_6$ from the elements (W as wire or powder) can be accomplished at temperatures as low as 200°C. Care has to be taken to remove traces of O-containing impurities (H$_2$O, oxides) from the metal, or else the WF$_6$ will be contaminated by WOF$_4$, see also [17, 18].

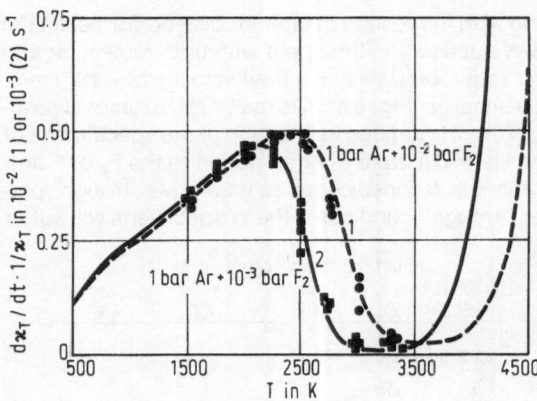

Fig. 7. Tungsten removal rate in an
Ar–F_2 flow of 40 cm/s, measured as the
relative change in the electrical conductivity κ [24].

The reaction products formed at low F_2 pressures ($\leq 7 \times 10^{-4}$ Torr) and high temperatures
were studied by line-of-sight mass spectrometry. A constant spectrum was only obtained at
very high temperatures where only F atoms are formed. Separation of WF_6 and WF_5 was barely
possible, but it could be established that the production of WF_5 increased at the expense
of WF_6 production as the temperature increased from 1200 to 1500 K [19, pp. 170/2], [20].
For difficulties encountered in the mass spectrometric distinction of WF_6 and WF_5, see also
[21, p. 223]. Quasi-equilibrium model calculations based on available thermodynamic data for
W and W fluorides, and assuming WF_6 [22] or WF_6 and WF [10] as W-containing reaction
products, did not satisfactorily reproduce the above experimental results of [10] for the
reaction kinetics of solid tungsten with atomic and molecular fluorine, see also [23]. By
adapting calculated W removal rates to those experimentally observed (see above), the
thermodynamic data of lower valent W fluorides involved in the reaction were derived. For
these compounds, only divergent and unreliable specifications were available in the literature,
while the thermodynamic properties of WF_6 and WF_5 were well established. On this new data
basis, the equilibrium gas phase compositions for the W(s)–F system were calculated. For
results obtained for a total effective F_2 pressure $\Sigma p(F_2) = 10^{-2}$ bar [24, 25], see Fig. 24, p. 48.

Reactions with Fluorine–Oxygen Mixtures

Reactions in the W–F–O system were studied by measuring the W removal rate from a hot
tungsten filament in a gas flow of 1 bar Ar with additions of 2×10^{-3} bar O_2 and (2.2 to 8.8)$\times$
10^{-3} bar NF_3 for the fluorine supply. The $O_2 : NF_3$ ratios in the gas flow corresponded to $O_2 : F_2$
ratios of 1:4, 1:3, 1:2, and 1:1 at the filament and were chosen to establish the effect of the
individual W oxide fluorides on the reaction. Equilibrium gas-phase compositions in the W–F–O
system derived from the measurements are represented in Fig. 29 and Fig. 30, p. 52 [24], see
also "Tungsten" Suppl. Vol. A 7, 1987, p. 72.

References:

[1] Hauffe, K. (Z. Werkstofftech. **15** [1984] 427/35).
[2] McInnis, M. B. (Tungsten Proc. 2nd Int. Tungsten Symp., San Francisco 1982, pp. 41/51;
    C.A. **98** [1983] No. 200593).
[3] Lugscheider, E.; Eck, R.; Ettmayer, P. (Radex-Rundsch. **1983** No. 1/2, pp. 52/84).
[4] Bachmann, W. T. (Mater. Design Eng. **64** No. 6 [1966] 106/8).
[5] Agte, C.; Vacek, J. (Wolfram und Molybdän, Akad.-Verlag., Berlin 1959, pp. 214/34).

[6] Iwasaki, M.; Yahata, T.; Suzuki, K.; Oshima, K. (Kogyo Kagaku Zasshi **65** [1962] 1165/7 from C.A. **58** [1963] 3092).

[7] Greenberg, E.; Settle, J. (ANL-5858 [1959] 1/132, 81/6; N.S.A. **14** [1960] No. 4497).

[8] Vincent, L. M. (Bull. Inf. Sci. Tech. Commis. Energ. At. [Fr.] No. 161 [1971] 17/31).

[9] Metlay, M.; Kimball, G. E. (J. Chem. Phys. **16** [1948] 779/81).

[10] Rosner, D. E.; Allendorf, H. D. (J. Phys. Chem. **75** [1971] 308/17).

[11] Rosner, D. E.; Allendorf, H. D. (AD-703881 [1970] 1/22; C.A. **73** [1970] No. 91891).

[12] Rosner, D. E.; Nordine, P. C. (AD-A022582 [1975] 1/16; C.A. **85** [1976] No. 113078).

[13] Johnson, R. L.; Siegel, B. (J. Inorg. Nucl. Chem. **31** [1969] 955/63).

[14] Byalobzheskii, A. V.; Tsirlin, M. S.; Krasilov, B. I. (Vysokotemperaturnaya Korroziya i Zashchita Sverkhtugoplavkikh Metallov [High-Temperature Corrosion and Protection of Refractory Metals], Moscow 1977, pp. 1/224; C.A. **88** [1978] No. 179375).

[15] Barber, E. J.; Cady, G. H. (J. Phys. Chem. **60** [1956] 505/6).

[16] Meinert, H.; Friedrich, L. (Z. Chem. **15** [1975] 411/2).

[17] Priest, H. F. (Inorg. Synth. **3** [1950] 171/83, 181/3).

[18] Brauer, G. (Handbuch der Präparativen Anorganischen Chemie, Vol. 1, Enke, Stuttgart 1975, pp. 1/608, 267/8).

[19] Philippart, J. L.; Caradec, J. Y.; Weber, B.; Cassuto, A. (Proc. Electrochem. Soc. **77-5** [1977] 169/80; C.A. **89** [1978] No. 66033).

[20] Philippart, J. L.; Caradec, J. Y.; Weber, B.; Cassuto, A. (J. Electrochem. Soc. **125** [1978] 162/6).

[21] Falconer, W. E.; Jones, G. R.; Sunder, W. A.; Vasile, M. S.; Muenter, A. A.; Dyke, T. R.; Klemperer, W. (J. Fluorine Chem. **4** [1974] 213/34).

[22] Abbot, P. C.; Stickney, R. E. (J. Phys. Chem. **76** [1972] 2930).

[23] Rosner, D. E.; Nordine, P. C. (J. Phys. Chem. **76** [1972] 2930/1).

[24] Dittmer, G.; Klopfer, A.; Schröder, J. (Philips Res. Rep. **32** [1977] 341/64).

[25] Dittmer, G.; Klopfer, A.; Ross, D. S.; Schröder, J. (J. Chem. Soc. Chem. Commun. **1973** 846/7).

[26] Daniel, P. L.; Rapp, R. A. (in: Fontana, M. G.; Staehle, R. W.; Advances in Corrosion Science and Technology, Vol. 5, Plenum, New York 1975, p. 55 from [27]).

[27] Hess, D. W. (Solid State Technol. **31** [1988] 97/103).

2.2 Chlorine

General

The attack of Cl_2 on W starts at 250 to 300°C [1, p. 60], [2, p. 220], [3, 4]. Ultrafine tungsten powder, obtained by hydrogen reduction of WCl_6, picks up some chlorine (weight gain several percent) from a 20:80 Cl_2–Ar gas mixture at room temperature. The chlorine can be removed (together with a small amount of W chlorides) by passing pure Ar over the sample at $\geq$300°C. With pure, hot Cl_2 the W powder reacts explosively [5, p. 269]. However, pressed pieces of W are reported to be 2.3 times faster converted into WCl_6 at 860 to 880°C as when the W is in an unpressed (powder) form [35].

Tungsten is less reactive to chlorine than molybdenum; the difference in the temperature for equal chlorination rates is $>$200 K. The attack by Cl_2 is highly anisotropic, i.e., the reactivity of W varies strongly with the crystal planes exposed at the surface [6].

The presence of oxygen greatly enhances the ablation of tungsten by chlorine at high temperatures.

As in the case of bromine (see p. 29), implantation of chlorine in subsurface regions is observed when the energy of Cl^+ ions impinging on a polycrystalline tungsten ribbon exceeds a certain threshold, which for chlorine lies at 440 eV. The incidence angle was 45°, and the target temperature 1800 to 2400 K [7], see also [8].

Products

A list of tungsten chlorides and oxide chlorides reported in the literature has been compiled [9, p. 46].

The main product of the reaction of W with Cl_2 at low and medium high temperatures (red heat) is the hexachloride WCl_6, see, e.g. [4, 10 to 13, 35]. In the presence of oxygen (impurity in the metal or in Cl_2) or moisture, oxide chlorides $WOCl_4$ and WO_2Cl_2 are also formed [4, 10, 14, 15], [16, p. 243]. Formation of WCl_4 is described [38]; thermodynamic calculations are presented for the optimal temperature of chlorination by Cl_2 and HCl. For the formation of WCl_4 by the reaction of tungsten with molecular and atomic chlorine at ≤1000 K and low pressures, see also pp. 18/9.

The free energy changes ΔG for five possible reactions of Cl_2 with W have been calculated for temperatures between 800 and 2000 K and pressures of 13.3 and 1333 Pa. The reactions considered are

$$Cl_2(g) + W(s) \rightarrow WCl_2(g) \qquad (I)$$
$$2\,Cl_2(g) + W(s) \rightarrow WCl_4(g) \qquad (II)$$
$$2.5\,Cl_2(g) + W(s) \rightarrow WCl_5(g) \qquad (III)$$
$$3\,Cl_2(g) + W(s) \rightarrow WCl_6(g) \qquad (IV)$$
$$Cl_2 \rightarrow 2\,Cl \qquad (V)$$

The initially negative ΔG values for the reactions II, III, and IV increase with T at both pressures, while the initially positive values for the reactions I and V decrease. Zero values for ΔG are reached for III and IV at >1000 to 1500 K, for I and V at ~1300 K (13.3 Pa) and 1700 to 1800 K (1333 Pa). The ΔG vs. T curve for II passes through zero at ~1800 K (13.3 Pa) or >2000 K (1333 Pa). The derived vapor pressure diagrams show that WCl_5 is the main product besides atomic Cl at ~1400 K and 13.3 Pa or ~1600 K and 1333 Pa. Dissociation of the pentachloride according to $WCl_5 \rightarrow W + 2.5\,Cl_2$ will occur at the walls of the reaction chamber at ~1200 K and 13.3 Pa [17]. Earlier graphs of calculated equilibrium vapor compositions in the W–Cl_2 system at 500 to 3000 K and 1, 10^{-3}, and 10^{-6} atm total pressure show WCl_2 to be the predominate gas phase component above 2000 K. The possible importance of WCl at high temperatures is suggested in the text of the paper but no data on this species are presented [18, p. 284].

The dichloride is also believed to be the main W–Cl species involved in the sublimation of tungsten in a chlorine atmosphere at 3000 K. The reactions

$$W(g) + 2\,Cl(g) \rightarrow WCl_2(g) \quad (VI) \text{ and}$$
$$W(g) + Cl_2(g) \rightarrow WCl_2(g) \quad (VII)$$

are assumed to proceed at low and high chlorine pressures, respectively. The enthalpies ΔH and equilibrium constants K for these reactions are calculated to be [19, p. 15]:

$$\Delta H = -246.1\,kcal/mol; \ K = p(WCl_2)/(p(W) \cdot p^2(Cl)) = 6.31 \times 10^5 \ atm^{-2} \qquad (VI)$$
$$\Delta H = -184.6\,kcal/mol; \ K = p(WCl_2)/(p(W) \cdot p(Cl_2)) = 5.7 \times 10^7 atm^{-1} \qquad (VII)$$

For equilibrium constants of the reactions $1/n\,W(s) + 1/2\,Cl_2 \rightarrow 1/n\,WCl_n(g)$ and $1/n\,W(s) + Cl \rightarrow 1/n\,WCl_n(g)$, see also [20]. For the gas phase composition in the W–Cl system at higher temperatures, see also pp. 61/4.

Kinetics

Tungsten chlorinates in a Cl_2–Ar stream (20% Cl_2, ~152 Torr) at 600 to 775°C at rates between ~10^{-7} and ~10^{-5} mol·cm^{-2}·min^{-1}. The least-squares equation for the chlorination rate v is $\log v = 2.65 - 8254\,T^{-1}$ with v in mol·cm^{-2}·min^{-1}. The rate depends approximately on the 0.6 power of the chlorine concentration. This dependency on the nearly half power of the Cl_2 pressure probably has to be attributed to the dissociation of the Cl_2 molecule at the metal surface. At the highest rates of chlorination measured, changes of total gas flow rates from 500 to 1000 mL/min (28-mm reaction tube) produced no change in the measured chlorination rates demonstrating that the true chemical reaction was rate-limiting. Upon chlorination all samples changed in appearance because of the anisotropic attack of chlorine. The polycrystalline samples developed ridges and troughs running in the direction in which they had been hot-rolled. Extensive chlorination of originally cylindrical single crystals produced two planar surfaces parallel to the {100} and {210} planes. While certain other crystal planes are not chlorinated, the {211} faces react rather rapidly. Electron and optical micrographs of chlorinated {100} and {320} planes reveal the formation of small tetrahexahedrons (also rounded square pyramids) with faces made up of {210} planes [6]. An activation energy of 43 kcal/mol and a chlorine concentration exponent of 0.58 to 0.60 had already been derived from earlier measurements on sheet samples at 617 to 776°C and chlorine contents of 5 to 35% in the Cl_2–Ar gas mixture. During all the chlorination runs the tungsten samples maintained a bright surface. Condensed chlorination products contained only the hexachloride. Intense gamma radiation produced no detectable change in the chlorination rate [21].

Early studies at low Cl_2 pressures showed that the velocity of the reaction, yielding volatile WCl_6, reaches a maximum at about 1500 K and becomes extremely small at higher temperatures. This is explained by the dissociation of some intermediate products which are assumed to be necessary steps in the formation of WCl_6. The chlorine leaving the W filaments used in the experiments at high temperatures was largely dissociated into atoms which were able to react with W at rather low temperatures [22, pp. 1154, 1162]. The occurence of a maximum in the chlorination rate at high temperatures was confirmed by later experiments. In an investigation [23], tungsten wires were reacted at 1500 to 2700 K with flowing Cl_2–He mixtures containing up to 10% Cl_2. The total pressure was 1 atm; the gas flow rate was chosen such (150 mL/min, reaction tube with 3 mm inside diameter) that further increase in velocity did not significantly affect the W loss rate; i.e., diffusion effects were no longer important and only pure chemical kinetics were involved. A maximum rate of W attack of ~1.7×10^{-4} mol·cm^{-2}·min^{-1} was found around 2100 K. The rate then drastically decreased; the rate constant at 2500 K being one fifth that at the maximum. A linear relationship existed between the tungsten reaction rate, and the reciprocal of the chlorine partial pressure. A reaction model was derived from the experimental data which involves a single site adsorption reaction, first order with respect to chlorine, followed by a rate-controlling surface reaction step. Adsorption enthalpy, activation energy of reaction, and pre-exponential factors were calculated for the reactions involved; they are of the Arrhenius type, except for transition temperature ranges within which the calculated thermodynamic values are no longer constant. Formation of WCl_6 is believed to occur at the W surface below 1500 K, formation of WCl_4 between 1750 and 2050 K, and formation of WCl_2 at 2250 to 2700 K. In the intermediate temperature ranges, a gradual transition between subsequent chlorides takes place. The chlorides formed at the W surface can be subjected to further reactions (e.g., disproportionation) in the gas phase [23].

As shown in **Fig. 8**, the reaction probabilities of tungsten with mono- and diatomic chlorine also have maxima at ~2000 to 2200 K. The reaction probability ε is defined as the ratio of the flux of removed W atoms (regardless of their chemical state of aggregation) to the flux of Cl or Cl_2 incident upon the surface. At higher temperatures (but below the sublimation threshold) the chlorination probabilities drop off sharply, obviously due to the evaporation of unreacted chlorine. The reaction order of W chlorination shows a complex behavior. Zero-order regions where the rates are independent of the reactant pressure are found for both the $W(s)/Cl(g)$ and $W(s)/Cl_2(g)$ reactions. However, even in these zero-order regions (with reactant pressure ~10^{-2} Torr) at a surface temperature of about 1860 K, the Cl atom reaction, probability for W atom removal, still exceeds that for Cl_2 by a factor of about 32 [24, pp. 714/5].

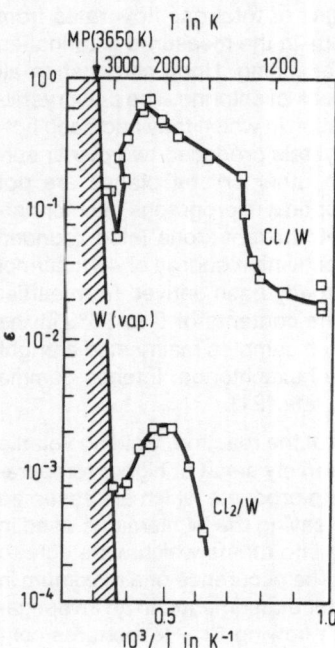

Fig. 8.   Reaction probability ε of W with Cl and Cl_2 as a function of the reciprocal temperature. Pressures (in Torr); $p(Cl) = 4.3 \times 10^{-3}$, $p(Cl_2) = 3.0 \times 10^{-2}$, $p(total) = 1$ (diluent usually Ar). The line W(vap) represents the apparent reaction probability contributed by pure W sublimation [24].

The reactions of atomic and molecular chlorine with tungsten were also investigated by modulated beam-mass spectrometric methods over the temperature range 300 to 1350 K at low pressures. The atomic beam was generated by a radio frequency plasma discharge. With both atomic and molecular beams, the main reaction product up to about 1000 K was WCl_4. The reaction probability with atomic chlorine was approximately a factor of ten higher than that obtained with molecular chlorine. In this case, the (apparent) reaction probability ε is the ratio of the product and reactant signals, which are corrected for the ionization efficiencies of the mass spectrometer. The ε values for molecular Cl_2 were in the range of about 10^{-5} to 10^{-3} and dependent on temperature and beam intensity, and slightly on the modulation frequency. The reaction was nonlinear with respect to Cl_2 intensity at low beam fluxes but approached linearity at high beam intensities (above ~2×10^{16} molecules·cm^{-2}·s^{-1} for 393 K and a modulation frequency of 20 Hz). Above 1000 K, the main reaction product was atomic chlorine. For Cl, the reaction probability increased rapidly with temperature; at 1300 K nearly complete dissociation of Cl_2 was observed. A kinetic model based on the Eley-Rideal mechanism was proposed, which satisfactorily described the experimental results and also accounted for the trends in the etch rate data in a Cl/Cl$_2$ flow at higher pressures described below [28, 31]. At Cl$_2$ pressures equivalent to the beam intensities used in [31], the "quasi-equilibrium" model [32] predicts WCl$_4$ as the main product from 300 to ~800 K with Cl appearing above 1000 K. This is in good agreement with the molecular beam results. Between 800 to 1000 K, WCl$_2$ should be formed, but is not found experimentally [31]. Using a modulated molecular beam technique and monitoring only the scattered Cl signal from a tungsten surface, concluded that the interaction consists of dissociative adsorption followed by strong bulk diffusion [33]. In contrast, no indication of a diffusion process was observed in [31].

The etching of thin W films on oxidized Si wafers in a Cl$_2$ plasma discharge was studied under various experimental conditions. These ranged from 0.1 to 1 Torr in pressure, 30 to 200°C in temperature, 0.2 to 1.0 W/cm^2 in power density, and 3 to 200 sccm in Cl$_2$ flow rate. Etch rates varied from below 10 to 90 nm/min. An increase in power, pressure, or gas residence time increased the etch rate. The plasma etching process evidently depends, in addition to the ion bombardment energy and flux, critically on the concentration of Cl atoms. Molecular Cl$_2$ in the absence of a discharge did not etch W under the above conditions. A drop in etch rate below a certain temperature (~60°C for 0.2 Torr pressure, 100 sccm Cl$_2$ flow rate, and 100 W power) suggests a threshold for the chemical reaction step or a change in the etching mechanism. If an Arrhenius rate expression is assumed, the data above the threshold temperature yield an effective activation energy of 0.12 eV/molecule. Small additions (0 to 5%) of BCl$_3$ to the chlorine discharge increased the etch rate and improved the reproducibility. When the samples were positioned downstream from the Cl$_2$ discharge, etching proceeded solely by chemical reaction of the W film with Cl atoms. For a pressure of 0.2 Torr and a 100 sccm Cl$_2$ flow rate, corresponding to a Cl atom mole fraction of 0.5 for the atomic etching, downstream and in-plasma etch rates were approximately equal at 110°C, but the Cl atom etch rate dropped more rapidly than the in-plasma etch rate as the temperature was decreased. If an Arrhenius temperature dependence was assumed, the apparent activation energies for the reactions were 0.1 and 0.3 eV/molecule, for plasma etching and atom etching, respectively. The chemical reaction between Cl atoms and the W films was proportional to the gas-phase Cl atom mole fraction [26 to 29], see also [30].

For the laser-induced etching of thin W films on fused silica in a Cl$_2$ atmosphere, see [34, 36, 37].

The rate of conversion of pressed pieces of tungsten into WCl$_6$ (containing <0.5 wt% O) in a Cl$_2$ stream at high temperatures was 2.3 times faster than the rate when the W was in an unpressed form. Thus, W powder was mechanically compacted in a ram and die press. The two resulting bars (each 1/2×1/2×18 in and weighing 2.5 kg) were loaded directly into a quartz tube resting in a horizontal tube furnace and heated to 860 to 880°C. Chlorine was passed at 1 L/min over the W bars. After 26 h, the bars were completely consumed which equated to a WCl$_6$ production rate of 415 g/h [35].

Reactions with Chlorine–Oxygen Mixtures

As already mentioned on p. 16, the combined attack of Cl$_2$ and O$_2$ results in the formation of tungsten oxide chlorides, e.g., WO$_2$Cl$_2$ and WOCl$_4$ [14 to 16, 24, 25].

The W removal rate in Cl$_2$–O$_2$ mixtures at high temperatures is faster than the sum of the rates in the pure gases reacting separately. A series of measurements was carried out with pure W wires in flowing Cl$_2$–O$_2$–Ar mixtures in which the sum of the Cl$_2$ and O$_2$ pressures was kept constant at 10^{-1} Torr. The total pressure was 1 Torr. Results at a surface temperature of 1473 K are shown in **Fig. 9**, together with the tungsten removal rates that would be expected if previous data for pure Cl$_2$ and pure O$_2$ are simply combined in accord with their prevailing concentrations (under "no-coupling conditions"). As can be seen, the actual removal rates exceed the no-coupling predictions under these conditions by up to twentyfold. Higher surface temperatures suppress the magnitude of the peak enhancement and cause the maximum coupling effect to occur at higher Cl_2/O_2 ratios (e.g., at 2155 K the maximum enhancement is only about twofold and occurs at Cl_2/O_2 ratios in excess of 7:1) and become less sharply peaked. The temperature dependence of the W removal rate at $p(Cl_2)=1\times10^{-2}$ and $p(O_2)=2\times10^{-2}$ Torr (total pressure 1 Torr) is depicted in **Fig. 10**. Time-independent removal rates were observed at temperatures as low as 1130 K, whereas in the absence of Cl_2, semiprotective oxide films were already noticeable at 1450 K and below. At constant O_2 pressure, the apparent reaction order with respect to chlorine passes from large positive values to large negative values beyond the rate maximum as the Cl_2 pressure increases. The reaction order for $p(Cl_2) \sim 0.6\times10^{-1}$ Torr is about 1.3. It is suggested that the thermodynamically favored (see p. 22) formation and rapid desorption of volatile WO_2Cl_2 is responsible for the enhancement of the W removal rate in Cl_2-O_2 mixtures [14], also see [24].

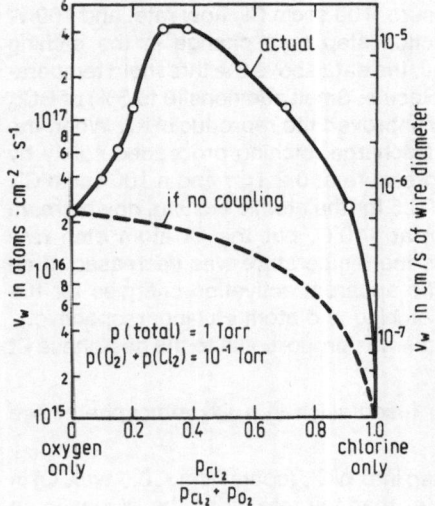

Fig. 9.   Tungsten removal rate v_W in Cl_2-O_2 mixtures at 1473 K as a function of the $p(Cl_2)/(p(Cl_2)+p(O_2))$ ratio [14].

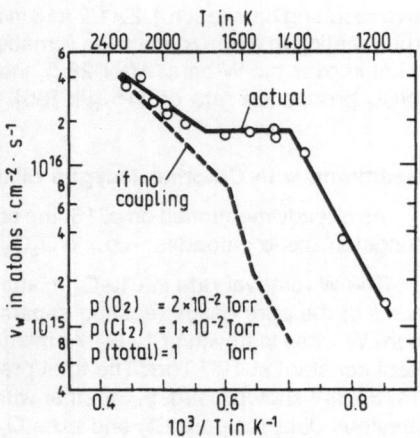

Fig. 10.   Tungsten removal rate v_W in Cl_2-O_2 mixtures as a function of the surface temperature [14].

Mass spectrometric investigations of the volatile reaction products formed on W ribbons in Cl_2–O_2 mixtures indeed showed that WO_2Cl_2 is the major product over wide ranges of temperature and pressure. Volatile oxides, $WO_2(g)$ and $WO_3(g)$ are produced as in pure oxygen, but at somewhat lower rates. Some atomic oxygen and possibly atomic chlorine are also formed. Ions indicating the occurrence of $WOCl_4$, binary W chlorides, or chlorine oxides were not detected in the mass spectrum in the temperature range of 1200 to 2400 K and total pressure range of 10^{-6} to 5×10^{-4} Torr studied. A typical plot of product desorption rates vs. tungsten surface temperature is shown in **Fig. 11**. The WO_2Cl_2 desorption rate data give a nonlinear Arrhenius plot. Its slope increases with temperature approaching 45 kcal/mol at 2400 K, a lower limit for the activation energy of the rate limiting formation step. The results in Fig. 11 suggest strongly that the surface processes leading to $WO_2Cl_2(g)$, $WO_2(g)$, or $WO_3(g)$ are competitive, and that the three products have a common precursor formed from adsorbed oxygen and chlorine. This is also apparent in the behavior of the major product desorption rates when gas pressures are varied. At sufficiently low pressures ($< 5 \times 10^{-5}$ Torr) all three products form at rates that are first order in O_2 pressure. The dependence of the rates on the Cl_2 pressure is different for WO_2Cl_2 and the oxides. The rate of WO_2Cl_2 release increases linearly with the Cl_2 pressure, while the oxide evolution rates show a slight decrease with increasing $p(Cl_2)$. As the total pressure is raised above some high value ($> 5 \times 10^{-4}$ Torr), the product desorption rates fall rapidly from first to zero-order pressure dependence for any given surface temperature. The total pressure at which the change in reaction order occurs depends on the surface temperature and the Cl_2/O_2 ratio. As the temperature is raised, the first-order dependence continues to higher total pressures. Chlorine is much less effective than oxygen in causing the decrease in pressure dependence. For example, at 1850 K with $p(Cl_2)$ fixed at 2.5×10^{-4} Torr, the evolution rates of both $WO_2(g)$ and $WO_2Cl_2(g)$ are linear with

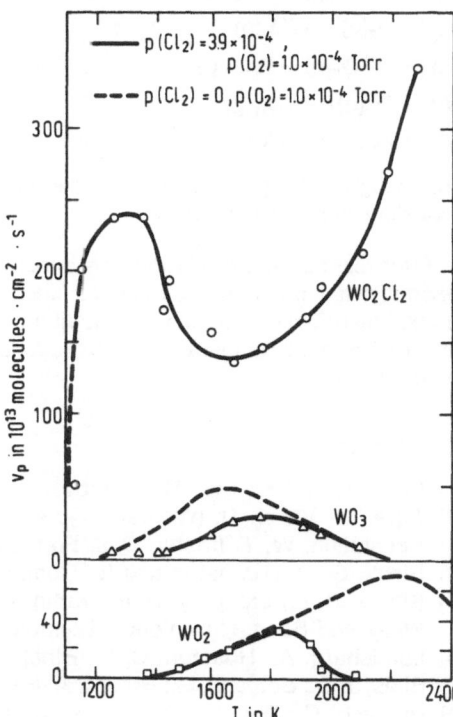

Fig. 11.   Product desorption rates v_p in Cl_2–O_2 mixtures as a function of the tungsten surface temperature [15].

$p(O_2)$ up to 5×10^{-4} Torr; on the other hand, with $p(O_2)$ held at 3.0×10^{-4} Torr, both product evolution rates are linear in $p(Cl_2)$ to $>7 \times 10^{-4}$ Torr. Obviously, at higher temperatures, and consequently higher product release rates, a larger Cl_2/O_2 flux is necessary to saturate the surface and, since oxygen sticks to the surface more effectively than chlorine, a higher partial pressure of Cl_2 is required to saturate the surface and cause a decrease in reaction order [15].

Dissociative adsorption of a fraction of both incident Cl_2 and O_2 molecules as the first reaction step was also assumed [14]. The adsorbed Cl and O atoms (the latter being more strongly bound) are believed to promote collectively the evaporation of W atoms by WO_2Cl_2 formation/desorption, while separately they may leave the surface without enticing W atoms to join them. Moreover, the high volatility of WO_2Cl_2 is assumed to help clear the surface, thereby raising the respective O_2 and Cl_2 sticking probabilities. At, e.g., 1515 K, $p(Cl_2) = 3.3 \times 10^{-2}$ Torr, and $p(O_2) = 0.67 \times 10^{-1}$ Torr, the net effect is to raise the fraction of O_2 encounters leading to W atom removal from about 1/500 in the absence of Cl_2 to about 1/25 in the presence of Cl_2 [14]. According to [15], the $WO_2Cl_2(g)$ formation rate approaches 10% of the reagent gas impact rate at 2000 K.

Free energy changes ΔG in the range from 1500 to 2500 K that are based on JANAF thermochemical data were compiled for several possible reactions in the $W-Cl_2-O_2$ system [14]:

reaction	$-\Delta G$ in kcal/mol at		
	1500 K	2000 K	2500 K
$W(s) + O_2(g) + Cl_2(g) \rightarrow WO_2Cl_2(g)$	124.312	106.962	89.618
$W(s) + 1/2\,O_2(g) + 2\,Cl_2(g) \rightarrow WOCl_4(g)$	79.582	53.737	28.071
$W(s) + O_2(g) \rightarrow WO_2(g)$	−9.105	−5.886	−2.983
$W(s) + 3/2\,O_2(g) \rightarrow WO_3(g)$	50.833	44.391	37.809
$W(s) + Cl_2(g) \rightarrow WCl_2(g)$	−23.321	−17.878	−12.752
$W(s) + 2\,Cl_2(g) \rightarrow WCl_4(g)$	34.419	18.854	3.309

According to [24, p. 716], the formation of chlorine oxides also has to be considered when regarding thermochemical data.

Prior dissociation of chlorine produces enhancements of the $W(s)-O_2(g)$ (and $W(s)-O(g)$) reaction rate close to those found for undissociated Cl_2, but at lower chlorine : oxygen atomic ratios. The effect of $Cl(g)$ or $Cl_2(g)$ addition on the attack of tungsten by atomic oxygen is much less marked at all temperature levels (e.g., at 1515 K the maximum enhancement is only about twofold) [14].

References:

[1] Lugscheider, E.; Eck, R.; Ettmayer, P. (Radex-Rundsch. **1983** 52/84).
[2] Agte, C.; Vacek, J. (Wolfram und Molybdän, Akad. Verlag, Berlin 1959, pp. 214/34).
[3] Bachmann, W. T. (Mater. Des. Eng. **64** No. 6 [1966] 106/8).
[4] Rieck, G. D. (Tungsten and Its Compounds, Pergamon, Oxford 1967, pp. 1/154, 43).
[5] Ripley, R. L.; Lamprey, H. (in: Kuhn, W. E.; Lamprey, H.; Sheer, C.; Ultrafine Particles, Wiley and Sons, New York – London – Sydney 1963, pp. 262/7).
[6] Landsberg, A.; Hoatson, C. L.; Block, F. E. (J. Elektrochem. Soc. **118** [1971] 1331/6).
[7] Blais, J. C.; Bolbach, G.; Brunot, A. (Int. J. Mass. Spectrom. Ion Phys. **47** [1983] 269/73).
[8] Bolbach, G.; Blais, J. C.; Brunot, A.; Marilier, A. (Surf. Sci. **90** [1979] 65/77).

[9] McInnis, M. B. (Tungsten Proc. 2nd Int. Tungsten Symp., San Francisco 1982, pp. 41/51; C.A. **98** [1983] No. 200593).

[10] Weiß, L. (Z. Anorg. Allg. Chem. **65** [1910] 279/43, 340).

[11] Defacqz, E. (Ann. Chim. Phys. [7] **22** [1901] 238/88).

[12] Eméleus, H. J.; Gutmann, V. (J. Chem. Soc. **1950** 2115/8).

[13] McCarley, R. E.; Brown, T. M. (Inorg. Chem. **3** [1964] 1232/6).

[14] Rosner, D. E.; Allendorf, H. D. (AIAA J. **5** [1967] 1489/91).

[15] McKinley, J. D. (React. Solids Proc. 6th Int. Symposium, Schenectady, N.Y., 1968 [1969], pp. 345/51).

[16] Rosner, D. E.; Allendorf, H. D. (in: Belton, G. R.; Heterogeneous Kinetics at Elevated Temperatures, New York 1970, pp. 231/51).

[17] Prokoshkin, D. A.; Tret'yakov, V. I.; Demin, Yu. N. (Izv. Vyssh. Uchebn. Zaved. Mashinostr. **1982** No. 3, pp. 84/7; C.A. **96** [1982] No. 169769).

[18] Brewer, L.; Bromley, L. A.; Gilles, P. W.; Lofgren, N. L. (Nat. Nucl. Energy Ser. Div. IV B **19** [1950] 276/311).

[19] Tower, L. K. (NASA-TN-D 1194 [1962] 1/45; C.A. **56** [1962] 13911).

[20] T'Jampens, G. R.; van de Weijer, M. H. A. (Philips Tech. Rev. **27** [1966] 173/9; Philips Tech. Rundsch. **27** [1966] 165/71).

[21] Landsberg, A.; Block, F. E. (Bur. Mines Rep. Invest. No. 6649 [1965] 1/26).

[22] Langmuir, I. (J. Am. Chem. Soc. **37** [1915] 1139/71).

[23] Meubus, P. (in: Foroulis, Z. A.; Pettit, F. S.; Properties of High Temperature Alloys, Met. Soc. AIME, Princeton 1976, pp. 395/410; C.A. **89** [1978] No. 78852).

[24] Rosner, D. E.; Allendorf, H. D. (Proc. 3rd Int. Symp. High Temp. Technol., Asilomar, Calif., 1967 [1969], pp. 707/19).

[25] Colton, R.; Tomkins, I. B. (Austral. J. Chem. **21** [1968] 1975/9).

[26] Fischl, D. S.; Hess, D. W. (J. Electrochem. Soc. **134** [1987] 2265/9).

[27] Fischl, D. S.; Rodrigues, G. W.; Hess, D. W. (J. Electrochem. Soc. **135** [1988] 2016/9).

[28] Fischl, D. S. (Diss. Univ. Berkeley, Calif., 1988, 163 pp. from Diss. Abstr. Int. B **49** [1989] 4923).

[29] Hess, D. W. (Proc. Jpn. Symp. Plasma Chem. **1** [1988] 35/41; C.A. **111** [1989] No. 125014).

[30] Hess, D. W. (Solid State Technol. **31** No. 4 [1988] 97/103 from C.A. **109** [1988] No. 64881).

[31] Balooch, M.; Fischl, D. S.; Olander, D. R.; Siekhaus, W. J. (J. Elektrochem. Soc. **135** [1988] 2090/5).

[32] Batty, J. C.; Stickney, R. E. (J. Chem. Phys. **51** [1969] 4475/84).

[33] Prince, R. H.; Lampert, R. M. (Chem. Phys. Lett. **67** [1979] 388/92).

[34] Rothschild, M.; Sedlacek, J. H. C.; Ehrlich, D. J. (Appl. Phys. Lett. **49** [1986] 1554/6).

[35] Munn, R. W.; McClintic, R. P.; Reilly, K. T. (U.S. 4803062 [1988/89] from C.A. **110** [1989] No. 157153).

[36] Rothschild, M.; Sedlacek, J. H. C.; Black, J. G.; Ehrlich, D. J. (J. Vac. Sci. Technol. [2] B **5** [1987] 414/8).

[37] Sesselmann, W.; Chuang, T. J. (J. Vac. Sci. Technol. [2] B **3** [1985] 1507/12).

[38] Thomas, N.; Blanquet, E.; Vahlas, C.; et al. (Mater. Res. Soc. Symp. Proc. **204** [1991] 451/6 from C.A. **115** [1991] No. 269269).

2.3 Bromine

General

Tungsten reacts with bromine at red heat [1] or bright red heat [2, pp. 220, 223] to form bromides. The reaction starts at ~450 to 500 K [3], [4, p. 60]. Only superficial attack of W heated in Br_2, presumably due to the presence of oxygen impurities, was noted earlier [5].

Products

Pure tungsten reacts with pure Br_2 to form compounds WBr_x (x = 2, 3(?), 4, 5, 6), [6, p. 16]. In the presence of O impurities or moisture, oxide chlorides, such as $WOBr_4$ or WO_2Br_2 are also formed as in the case of the other halogens. A list of tungsten bromides and oxide bromides reported in the literature has been compiled [7, p. 46].

The reaction of tungsten metal with gaseous bromine at elevated temperatures can be used to prepare WBr_5. For this purpose, Br_2 in an N_2 carrier gas stream was passed over the metal at 450 to 500°C [3] or 600°C [8]. Direct combination of W with large amounts of Br_2 in a closed system at $\geq$500°C yielded a mixture of WBr_5 and WBr_6, from which WBr_5 could be obtained by sublimation (the WBr_6 disproportionates during sublimation into WBr_5 and Br_2) [9 to 11]. For the formation of WBr_5 from tungsten and bromine, see also the early studies of [12] and [13, p. 362].

W_6Br_{12} (WBr_2) could be prepared from stoichiometric amounts of W and Br_2 in a sealed tube kept in a temperature gradient of 760 to 560°C [14, pp. 309/10].

According to thermodynamic calculations, WBr_6 might be prepared by the action of Br_2 under pressure upon W metal at temperatures around 400 K, provided that the rates were fast enough. At this temperature the pressure of Br_2 would have to be about 7.5 atm. The calculations also show that WBr_4 is the main W-containing gas-phase component in the W–Br system over a wide range of temperatures at total pressures of 10^{-6} to 1 atm. WBr_6 and WBr_3 are never important species of the gas phase. Vapor compositions for total pressures of 10^{-3} and 1 atm are graphically represented [30]. For calculated gas-phase compositions in the W(s)–Br_2 system at high temperatures, see pp. 36, 80/1.

The predominance of WBr_4 in the gas phase was experimentally confirmed by studies of the W + Br_2 reaction in a Knudsen double cell connected to a mass spectrometer. The Knudsen cell was filled with W wool; the bromine was in a separate compartment. At Br_2 pressures between 6 and 50 Torr in the reaction cell and temperatures between 1000 and 1500 K only WBr_4(g) was formed. Up to about 1000 K, no reaction products could be detected in the gaseous phase. The intensities of the ions originating from WBr_4 increased up to about 1270 K and then decreased. Above 1500 K no ions could be detected. At higher pressures W_2Br_8 also appeared in the gas phase via the secondary reaction $2\,WBr_4(g) \rightarrow W_2Br_8(g) + Br_2(g)$ [15], see also [16, 17]. Later measurements of the intensity of the bromine ions effusing from a W-filled Knudsen cell indicated that the reaction $W(s) + 2\,Br_2(g) \rightarrow WBr_4(g)$ starts at ~600 K when the initial bromine pressure in the reaction cell is adjusted to 6.7 Pa (~5×10^{-2} Torr) by a glass capillary connecting the cell to a bromine reservoir held at constant pressure. The consumption of bromine attains a maximum value at 750 K, where the reaction rate is high enough for the establishment of the thermodynamic equilibrium of the reaction. At higher temperatures the bromine consumption, i.e., the reaction rate, decreases due to the exothermic nature of the reaction between tungsten and bromine [18]. For a discussion of the gas-phase species probably formed in the reaction of tungsten with bromine at elevated temperatures, see also [9].

Kinetics and Thermodynamics of the Reactions

The kinetics of the $W + Br_2$ reaction in the temperature range between 300 and 850 K depend strongly on both the surface conditions of the metal and the purity of the bromine vapor. The apparent activation energy for the reaction with pure bromine is 169 kJ/mol under the above conditions and only 91 kJ/mol for bromine containing 1% oxygen. The order of the reaction $W(s) + 2Br_2 \rightarrow WBr_4(g)$ is 0.7 with respect to $p(Br_2)$ at 13.3 Pa and 650 K [18]. In a subsequent publication the reaction rate at <10 Pa and 400 to 800 K is given by the equation

$$v(Br_2) \text{ (in } mol \cdot cm^{-2} \cdot s^{-1} \cdot Pa^{-1}) = 3.7 \cdot p(Br_2)^{0.7} \cdot exp[-144/RT],$$

where $v(Br_2)$ is the Br_2 consumption rate and the activation energy is in kJ/mol. From the theory of absolute reaction rates assuming an Eley-Rideal mechanism, a reaction rate at 600 K of 4.5×10^{-12} $mol \cdot s^{-1} \cdot cm^{-2} \cdot Pa^{-1}$ was derived, which is in good agreement with the experimental value of 9.3×10^{-13} $mol \cdot s^{-1} \cdot cm^{-2} \cdot Pa^{-1}$ [19]. The dependence of the reaction (ablation) rates of polycrystalline W filaments on the Br_2 partial pressure in an Ar stream is shown in **Fig. 12**. The temperatures and Br_2 pressures range form 975 to 1448 K and. 3×10^{-3} to 3×10^3 Pa, respectively; the total pressure is 1 atm. No essential difference was noted between the results obtained for two different types of lamp filament quality. When atomic bromine is assumed as the active species, the tungsten reaction rate $v(W)$ is independent of temperature. The $v(W)$ vs. $p(Br)$ dependence is depicted in **Fig. 13**. A least-squares treatment of the data in the interval 975 to 1282 K yields the relation

$$v(W) \text{ (in } mol \cdot cm^{-2} \cdot s^{-1}) = 4.7 \times 10^{-13} \, p(Br) + 4.0 \times 10^{-15} \, p(Br)^2,$$

when $p(Br)$ is measured in Pa. It is proposed that the rate-limiting step in the reaction of tungsten with pure bromine is the formation of a tungsten-bromine surface complex by an Eley-Rideal mechanism [20].

The reaction between W and Br_2 in a static system was studied between 1063 and 1213 K with Br_2 concentrations of 0.5 to 1.5 mol/m^3 (10 to 25 Torr). A quartz spring was utilised to record the tungsten foil weight loss, which is linear with time at all temperatures and pressures of bromine. Thus the reaction is zero-order with respect to tungsten. The reaction with respect to bromine varies in order from 1.70 to 1.16 with an increase in temperature. The rate constant is given by

$$k \text{ (in } mol^{-1/2} \cdot m^{5/2} \cdot s^{-1}) = 1.12 \times 10^{-9} \cdot exp[(-82.7 \pm 4.0)/RT]$$

if a mean order of 3/2 is assumed; the activation energy is in kJ/mol. The species produced at the W surface are probably WBr_2 and WBr_5, the former prevailing at the highest temperatures studied. Their formation is attributed to the stepwise attack of atomic bromine on the tungsten. The major contribution to the activation energy is from adsorption of atomic Br onto, and desorption of the products from the W surface [9].

On exposure of tungsten to bromine gas in a flow system at relatively high temperatures (see above), heavy local erosion of the tungsten samples was observed. This consists of the formation of hemispherical pits (craters) in the tungsten surface. The influence of the experimental parameters, i.e., reaction time, temperature, and Br_2 partial pressure, and of the thermal pretreatment of the W samples on pit formation was investigated. From the results and supporting experiments using Scanning Auger Microscope (SAM) techniques, it is proposed that a local impurity in the tungsten, the nature of which remained unidentified in the SAM experiments, acts as a catalyst for the $W + Br_2$ reaction. The growth rate of the pits becomes zero after some time due to consumption or poisoning of the catalyst and/or dilution when the pit area is expanding. An oxidizing thermal pretreatment at ~1300 K considerably reduces the number of pits formed. However, this protective action is not permanent and extends only to a layer of limited thickness [21].

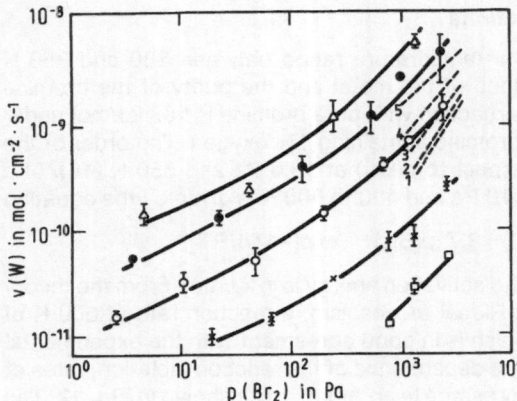

Fig. 12.   Tungsten reaction rate v(W) as a function of partial Br_2 pressure $p(Br_2)$ at 975 K (□), 1067 K (×), 1166 K (o), 1282 K (•), and 1448 K (△) [20]. The broken lines represent measurements of [9] at 1063 K (1), 1083 K (2), 1113 K (3), 1163 K (4), and 1213 K (5).

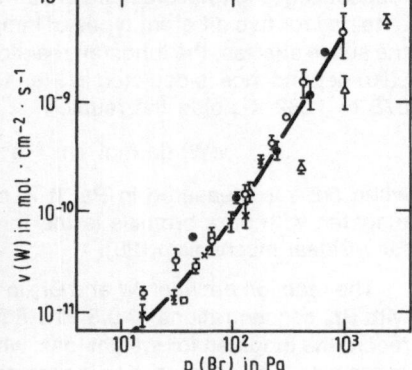

Fig. 13.   Tungsten reaction rate v(W) as a function of atomic bromine pressure p(Br) at 975 K (□), 1067 K (×), 1166 K (o), 1282 K (•), and 1448 K (△). The drawn curve obeys the equation given on p. 25 [20].

From the mass-spectrometric ion intensities, the following average values were obtained for the second law heat, free energy change, and entropy change of the reaction W(s) + $2Br_2(g) \rightarrow WBr_4(g)$ at 1317 K and at initial Br_2 pressures of 1×10^{-2} to 8.7×10^{-2} Torr: $\Delta H^{\circ}_{1317} = -36.7 \pm 1.7$ kcal/mol; $\Delta G^{\circ}_{1317} = -8.6 \pm 0.6$ kcal/mol; $\Delta S^{\circ}_{1317} = -21.3 \pm 1$ cal·mol^{-1}·K^{-1}. Under the experimental conditions prevailing in the experiments (see p. 24), the reaction equilibrium was established above 1270 K [15].

Influence of Additions and Impurities

A tungsten rod exposed for 4 d to a mixture of water and liquid bromine showed no apparent reaction and did not change its weight [22]. A weight loss of 2.9 mg·dm^{-2}·d^{-1} was found in liquid bromine saturated with H_2O after 170 h at 20°C [23, p. 249].

For the effect of addition of noble gases and nitrogen on the kinetics of the reaction between tungsten and gaseous bromine, see below. In the presence of O impurities or moisture, oxide chlorides, such as $WOBr_4$ or WO_2Br_2, are formed as in the case of the other halogens.

A high bromine-to-oxygen ratio in an N_2 carrier gas stream passing heated tungsten at 250 to 300°C resulted in the formation of large quantities of the purple-black $WOBr_4$ and only small amounts of WO_2Br_2. Low bromine-to-oxygen ratios yielded flaky orange WO_2Br_2 with some

$WOBr_4$. The main products can be easily separated from the side products by sublimation in the gas stream, the $WOBr_4$ being more volatile [3, 24]. Tungsten wool containing some WO_3 reacts with gaseous Br_2 at 400°C to give $WOBr_4$, which crystallizes in dark brown needles [17].

The formation of WO_2Br_2 at the expense of $WOBr_4$ is also promoted by increasing temperatures. A reddish brown wall deposit of WO_2Br_2 was observed downstream of the reaction zone when an N_2 stream containing 0.3 to 2.7 Torr Br_2 and 4.7×10^{-4} to 4.3×10^{-2} Torr O_2 passed over W strips heated to 600 to 950°C. At the highest temperatures and O_2 pressures, a less volatile yellow deposit assumed to be WO_3 was observed near the reaction zone. There was no evidence of $WOBr_4$ [25]. When repeating these experiments under essentially the same conditions, a yellow deposit of (presumably) WO_3 within the furnace zone, a yellow-brown deposit of WO_2Br_2 adjacent to the furnace, and red crystals of WO_2Br_2 were obtained further downstream. With He as the carrier, there was less crystalline WO_2Br_2 and considerably more WO_3 in the hot zone. Presumably, higher diffusion rates in He allow more $WO_2Br_2(g)$ to reach the wall of the reaction tube and undergo decomposition [26]. After gas-flow experiments with Ar as carrier gas (at atmospheric pressure) at 975 to 1448 K, the material that condensed in the cooler parts of the reaction tube showed W:Br ratios in the range of 1:2 to 1:2.8, according to chemical analyses, and apparently consisted of mixtures of WO_2Br_2 and WBr_4. The presence of WO_2Br_2 as a dominating species in the reaction product was demonstrated by X-ray diffraction. The Ar carrier gas was at atmospheric pressure, with bromine and oxygen at 3×10^{-3} to 3×10^3 and 6×10^{-2} to 30 Pa ($1\,Pa = 0.75 \times 10^{-2}$ Torr), respectively. Oxidation phenomena were found in cases where a relatively high $O_2:Br_2$ ratio was sustained during the experiments. These phenomena ranged from needle-like crystals in pits, via veils extending to some distance around the W wires studied, to surfaces fully blanketed with oxide nodules [20]. Formation of orange-colored WO_2Br_2 was also observed in flowing Br_2–N_2 mixtures at 1163 K, when small amounts of O_2 were present [9]. According to mass spectrometric studies, formation of gaseous WO_2Br_2 occurs when previously oxidized tungsten samples are exposed to Br_2 at 6.7 Pa in a Knudsen cell. Measurements of the intensity of the Br_2^+ ions leaving the cell indicate that the reaction starts at ~400 K and is fast enough at 600 K for total consumption of the bromine entering the reaction through a capillary from a constant pressure reservoir [18].

First systematic studies of the kinetics of the $W + Br_2 + O_2$ reaction were performed in a microbalance-flow system with an N_2 carrier gas as described above at 600 to 950°C. Initial pressures of Br_2 and O_2 were varied between 0.3 and 2.7 Torr and 10^{-4} and 10^{-2} Torr, respectively. The reaction was found to be zero-order with respect to $p(Br_2)$ and first-order with respect to oxygen. An empirical rate equation with an apparent activation energy of 31 kcal/mol (130 kJ/mol) was obtained and compared with similar rate equations for tungsten oxidation given in the literature for comparable oxygen pressures but at temperatures above 1000°C where the oxides are volatile. It was concluded that the rate-limiting step was tungsten oxidation followed by the rapid formation of volatile WO_2Br_2 which precluded the formation of an oxide surface layer. The favorable correlation with tungsten oxidation rate data at low total pressures suggested that the N_2 at atmospheric pressure did not affect the reaction kinetics [25]. This conclusion is not consistent with the known strong chemisorption of N_2 on W in this temperature region and motivated a continuation of this work [26]. Rate measurements at 700 to 900°C were carried out in Ar, He, and N_2 containing 0.01 to 3 Torr bromine and 10^{-3} to 10^{-2} Torr oxygen. The reaction rate of W was linear with oxygen, but was a complicated function of both the bromine pressure and the carrier gas. Only under certain conditions was zero-order dependence on $p(Br_2)$ observed. Except at the lowest Br_2 pressures, adsorbed bromine inhibited the reaction and the rate was higher in N_2 than in Ar or He where the rate was the same. These observations are explained by the two-layer adsorption model proposed by Schissel and Trulson [27] for tungsten oxidation. It is assumed that in the first layer oxygen is selectively adsorbed as atoms, while in the second layer bromine and oxygen are coadsorbed and at saturation compete for adsorption sites. Higher Br_2 pressures and lower temperatures promote bromine adsorption in the second layer at the expense of oxygen adsorption and thus reduce the overall reaction rate. The observed increase in rate with increased O_2 pressure indicates that oxygen can successfully compete with a large excess of bromine for adsorption sites. Even with a large $Br_2:O_2$ ratio, a change in O_2 pressure during a run resulted in an immediate corresponding change in reaction rate. Nitrogen possibly can displace adsorbed bromine or, when adsorbed, can be displaced by oxygen but not by bromine. In each case, the adsorbed oxygen concentration will be higher in the presence of N_2 carrier gas, which explains the higher reaction rates observed [26].

The simple reaction mechanism proposed in [25] with oxide formation as rate-determining step was questioned [18]. The activation energy of only 91 kJ/mol found with Br_2 containing 1% oxygen (see p. 25) is much below the 130 kJ/mol assumed for oxide formation. Furthermore, the appearance of ions from $WOBr_4$ and WBr_4 in addition to that from WO_2Br_2 in the mass spectra indicates clearly that the function of bromine is not restricted to removing the reaction product. The preliminary results obtained for 300 to 850 K show that the reaction kinetics in the $W+Br+O_2$ system is a rather complex function of the pressure of both gases and/or the surface characteristics of the metallic tungsten [18]. In a subsequent study by the same authors [19], the reaction rate in the presence of 10% O_2 was found to be lower than the reaction with pure bromine at the same temperature in the range of ~ 600 to 650 K at < 10 Pa. This was apparently in contradiction with other experimental evidence [20, 25], but could be explained by the dominant adsorption of bromine at the lower temperatures used in the experiment. Also, the reaction of tungsten with pure bromine at 600 K was an order of magnitude faster than the reaction with pure oxygen, while the reaction of bromine with an oxidized tungsten surface proceeded at significantly lower temperatures than the reaction with pure tungsten (see a figure in the paper which compares the rates of the $W+Br_2$, $W+(Br_2+10\% O_2)$, $W+O_2$, and $W_xO_y+Br_2$ reactions [19].

The ablation rate v_W of polycrystalline tungsten in bromine and oxygen at temperatures of 975 to 1448 K was measured in gas-flow experiments with Ar as the carrier gas at atmospheric pressure. The Br_2 and O_2 pressures ranged between 3×10^{-3} and 3×10^3 Torr and between 6×10^{-2} and 30 Pa, respectively. Data for 1166 K are presented in **Fig. 14**; additional plots for 1067 and 1282 K are given in the paper. The results show that the overall reaction in the

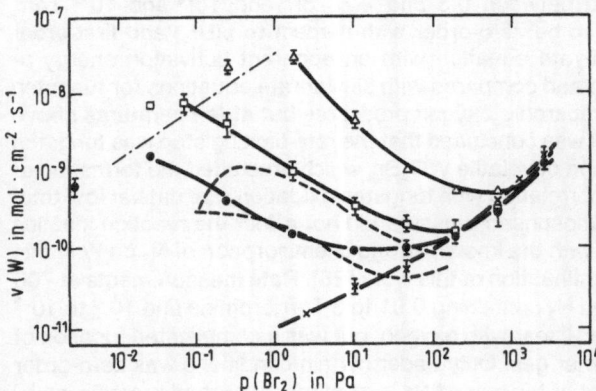

Fig. 14. Tungsten reaction rate v(W) at 1166 K as a function of partial Br_2 pressure at various O_2 pressures: $p(O_2) = 0$ (x), 0.25 (•), 1 (□) and 8 Pa (△). Left of the dash-dotted line oxide formation takes place [20]. The broken curves represent measurements at 1173 K in He [26]. The lower one gives results at an O_2 pressure of 0.27 Pa, the upper one at 1.15 Pa.

$W+Br_2+O_2$ system has to be described by at least two different types of reaction. The first one, called the W–Br reaction, is found in its pure form in the $W–Br_2$ system (see p. 25). The second reaction, called the W–O reaction, is distinguished by a rate which is more or less proportional to the oxygen partial pressure and, in contrast to the W–Br reaction, decreases with increasing bromine pressure:

$$v_w(W\text{–}O) = k(W\text{–}O) \cdot p(O_2)^f \cdot p(Br_2)^g.$$

The temperature-dependent parameters k(W–O), f, and g have the following values when v_w is in $mol \cdot cm^{-2} \cdot s^{-1}$ and $p(O_2)$ and $p(Br_2)$ are in Pa:

T in K	k(W–O)	f	g		T in K	k(W–O)	f	g
1067	9.8	1.55	−0.92		1282	23	1.05	−0.40
1166	12	1.20	−0.61					

It is proposed that the rate-limiting step in the W–Br in the W–O reactions is the formation of a tungsten-bromine surface complex by an Eley-Rideal mechanism, and the formation of a tungsten-oxygen surface complex by a Langmuir-Hinselwood mechanism, respectively [20].

Bombardment Effects

The desorption kinetics of bromine from a polycrystalline tungsten ribbon was studied at low coverages ($\Theta \leq 10^{-2}$ monolayer) and high temperatures (1800 to 2400 K), using a pulsed ionic beam method. A Br^+ beam impinged on the W surface at an incidence angle of 45°; the intensity of the negative ions leaving the surface was recorded. At low incidence energies, $E_i \leq 500$ eV, a first-order kinetics corresponding to the desorption of adsorbed Br from the surface was observed. In the range $500 < E_i < 2500$ eV, the desorption kinetics was more complex due to the implantation of an important fraction of the incident ions in subsurface regions (<100 Å) of the tungsten and their subsequent diffusion towards the surface [28]. The threshold corresponding to the appearance of complex kinetics was later located at 360 eV. An analysis of the kinetic results based on a one-dimensional diffusion model led to the conclusions that pure diffusion kinetics will occur when the mean adsorption lifetime of the Br is smaller than the mean diffusion time of the implanted atoms towards the surface and the diffusion of Br from the bulk of the surface is negligible [29].

References:

[1] Bachmann, W. T. (Mater. Des. Eng. **64** No. 6 [1966] 106/8).

[2] Agte, C.; Vacek, J. (Wolfram und Molybdän, Akad.-Verlag, Berlin 1959, pp. 214/34).

[3] Colton, R.; Tomkin, I. B. (Austral. J. Chem. **19** [1966] 759/63).

[4] Lugscheider, E.; Eck, R.; Ettmayer, P. (Radex-Rundsch. **1983** No. 1/2, pp. 52/84).

[5] Weiß, L. (Z. Anorg. Allg. Chem. **65** [1910] 279/340, 339/40).

[6] Byalobzheskii, A. V.; Tsirlin, M. S.; Krasilov, B. I. (Vysokotemperaturnaya Korroziya i Zashchita Sverkhtugoplavkikh Metallov [High-Temperature Corrosion and Protection of Refractory Metals], Atomizdat, Moscow 1977, pp. 1/224; C. A. **88** [1978] No. 179375).

[7] McInnis, M. B. (Tungsten Proc. 2nd Int. Tungsten Symp. 1982, pp. 41/51; C. A. **98** [1983] No. 200593).

[8] Emeléus, H. J.; Gutmann, V. (J. Chem. Soc. **1950** 2115/8).

[9] Goddard, V. W.; Pett, C. (J. Chem. Soc. Dalton Trans. **1975** 767/71).

[10] McCarley, R. E.; Brown, T. M. (Inorg. Chem. **3** [1964] 1232/6).

[11] Shchukarev, S. A.; Kokovin, G. A. (Zh. Neorg. Khim. **9** [1964] 1309/15; Russ. J. Inorg. Chem. [Engl. Transl.] **9** [1964] 715/8).

[12] v. Borch, J. B. (J. Prakt. Chem. **54** [1851] 254/60).

[13] Roscoe, H. E. (Justus Liebigs Ann. Chem. **162** [1872] 349/68).

[14] Schäfer, H.; v. Schnering, H.-G.; Tillack, J.; et al. (Z. Anorg. Allg. Chem. **353** [1967] 281/310).

[15] Popović, A.; Marsel, J.; Kaposi, O. (J. Inorg. Nucl. Chem. **41** [1979] 1289/94).

[16] Kaposi, O.; Popović, A.; Marsel, J. (Adv. Mass. Spectrom. **A 7** [1978] 631/5).

[17] Kaposi, O.; Popović, A.; Marsel, J. (J. Inorg. Nucl. Chem. **39** [1977] 1809/15).

[18] Marsel, J.; Popović, A.; Susić, R.; Kaposi, O. (Vestn. Slov. Kem. Drus. **28** [1981] 11/20; C. A. **94** [1981] No. 181463).

[19] Popović, A.; Susić, R.; Kaposi, O.; Marsel, J. (Int. J. Mass. Spectrom. Ion Phys. **47** [1983] 277/80).

[20] Rouweler, G.; de Maagt, B. J. (Philips J. Res. **33** [1978] 1/19).

[21] Rouweler, G. C. J.; de Maagt, B. J. (Philips J. Res. **36** [1981] 195/209).

[22] Theobald, L. S.; Stern, J. P. (Analyst. [London] **77** [1952] 99/101).

[23] Ishikawa, H.; Ishii, E.; Uehara, I.; Nakane, M. (Osaka Kogyo Gijutsu Shikensho Kiho **33** No. 3 [1982] 244/52; C. A. **98** [1983] No. 130286).

[24] Colton, R.; Tomkins, I. B. (Austral. J. Chem. **21** [1968] 1975/9).

[25] Zubler, E. G. (J. Phys. Chem. **74** [1970] 2479/84).

[26] Zubler, E. G. (J. Phys. Chem. **76** [1972] 320/2).

[27] Schissel, P. O.; Trulson, O. C. (J. Chem. Phys. **43** [1965] 737/43).

[28] Bolbach, G.; Blais, J. C.; Brunot, A.; Marilier, A. (Surf. Sci. **90** [1979] 65/77).

[29] Blais, J. C.; Bolbach, G.; Brunot, A. (Int. J. Mass. Spectrom. Ion Phys. **47** [1983] 269/73).

[30] Brewer, L.; Bromley, L. A.; Gilles, P. W.; Lofgren, N. L. (Natl. Nucl. Energy Ser. Div. IV B **19** [1950] 276/311).

2.4 Iodine

General. Products

Tungsten is stable against iodine vapor up to $\sim 500°C$ [1, p. 60], or up to 700°C [2]. Iodides are formed at red heat [3], [6, p. 366] or bright red heat [4, pp. 222/3]; see also [5].

In an early study, tungsten heated in I_2 vapor did not show severe attack; some visible sample changes were ascribed to oxidation by impurities [7]. This seems to be confirmed by more recent observations. Upon passing pure, dry I_2 vapor (0.285 Torr) over coiled W filaments at 530, 720, or 900°C, the only evidence of a reaction was a faint, gray film forming on the cool walls during the first few seconds. Thereafter, the film no longer grew or visibly changed, even when the iodine vapor was exposed to a high-frequency discharge. The tungsten filaments did not show any change. The same behavior of the W filaments was observed when they were exposed at room temperature to 700 Torr of O_2 prior to reaction with I_2. The quantity of the gray solid was in no case large enough for analysis, but the product was in all likelihood WO_2I_2. No evidence for the formation of a stable W iodide was obtained under the conditions of these experiments [8]. These findings are also in agreement with the results of mass spectrometric studies of tungsten-iodine reactions, which did not furnish any evidence for W iodide formation between 300 and 2200 K at I_2 pressures of 10^{-6} to 10^{-4} Torr [9]. Also W iodides did not form when oxygen was purposely excluded from reaction systems in which iodine of 1 to 500 Torr and tungsten were brought together in the 700 to 1000°C range [10]; see also [11].

Formation of a nonvolatile iodide was reported to occur at temperatures as low as 44°C [12]. Tungsten filaments of 5×10^{-3} cm diameter were exposed to I_2 vapor of low pressure ($\sim 10^{-2}$ Torr) in a glass bulb. Formation of the iodide was monitored by manometric and resistivity measurements. The iodide formed up to 1445 K and decomposed above 1500 K [12, pp. 398/401]. Formation of WI_2, that became gaseous above 250°C, and decomposed above 1400°C was assumed to explain the processes in I_2-containing (~1 μmol) incandescent lamps [13]; see also p. 103. According to [14], WI_2 may be prepared by the action of gaseous iodine on tungsten metal at about 1000 K. Thermodynamic calculations show that the maximum stability of gaseous WI_2 in the W–I_2 system is to be expected in the temperature region around 1400 K, at which it has a pressure of about 10^{-3} atm when the total pressure is 1 atm. The iodine pressures necessary to maintain a constant pressure of 10^{-5} atm of gaseous WI_2, or to just start formation of solid WI_2, are given in a log $(p(I) + p(I_2))$ vs. T plot for ~400 to 2000 K; the I_2 and I contents of the gas phase over W in this temperature range at 1 atm total pressure are given in another figure in the paper. Higher iodides (WI_3, WI_4, WI_5) are not stable in the solid state and do not represent important constituents of the gas phase [14]. The calculations [14] (and later ones [15] based on the same thermodynamic data), which suggest that WI_x compounds can be formed in O_2-free quartz-iodine lamps, are inconsistent with the above experimental results [8 to 10]. It was suggested that WI_x species are formed from the elements only at high iodine pressures (4 atm) and low temperatures (300°C). Neither of these conditions exists in quartz-iodine lamps [10]. According to [8], the disagreement among these findings is not surprising, since the validity of the thermodynamic data basis used [14] has been seriously questioned [16].

When W is locally heated by a laser beam in I_2 vapor, the metal is etched along the path of the beam by evaporation of volatile WI_2 [17].

When finely divided tungsten powder is heated with iodine in amounts corresponding to the stoichiometry of the diiodide in a sealed tube at a temperature gradient of 450°C (W) to 300°C (liquid iodine), a higher iodide is formed in the colder zone which decomposes to W_6I_{12} (= WI_2) at ~500 to 600°C [18, pp. 309/10].

WI_3 was obtained from the elements (excess of iodine) in a closed reaction tube heated in a temperature gradient of 500 to 350°C. The tungsten in the form of an O-free, flaky powder with a highly active surface was contained in the lower, hotter end of the reaction tube which had been brought into an inclined position; the iodine condensed in the upper, colder end and refluxed to the W. The WI_3 formed in the 350°C zone. The product yield after 3 d was 15%, related to the W input [19], [11, p. 19].

Atomic iodine did not attack the hot regions of W filaments, but reacted readily with the cooler regions to form a brown deposit on the walls of the glass reaction vessel; thus, a volatile iodide must have formed, which is different from that obtained with molecular I_2 (see above) [12, p. 402].

A list of W iodides and oxide iodides reported in the literature has been compiled [20, p. 46].

Kinetics and Thermodynamics of the Reactions

According to gravimetric measurements, the loss rate of W in I_2 vapor (~7400 Torr) increases from 0.2×10^{-3} mg·mm^{-2}·h^{-1} at 700°C to 8×10^{-3} mg·mm^{-2}·h^{-1} at 1000°C. The test time was 5 h in all experiments [2]. For the kinetics of formation of the nonvolatile iodide produced with molecular I_2, specifically at 810 K and 0.0254 Torr initial pressure, see [12].

The equilibrium W(s) + 2 I (g) ↔ WI_2 (g) was studied at ~900 to 1400 K. Tungsten powder or tiny wire spirals (99.85 or 99.98% purity) were reacted with a flow of purified Ar saturated with I_2 at 15, 17, or 20°C. The W powder was mixed with crushed quartz to avoid caking. The temperature dependence of the equilibrium constant could be described by

$$\log K = -(1.262 \pm 0.087) + (4215 \pm 100)/T \quad (940 \text{ K} < T < 1340 \text{ K}).$$

The changes of enthalpy and entropy for the reaction at 1100 K are:

$$\Delta H° = -19.3 \pm 0.5 \text{ kcal/mol}; \quad \Delta S° = -5.8 \pm 0.4 \text{ cal} \cdot \text{mol}^{-1} \cdot \text{K}^{-1},$$

[21]. For calculated vapor phase compositions, see p. 103.

Influence of Additions and Impurities

For the influence of hydrogen, introduced as HI, on the tungsten–iodine reactions, see "Tungsten" Suppl. Vol. A 7, 1987, pp. 79/80.

While oxygen adsorbed at room temperature does not noticeably influence the behavior of tungsten against iodine vapor (see p. 30), oxygen added to the I_2 in a flow system appreciably enhances the attack on W. An Ar–0.01% O_2 mixture was passed over solid iodine at 24°C (iodine vapor pressure 0.285 Torr), and the resulting Ar–O_2–I_2 mixture was in turn flowed over the hot tungsten coils at 760°C. The flow rate was approximately 10 L/h; the diameter of the quartz reaction tube is not given. Bright blue solid tungsten oxide collected on the tube wall just outside the furnace. Gray, shiny solid WO_2I_2 condensed further away from the oven, its quantity increasing with time. Inspection of the tungsten coils after a reaction period of 5 h revealed that they were coated with a dull black solid and that their cross-sectional area had decreased. The nature of the black solid was not identified. Experiments at 530 and 900°C produced essentially the same results. Gross changes in the results also did not occur when a high-frequency discharge was applied to the flowing gases (at 760°C). In all cases a relatively large quantity of sublimable, gray solid was formed when oxygen was present in the flowing gas. Formation of WO_2I_2, sometimes accompanied by some W oxide production, was also observed in quartz-iodine lamps containing 1 to 3 Torr O_2. The formation of the oxide iodide was particularly pronounced in a special lamp atmosphere of (in Torr) 700 Ar, 12 I_2, and 3 O_2. The filament temperatures in this case ranged between 2730°C (wall temperature 625°C) and 1340°C (wall temperature 225°C) [8]. The latter results confirm observations [22], which found that in presence of oxygen or oxygen donor compounds, tungsten reacts with iodine upon heating in a sealed quartz tube according to

$$W + O_2 + I_2 \rightarrow WO_2I_2.$$

With H_2O as the oxygen donor, the reaction is either

$$W + 2H_2O + 3I_2 \rightarrow WO_2I_2 + 4HI \text{ or}$$
$$W + 2H_2O + I_2 \rightarrow WO_2I_2 + 2H_2.$$

In the presence of H_2O, a tungsten transport with I_2 is noted to the hot end of a gradient 800 to 1000°C, which again is assumed to proceed via formation and decomposition of gaseous WO_2I_2 [10, 11].

Bombardment Effects and Diffusion

As in the case of bromine (see p. 29), implantation of iodine in subsurface regions is observed when the energy of I^+ ions impinging on a polycrystalline tungsten ribbon exceeds 500 eV. The incidence angle was again 45°, the target temperature 1800 to 2400 K [23]; see also [24].

Irradiation of tungsten single crystals with fission products of enriched uranium at ≤100°C to a dose of 1.2×10^{13} fragments per cm² caused implantation of ^{131}I and ^{103}Ru into the metal lattice. Twenty days after the irradiation, the distribution of the radioisotopes was determined after a 20 h diffusion annealing treatment at 2000°C and 4×10^{-5} Pa. In contrast to ^{103}Ru, the ^{131}I diffused against the concentration gradient, i.e., a process took place analogous to the segregation of iodine at the surface. This probably has to be attributed to an inverse Kirkendall effect, i.e., displacement of the ^{131}I atoms in a direction opposite to the flow of excess vacancies due to the presence of pores [25].

Tungsten disks (fluoride quality), one impregnated with ^{131}I, the other clean, were pressed together with 350 atm and heated for 3 h to 1873 K$<$T$<$2123 K. From the distribution of the ^{131}I atoms, the diffusion coefficient of I in W was calculated to be 4.25×10^{-1} cm^2/s and the activation energy of diffusion 49.6 kcal/mol [26, pp. 577/8, 584].

References:

[1] Lugscheider, E.; Eck, R.; Ettmayer, P. (Radex-Rundsch. **1983** No. 1/2, pp. 52/84).

[2] Abakumov, A. S.; Malyshev, M. L. (Zh. Neorg. Khim. **23** [1978] 2601/5; Russ. J. Inorg. Chem. [Engl. Trans.] **23** [1978] 1442/4).

[3] Bachmann, W. T. (Mater. Des. Eng. **64** No. 6 [1966] 106/8).

[4] Agte, C.; Vacek, J. (Wolfram und Molybdän, Akad.-Verlag, Berlin 1959, pp. 214/34).

[5] Riche, A. (Ann. Chim. Phys. [3] **50** [1857] 5/80, 14).

[6] Roscoe, H. E. (Justus Liebigs Ann. Chem. **162** [1872] 349/68).

[7] Weiß, L. (Z. Anorg. Allg. Chem. **65** [1910] 279/340, 339/40).

[8] McHale, J. J. (Illum. Eng. [New York] **66** [1971] 280/4, discussion pp. 284/6).

[9] McCarroll, B. (J. Chem. Phys. **47** [1967] 5077/82).

[10] Dettingmeijer, J. H.; Tillack, J.; Schäfer, H. (Z. Anorg. Allg. Chem. **369** [1969] 161/77).

[11] Schäfer, H.; Grofe, T.; Trenkel, M. (J. Solid State Chem. **8** [1973] 14/28).

[12] van Praagh, G.; Rideal, E. K. (Proc. R. Soc. [London] A **134** [1932] 385/404).

[13] Schilling, W. (ETZ Elektrotech. Z. Ausg. B **13** [1961] 485/7).

[14] Brewer, L.; Bromley, L. A.; Gilles, P. W.; Lofgren, N. L. (Natl. Nucl. Energy Ser. Div. IV B **19** [1950] 276/311).

[15] Kopelman, B.; van Wormer, K. A., Jr. (Illum. Eng. [New York] **63** [1968] 176/82).

[16] Shchukarev, S. A.; Kokovin, G. A. (Zh. Neorg. Khim. **9** [1964] 1309/15; Russ. J. Inorg. Chem. [Engl. Transl.] **9** [1964] 715/8).

[17] Solomon, R.; Müller, L. F.; Varian Associates (U.S. 3364087 [1964/68]; C.A. **68** [1968] No. 63746).

[18] Schäfer, H.; von Schneering, H.-G.; Tillack, J.; et al. (Z. Anorg. Allg. Chem. **353** [1967] 281/310).

[19] Schäfer, H.; Schulz, H.-G. (Z. Anorg. Allg. Chem. **516** [1984] 196/200).

[20] McInnis, M. B. (Tungsten Proc. 2nd Int. Tungsten Symposium, San Francisco 1982, pp. 41/51; C.A. **98** [1983] No. 200593).

[21] Bartovska, L.; Cerny, C.; Bartovsky, T. (Sb. Vys. Sk. Chem. Technol. Praze, Fys. Chem. N1 [1974] 7/44; C.A. **85** [1976] No. 10936).

[22] Tillack, J. (Z. Anorg. Allg. Chem. **357** [1968] 11/24).

[23] Bolbach, G.; Blais, J. C.; Brunot, A.; Marilier, A. (Surf. Sci. **90** [1979] 65/77).

[24] Blais, J. C.; Bolbach, G.; Brunot, A. (Int. J. Mass. Spectrom. Ion Phys. **47** [1983] 269/73).

[25] Bekmukhambetov, E. S.; Daukeev, D. K.; Zhotabaev, Zh. R.; Musurmankulov, R. T. (At. Energ. **53** No. 4 [1982] 265/6; Soc. At. Energy [Engl. Transl.] **53** [1982] 727/8).

[26] Yang, L.; Hudson, R. G. (EUR-4210 [1968] 575/88).

2.5 Transport Reactions in Tungsten-Halogen Lamps

2.5.1 General

The present chapter describes processes by which metallic tungsten is transported in a temperature gradient via the gas phase to another place in an apparatus and deposits there in elemental form. The transport is not accomplished by simple evaporation and condensation of the tungsten, but by chemical reactions in which volatile halides or oxide halides of W are formed at the original site and decompose again at the second site (or temperature) to elemental W and halogens or halogen compounds. The process occurs at temperatures at which the vapor pressure of elemental W is still negligible. Transport reactions of this type play an important role in purification processes for W based on the classical van Arkel-de Boer method and in various chemical vapor deposition processes for metallic W described in "Wolfram" Erg.-Bd. A 1, 1979, pp. 108/12, 191/205. These reactions are also of paramount importance for the performance of modern tungsten-halogen lamps. The following sections are mainly concerned with the treatment of such reactions. The general description of the features of tungsten-halogen lamps and the nature of the chemical processes occurring in them is based on review articles and summarizing introductions to the entire literature compiled in the reference list on pp. 45/6.

Features of Tungsten-Halogen Lamps

Tungsten-halogen lamps are compact light sources which utilize regenerative actions of halogens to prevent tungsten deposition on lamp walls as well as to sustain long life for the filament. The process involves the reaction of tungsten evaporated from the hot filament with halogen near the relatively cool bulb wall to form gaseous halides or oxides halides. These dissociate again at or near the hot filament returning lost W to the filament and releasing halogen for further regenerative reaction cycles. The possibility of returning evaporated tungsten from the lamp envelope to the filament by use of a regenerative cycle was demonstrated as early as 1915 [1]. However, the high bulb temperatures required for satisfactory operation of the cycles known at that time prevented its use in practical lamps until silica lamp envelopes became both technically and economically possible as a result of the development of the high-pressure mercury-vapor lamp [2]. It was not until 1959 that lamps based on a regenerative iodine cycle were introduced [3 to 5]. In 1966 these lamps were further improved by using bromine instead of iodine [6, 7, 32]. While iodine is primarily added in elemental form (in spite of the drawbacks described on p. 102), bromine is almost exclusively introduced in the form of alkylbromides or HBr, which, however, creates difficulties because of its aggressiveness. The use of chlorine and fluorine additions, which on theoretical grounds should yield optimum lamp performance, in general has not passed the experimental stage because of the chemical reactivity of these lighter halogens and their compounds. Chlorine, added in the form of suitable compounds, is a possible choice for lamps whose bulb wall temperatures are so low that the bromine and iodine cycles do not function (low wattage lamps). Protective layers for the inner bulb wall have been proposed, but their practical benefit seems to await final confirmation.

The protection of the bulb wall, i.e., the prevention of bulb blackening, by the regenerative cycle allows the use of much smaller bulbs, which because of their increased mechanical strength can be filled with inert gas at a higher pressure. The small volume of such bulbs also permits the use of heavier inert gases, even though these are more expensive. Both measures reduce the evaporation of tungsten and thus permit a higher filament temperature resulting in higher luminous efficacy and/or longer lamp lives than in conventional incandescent lamps [2, 8, 28, 29].

Incandescent tungsten filaments normally fail at localized hot spots. At such a point a slightly higher temperature increases the rate tungsten evaporation. As the filament cross-section decreases, the hot-spot temperature increases further until the filament finally melts. The halogen cycle prevents an overall weight loss of the filament. However, since the gaseous halogen compounds formed generally dissociate at temperatures below the normal filament temperatures, the tungsten is not deposited preferentially onto the hottest spots of the filament and no improvement in lamp life attributable to the regenerative cycle is observed. The development of halogen lamps therefore always aimed at obtaining a lamp atmosphere containing the thermally most stable tungsten halogen compounds to approach a true regeneration of the filament. Under some conditions, erosion of tails (filament ends, leads) and supports can lead to premature failure of the lamps, while the filament coil is still intact. This is the case when the pressure of the tungsten halogen compounds in the medium temperature range is too high or, alternatively, the halogen or usually added halogen compound is too aggressive. The halogen will attack any solid tungsten which is below the dissociation (inversion) temperature of the tungsten halogen compound(s) formed. The deleterious effect can be remedied by careful dosing of the halogens or usual halogen compounds of by judicious limitation of the amount of free (otherwise unbound) halogen by an appropriate composition of the filling gas. In an ideal lamp the total effective tungsten pressure, which is the sum of the partial pressures of the gaseous tungsten compounds, each multiplied by a weighting factor equal to the number of W atoms per molecule of the compounds, should be as high (or somewhat higher) at the envelope wall as at the filament coil, to prevent bulb blackening, but it also should be as constant as possible within the range of temperatures found in the lamp up to the operating temperature of the tungsten coil [2, 8].

Linear lamps (such as used for floodlighting applications), in order to attain a long life of >2000 h, predominantly work with iodine, the chemically least active halogen, show another life-shortening phenomenon if they are burned in a position deviating more than $\pm 4°$ from the horizontal. Due to thermal diffusion and convection, the heavy iodine molecules separate within the gas mixture (iodine + inert gas) of the filling causing a halogen deficiency at the upper end of the lamp. This leads to bulb blackening and, in extreme cases, subsequent bulb fracture. Combinations of xenon with bromine and krypton with chlorine were shown to offer a solution to this problem of halogen separation. Both the halogens are added in the form of the molecule $(PNX_2)_n$ where X is the appropriate halogen atom. This has the additional beneficial effect that by dissociation of the halophosphonitrile, sufficient (atomic) halogen is available near the filament to combine with evaporated tungsten, while the remaining PN radical picks up molecular bromine near the relatively cold supports of the filament, thus preventing bromine attack at these points [9, 29].

The detrimental or beneficial effect of various characteristic impurities or intentionally introduced additional elements on the halogen cycle is described in the following section. For features of tungsten-halogen lamps, see also [25 to 27, 31].

Tungsten-halogen lamps are used in the illumination of airfields, sports arenas, and buildings. They are also used in automobile headlights, slide and film projectors, special optical instruments, and a number of other lighting equipments operating at low voltages [10, 11].

General Characteristics of the Transport Processes in Tungsten-Halogen Lamps

The tungsten-halogen cycle is a complex transport process with many variables, such as temperature, temperature distribution, gas composition, pressure, gas velocity distribution, system-geometry, etc., which can be analyzed in terms of thermodynamics, gas phase transport, or rate processes [12]. The chemical system involved is very complicated. There are always impurities such as oxygen, hydrogen, and carbon, which react to form many gaseous compounds. Seldom is a single compound or reaction responsible for the chemical transport phenomena.

Thermochemical Analyses

Understanding of the regenerative lamp chemistry has been approached primarily from thermodynamic considerations. Computer-aided calculation procedures were used which were based on the free-energy minimization of the system. In doing so, it has mostly been assumed that the reaction system was heterogeneous with respect to solid tungsten. Other basic assumptions are: (I) the chemical reactions are reversible, (II) thermodynamic equilibrium is achieved at the gas/solid interface, and (III) the reactions are not controlled by reaction kinetics and are fast compared with the diffusion processes [12]. The composition of the gas phase can then be calculated as a function of temperature and pressure, provided the nature of the gas phase components and their thermodynamic characteristics are known. Often no reliable thermodynamic data were or are available. In such cases there is no option but to use estimated values and then to verify and improve the estimates by matching them with equilibrium compositions found experimentally [8]. These thermochemical analyses have provided important insights into the conditions of favorable tungsten transport, the tungsten transporting species, and the role of common lamp impurities such as hydrogen, oxygen, and carbon [13].

Equilibrium gas phase compositions obtained for the binary $W(s)-X_2$ ($X = F$, Cl, Br) systems at initial halogen pressures $p^0 (X_2)$ of 10^{-3} atm are compared in **Fig. 15**. Compound formation with iodine at the operating temperatures of halogen lamps and iodine pressures of < 0.1 atm

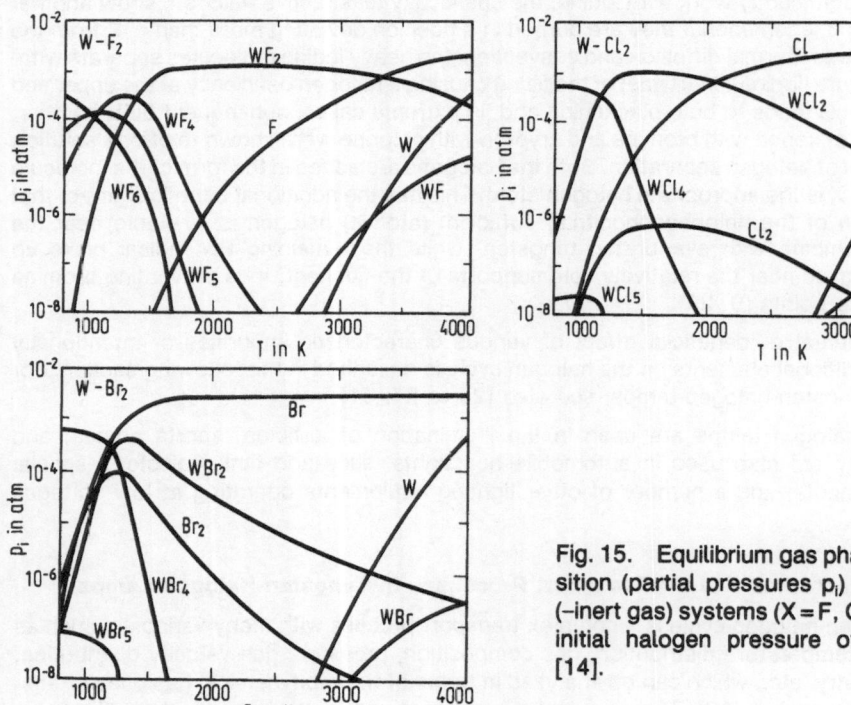

Fig. 15. Equilibrium gas phase composition (partial pressures p_i) of $W(s)-X_2$ (–inert gas) systems ($X = F$, Cl, Br) at an initial halogen pressure of 10^{-3} atm [14].

occurs only in the presence of oxygen (see p. 104). An appreciable formation of higher halides WX_6 and WX_5 is expected only in the $W-F_2$ system. The endothermal monohalides WX will appear only at the highest temperatures and even then only in minor amounts. The main W-containing components of the gas phase and those, responsible for the transport processes are the tetra- and dihalides, WX_4 and WX_2. At low temperatures W is transported from hot to cold sites by the reaction

$$2\,WX_2 \leftrightarrow WX_4 + W,$$

and in the W–F system additionally by the reactions

$$5\,WF_4 \leftrightarrow 4\,WF_5 + W \text{ and}$$
$$6\,WF_4 \leftrightarrow 4\,WF_6 + 2\,W.$$

In the range of medium temperatures, W transport from hot to cold proceeds via decomposition of the dihalides according to

$$WX_2 \leftrightarrow W + 2\,X.$$

At high temperatures, W is transported from hot to cold by simple physical evaporation and condensation [14].

If a single gaseous W compound is considered, tungsten will be transferred to the site (i.e., temperature) where the equilibrium pressure of the compound is lowest. The difference between the vapor pressure of the W compound at two different sites is a measure of the transport rate. The thermodynamics of the transport reactions therefore predict both direction and magnitude of the transport [15]. In the complex chemical system of the halogen lamp atmosphere, it is more convenient to treat the W-containing components of the gas (free tungsten and compounds) not separately but collectively. This is done by taking the sum of the partial pressures, each multiplied by a weighting factor equal to the number of W atoms per molecule of the compound in question. The summed pressure is called the total effective or global tungsten pressure and in the following is expressed by $\Sigma p(W)$ [8, 16]:

$$\Sigma p(W) = p(W) + p(WO_m) + p(WX_4) + p(WX_2) + x\,p(W_xO_yX_z) + \ldots$$

It is also termed the mass balance of tungsten in the gas phase. The mass balance of tungsten is often related to the initial concentration or the mass balance of the transporting agent (halogen and/or oxygen) by $W/(X_2(+O_2)) = \Sigma p(W)/(\Sigma p(X_2)(+\Sigma p(O_2)))$, the so-called mass balance ratio or "solubility" of tungsten in the gas phase [14]. The thermodynamic force driving the transport process in a certain (desired or undesired) direction is then the difference of the $\Sigma p(W)$ values between two considered sites or temperatures, e.g., W coil and bulb wall. W transport will occur from a site with a high value of $\Sigma p(W)$ to a site where this quantity is lower. Because of the high pressure of the inert gas, the mass transport takes place by diffusion [8, 16]. The temperature dependence of the mass balance ratios in the $W(s)-X_2$ systems calculated for different initial halogen pressures is shown in **Fig. 16**. The qualitative trend of the curves, in most cases showing maxima at lower temperatures and minima at higher temperatures, is similar for the individual $W-X_2$ systems. Chemical transport of W is possible over wide ranges of temperature, which, however, differ for the various X_2 and $p^0(X_2)$, and depend on the position of the maxima and minima. Depending on the slope of the W/X_2 vs. T curves, the direction of the transport is from hot to cold (positive slope) or from cold to hot (negative slope). In the region of the maxima, an attack on metal parts will occur, while in the regions around the minima, deposition of metallic tungsten from the gas phase has to be expected. Such effects can be observed as removal of W from the cold ends of the leads and growth of W needles and dendrites on supports and certain turns of the W coil in halogen lamps [14].

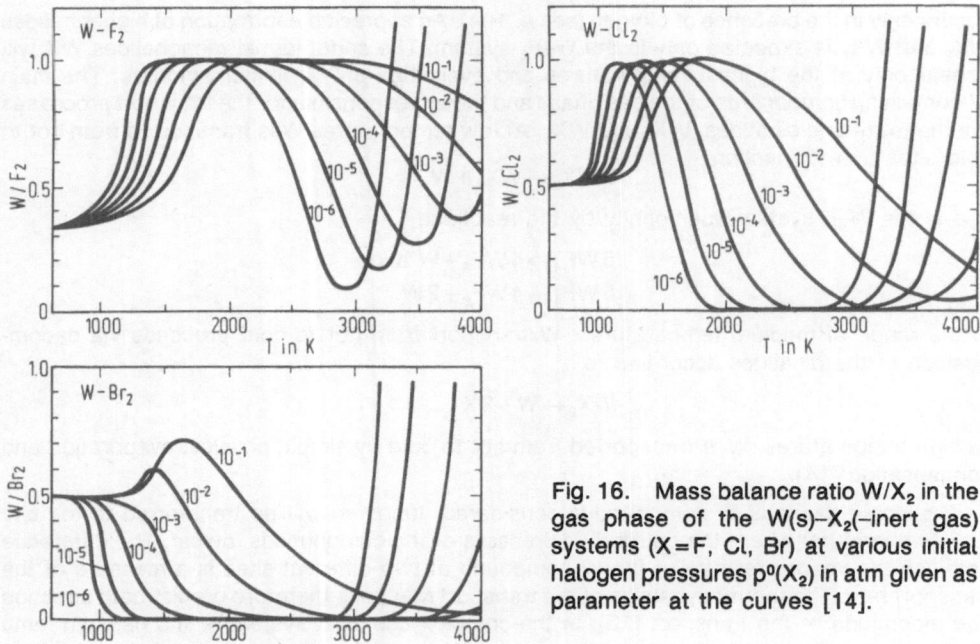

Fig. 16.   Mass balance ratio W/X_2 in the gas phase of the $W(s)–X_2(–inert\ gas)$ systems ($X = F$, Cl, Br) at various initial halogen pressures $p^0(X_2)$ in atm given as parameter at the curves [14].

In order to solve the problem of corrosion at colder tungsten parts in W–Br lamps, the use of hydrogen bromide instead of elementary bromine has been proposed. HBr is a stable compound and dissociates appreciably only near the filament. Thus, the pressure of free bromine in the colder regions of the lamp is low and this causes a decrease of $\Sigma p(W)$ or W/Br_2 in the range between 1400 and 2200 K; the attack on colder parts like filament tails is therefore much less [15, 17]. Exact dosing of the HBr (accomplished by use of, e.g., alkyl bromides), however, is required to avoid undesired side effects. The interference of hydrogen in the $W–X_2$ systems is very pronounced, as can be seen from **Fig. 17** and **Fig. 18** in which the temperature dependence of gas phase composition and mass balance ratio, respectively, are shown for $H_2 : Br_2 = 1:1$ and $p^0(X_2) = 10^{-3}$ atm or various other initial X_2 pressures. Halide formation at low and medium temperatures is drastically reduced, which leads to appreciably lower W/X_2 values in this region. The regimes of W transport from cold to hot are severely decreased; in some temperature ranges, there is even a reversal in the direction of the transport. The presence of hydrogen should, therefore, result in a strong deterioration or even complete suppression of the W transport from the wall back to the coil. This deleterious effect increases with an increasing proportion of hydrogen. Due to the reversed direction of transport, the material transport from hot to cold, on the other hand, is significantly enhanced. At relatively low hydrogen concentrations, only the needle-like or dendritic growth of W crystals at medium temperatures is intensified, while at higher hydrogen contents blackening of the bulbs is liable to occur. Due to the permanent presence of traces of oxygen and H_2O in commercial lamps, the mass balance of W at wall temperatures is raised by formation of W oxide halides to a level beyond that at coil temperatures, such that the problem of bulb blackening is overcome in most cases [14].

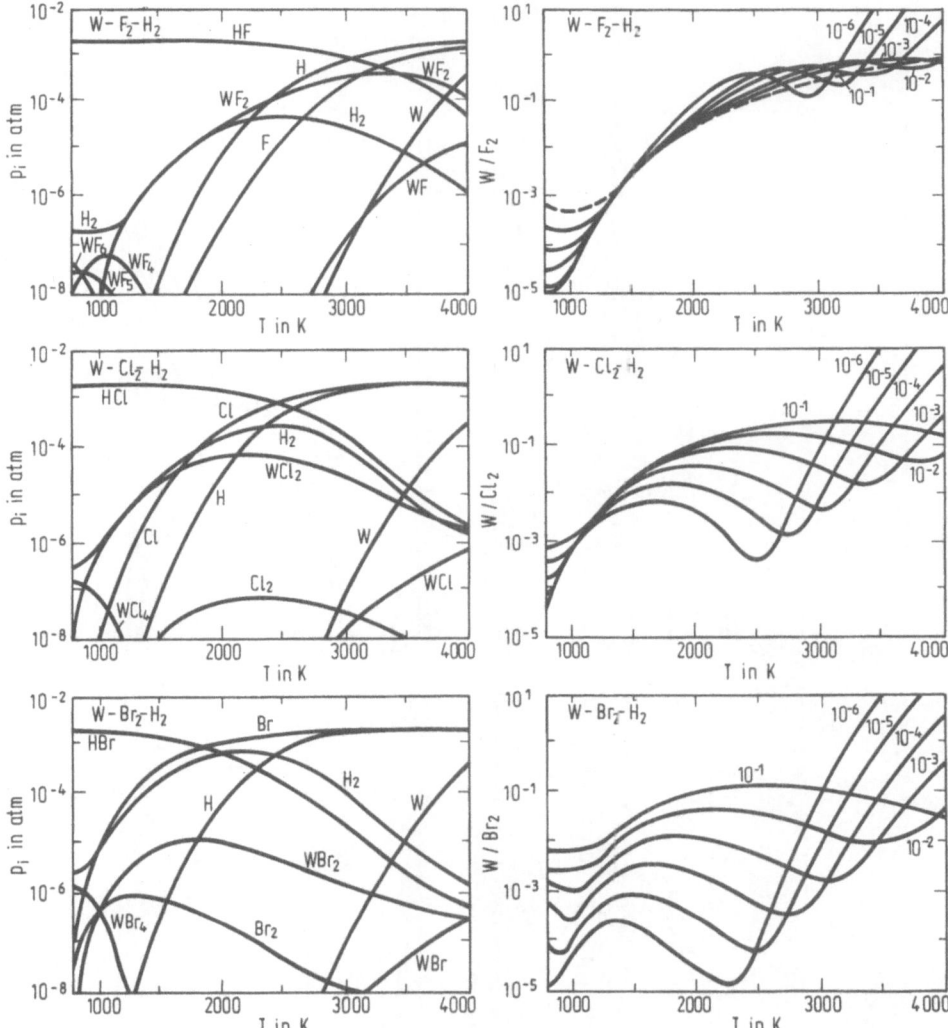

Fig. 17 (left). Equilibrium gas phase composition (partial pressures p_i) of the W(s)–X_2–H_2(–inert gas) systems (X = F, Cl, Br) for X_2:H_2=1:1 and initial halogen pressures $p^0(X_2)$=10^{-3} atm [14].

Fig. 18 (right). Mass balance ratio W/X_2 in the gas phase of the W(s)–X_2–H_2(–inert gas) systems at X_2:H_2=1:1 and various initial halogen pressures $p^0(X_2)$ in atm given as parameter at the curves [14].

Oxygen generally has an activating influence on the halogen cycle in incandescent lamps. By the formation of a large variety of oxygen compounds the mass balance of W is markedly increased at low and medium temperatures, and the transport processes are thus enhanced. The temperature dependence of the gas phase compositions in W–X_2–O_2 systems is depicted in **Fig. 19**, p. 40, for a halogen input pressure of 10^{-3} atm and a halogen to oxygen ratio of 1:1.

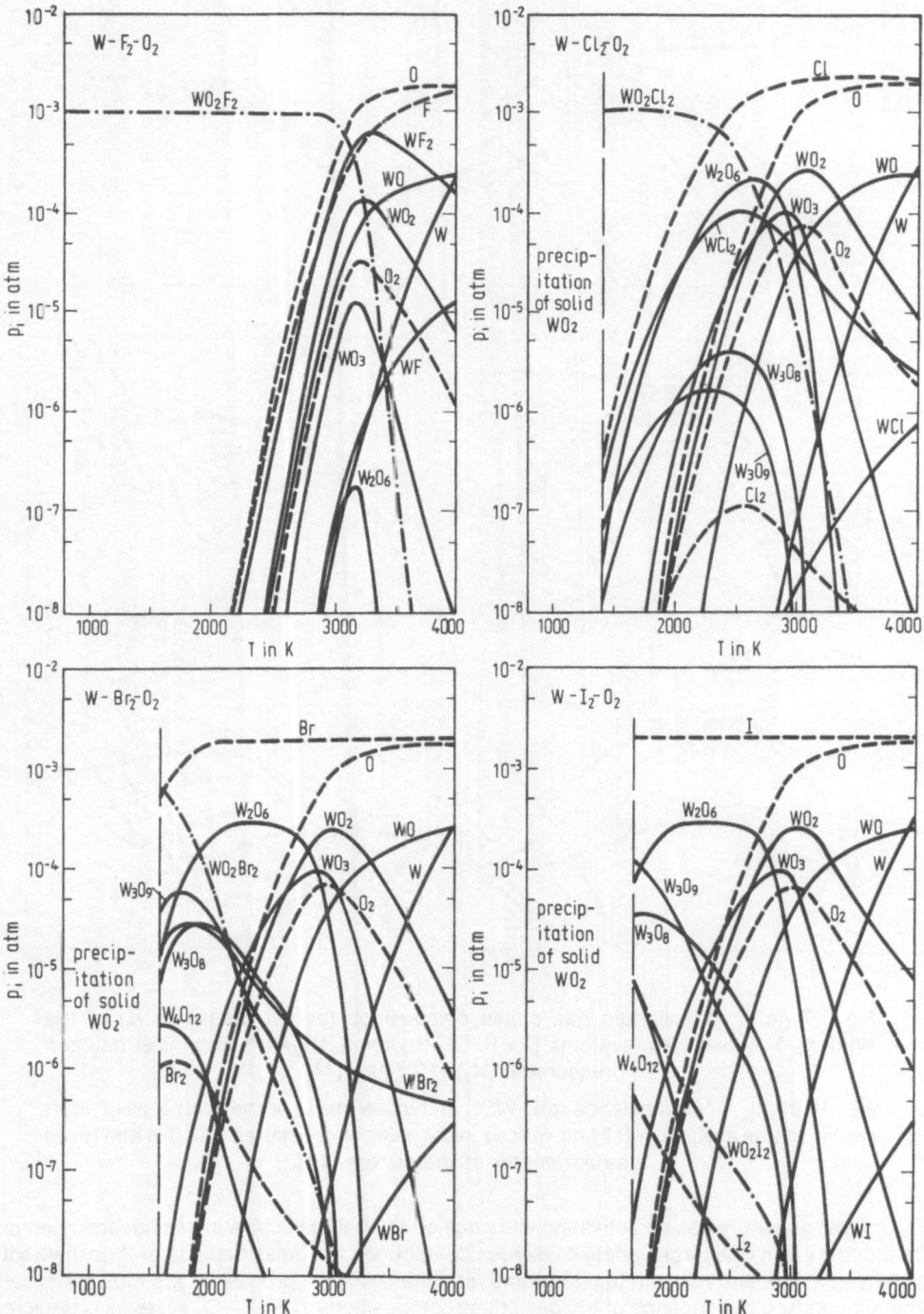

Fig. 19 (opposite page).
Equilibrium gas phase composition (partial pressures p_i)
of the W(s)–X_2–H_2(–inert gas) systems (X = F, Cl, Br, I) at
$X_2:O_2$ = 1:1 and initial halogen pressure $p^0(X_2)$ = 10^{-3} atm [14].

Most marked feature of these systems is the formation of mixed tungsten–oxygen–halogen compounds WO_2X_2 as the main components in the gas phase over a wide range of temperatures and the formation of tungstic oxides at higher temperatures. Furthermore, the precipitation of solid tungstic oxide is characteristic at temperatures around 1500 K, depending on the oxygen concentration. The thermal stability of W oxide halides decreases in the sequence

$$WO_2F_2 > WO_2Cl_2 > WO_2Br_2 > WO_2I_2,$$

and thus the extent to which the W–X_2 systems are influenced by the presence of oxygen. Although WO_2I_2 is the least stable of these compounds, its formation in the W–I_2 system is the precondition for a regenerative halogen cycle. With increasing X_2 or O_2 proportions, the gas phase compositions approach those in the binary W–X_2 and W–O_2 systems, respectively. At higher temperatures, depending on the thermal stability of the oxide halides, the phase relations of the W–X_2 and W–O_2 systems are established without mutual interference. The presence of oxygen, therefore, does not change the direction of the chemical transport processes; there is only an addition of the transport reactions. An appreciable influence of oxygen is, however, to be expected at medium temperatures, particularly in the W–F_2 and W–Cl_2 systems, in which a maximum of the W/($X_2 + O_2$) vs. T curves disappears with an increasing oxygen proportion. The corrosion of tungsten parts in this range is reduced and the region of W transport from cold to hot is extended to higher temperatures. The W transport in the direction from hot to cold is diminished and, finally, completely suppressed. At very high oxygens concentrations there will be a decrease in the solubility of tungsten in the vapor phase due to the precipitation of solid tungstic oxide. **Fig. 20**, p. 42, shows the temperature dependence of the mass balance ratios W/($X_2 + O_2$) for various $X_2:O_2$ ratios at an initial halogen pressure $p^0(X_2)$ of 10^{-4} atm and for $X_2:O_2$ = 1:1 and various $p^0(X_2)$. The dependence of the precipitation of solid WO_2 on the O_2 content is depicted in **Fig. 21**, p. 42 [14, 18].

In systems containing both hydrogen and oxygen, the gas phase equilibria are controlled by the competition between the formation of oxide halides on the one hand and hydrogen halides and water vapor on the other, the balance between these reactions depending on the temperature and the halogen present. At equimolar proportions of X_2 and O_2 and lower temperatures, oxygen is exclusively bound as WO_2F_2 in the case of fluorine, while in the case of chlorine, and even more so in the case of bromine, the formation of H_2O prevails; the formation of WO_2I_2 finally plays only a minor role. Hydrogen is mainly bound as HX. An increasing proportion of oxygen enhances the W transport. This is demonstrated by **Fig. 22** representing the temperature dependence of the mass balance ratios W/($X_2 + O_2$) at different $X_2:H_2:O_2$ ratios. At temperatures below ~1500 K, solid WO_2 is precipitated in these systems (except W–F_2–O_2–H_2) if the proportions of the halogen and oxygen are comparable. The vapor phase is thus deprived of oxygen, and the transport processes occurring become similar to those in the oxygen-free systems. When hydrogen and oxygen are added in the form of H_2O, the influence of oxygen prevails over that of hydrogen [14].

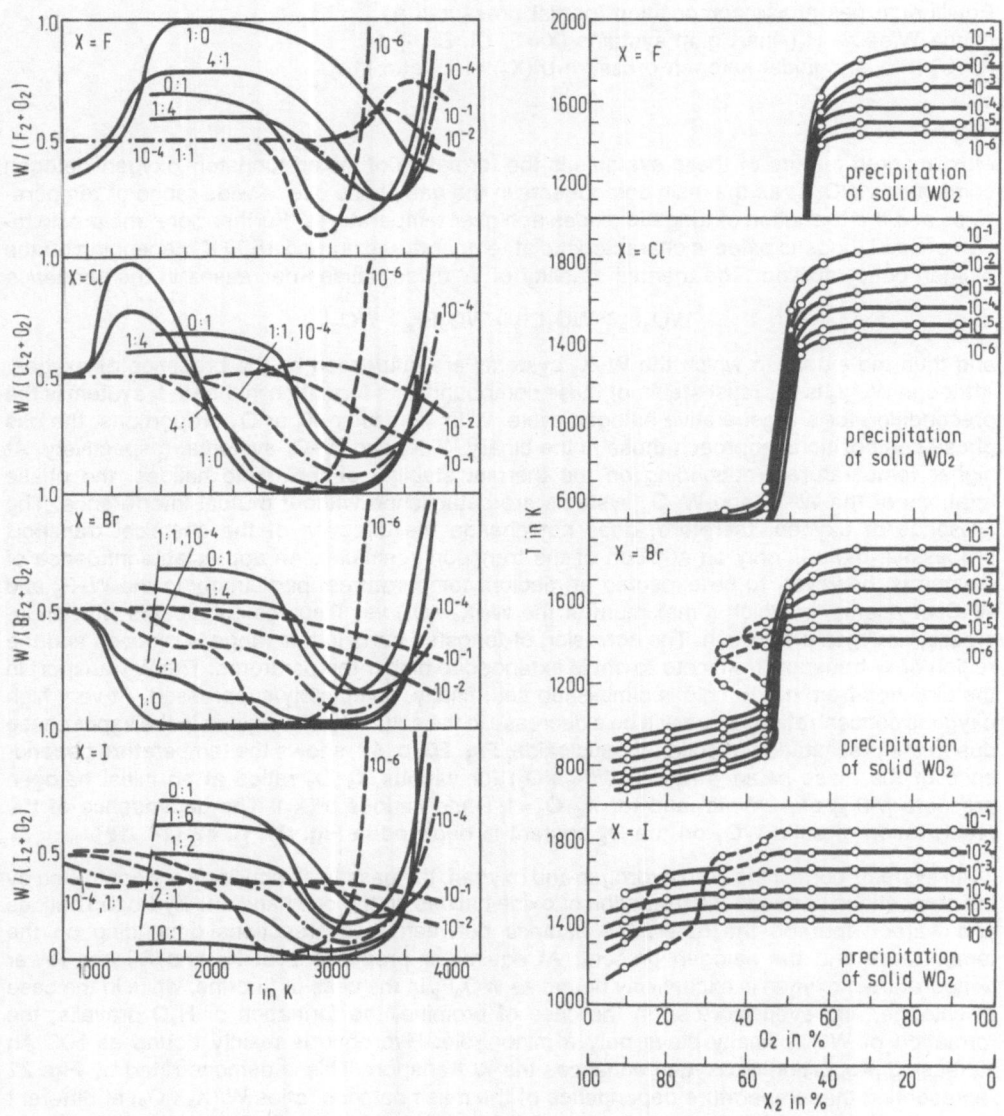

Fig. 20 (left).   Mass balance ratios $W/(X_2 + O_2)$ in the gas phase of the $W(s)–X_2–O_2$
(–inert gas) systems ($X = F$, Cl, Br, I) at $p^0(X_2) = 10^{-4}$ atm and various ratios $X_2 : O_2$ (solid
curves) or $X_2 : O_2 = 1:1$ and various $p^0(X_2)$ (in atm) (dashed curves). The dash-pointed
curves represent the temperature dependence for $X_2 : O_2 = 1:1$ and $p^0(X_2) = 10^{-4}$ atm [18].

Fig. 21 (right).   Dependence of the precipitation of solid WO_2 on the O_2 concentration
in $W(s)–X_2–O_2$(–inert gas) systems ($X = F$, Cl, Br, I). The initial halogen pressures $p^0(X_2)$
in atm are given as parameter at the curves [18].

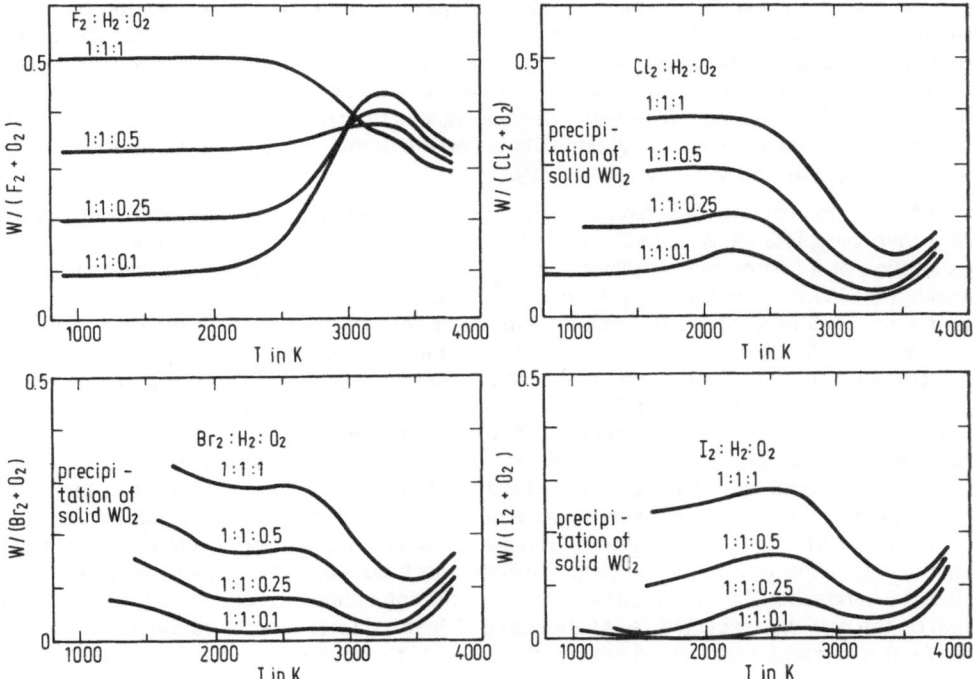

Fig. 22. Influence of oxygen on the temperature dependence of the mass balance ratios $W/(X_2 + O_2)$ in the gas phase of the $W-X_2-O_2-H_2(-inert\ gas)$ systems ($X = F$, Cl, Br, I) at $X_2:H_2 = 1:1$ (initial X_2 pressure 10^{-3} atm) [14].

The presence of carbon does not noticeably affect the gas phase composition and the W transport in oxygen-free systems. Carbon-containing compounds and radicals such as C_2H_2, CH, CH_2, CF, CF_2, etc., may be formed at medium and higher temperatures, but their concentrations are too low to influence the main phase relations. At temperatures below 3000 K deposition of solid C is observed, which can lead to the formation of W carbide resulting in an undesired embrittlement of the W filament. In contrast, carbon has a marked effect on systems containing oxygen. At low temperatures formation of CO_2, and at high temperatures formation of CO is found. The stability of CO is such that up to its stoichiometric composition all oxygen in extracted from the reaction systems at higher temperatures. Thus, carbon is an excellent getter for oxygen in halogen incandescent lamps. At lower temperatures, a competition between the formation of CO_2 and W oxide halides arises; deposition of solid C is possible. Added CO acts as an oxygen source in this temperature range and this together with its inertness at higher temperatures makes it an useful oxygen additive, which does not require meticulous dosing. It has been experimentally shown that for a good performance of any incandescent lamp, the use of CO is preferable to that of elementary oxygen, since an excess of the latter would enhance the transport of material in the high temperature region and lead to a shortening of lamp life due to W oxide formation, whereas an excess of CO results in no alteration of the transport behavior. The mass balance of W in the gas phase of carbon-containing $W-X_2-O_2$ systems (CO, CO_2 precursors) in this temperature range is determined by the oxygen concentration exceeding the stoichiometric 1:1 ratio. At lower temperatures (<1000 K), oxygen in any amount leads to an increase of the mass balance due to the formation of

W oxide halides. In systems containing water, oxygen is gettered by carbon and, therefore, the transport characteristics produced by hydrogen interference are found [12, 14]. The importance of CO added to the $W-Cl_2-O_2$ and $W-Br_2-O_2$ systems as an efficient getter for excess oxygen in the colder parts of corresponding lamps is also pointed out by [30]. The gettering action of the CO is assumed to prevent the oxide halides from decomposing into solid WO_3 and halogen. In the case of $W-I_2-O_2$ systems, however, the dominating species is reduced if I_2 and CO are present in equimolar amounts [30].

According to [16], the reasoning on which the usual thermodynamic approach employed in the above calculations is based has two serious pitfalls. The principal one is that the assumption of a homogeneous distribution of halogen, oxygen, and other chemically active species (not W) throughout the gas phase is by no means evident in a system as far from thermal equilibrium as an incandescent lamp with a temperature gradient of some 10^3 K/cm and a nonnegligible thermal convection. The other pitfall is more practical; if there are several solid components transported through the lamp (in particular, tungsten and carbon), then there appears to be no straightforward method to calculate the corresponding thermodynamical forces. In order to avoid such difficulties, a different calculation scheme was worked out which simulates more closely the actual situation in a well-working halogen incandescent lamp in steady state operation. It is presumed that neither tungsten nor carbon is transported in the radial direction. The approximations of a common diffusion coefficient for all gaseous components and local chemical equilibrium are used. Non-blackening is expressed by an inequality forbidding deposition of tungsten onto the bulb wall. The amount of carbon in a halogen incandescent lamp is usually assumed to be high enough to cause some carbon deposition on the bulb wall, acting as a reservoir of C. Chemical attack or growth on a given part of the W coil is characterized by the second derivative of the total effective pressure of W along the coil. For results obtained by applying the calculation scheme to $W-Br-O-C-H$ containing systems, see p. 98 and the paper [16]. The importance of an appropriate consideration of the condensed phases in thermodynamic calculations to avoid misleading conclusions has been pointed out [19, 20] and is demonstrated, also for $W-Br-O-C-H$ containing systems described on pp. 97/8.

The thermodynamic approach allowed the understanding and the prediction of transport phenomena in halogen incandescent lamps in many cases. Sometimes, however, it was found to be inadequate in explaining the observed lamp phenomena. The apparent failure in these cases may be due to a number of reasons, which are found in part in the invalidity of the basic assumptions as well as in the interference of complicating factors which cannot completely be taken into account in the simple models used. Gupta [19, 20] has summarized such reasons as follows: incomplete and/or incorrect thermodynamic treatment of the lamp system by exclusion of important yet unknown species and inclusion of nonexistent phases, lack of well-defined data, lamp geometry effects, filament impurities, and, perhaps most important, the questionable assumption of local thermal equilibrium, especially near the bulb walls. Existence of chemical equilibria implies that at the filament and wall, reaction rates are fast enough to establish equilibria with migrated gaseous species. Flow of species occurs via radial and thermal diffusion and convection. Extent of each mode of flow will depend on lamp geometry, temperature profiles, and pressures. At the filament temperatures of ~3000 K, reaction rates are likely to be fast enough to attain instant equilibrium; however, the same could not be assumed a priori at lamp wall temperatures which are in the range of ~1000 K. A disturbance of the thermal equilibrium at the bulb wall is, in particular, to be expected for systems in which hydrogen atoms and carbon monoxide appear as gas phase components, because H has an extreme mobility and CO an extraordinary kinetic stability [19, 20]. Emphasis was also placed on the point that the tungsten-halogen cycle in a strict sense does not represent an equilibrium system, but is a nonequilibrium system in a quasi-stationary state. It depends on the establishment of two different equilibria, which can very strongly influence each other: (I) establishment of the chemical reaction equilibrium in the direction of a state with local thermodynamic equilibrium and (II) establishment of the transport equilibrium in the direction of a state of minimum energy for the system as a whole, i.e., in this case a heterogeneous transport equilibrium in a state with a minimum value of the mass balance of tungsten [14].

For thermodynamic analyses of the transport processes in tungsten-halogen lamps, see also [21 to 24].

Kinetic Analyses

In view of the above limitations of thermodynamic arguments, a modified lamp model based primarily on kinetic considerations has been proposed which is based on rates of various processes and mass flow within the lamp. The migration of W atoms was calculated assuming a simple diffusion model. The blackening criterion under this approach assumes the form

$$j(W) > [r]_b.$$

Here, $j(W)$ is defined as the W atom flux reaching the wall; it depends on the filament temperature and gas phase reaction rates. $[r]_b$, the wall reaction rate of the W atoms, is controlled by wall temperature and reactant concentrations. The importance of gas phase and wall reaction rates is evident from these calculations, although these are also limited by the lack of appropriate kinetic data on wall reactions. Limitations to the simple model used are also imposed by the fact that convective and thermal diffusion flows are also prevalent in an operating lamp. However, the general conclusion can be drawn that thermodynamic calculations, if performed correctly, can give valuable insight into transport processes occurring near the filament, while kinetic considerations are far more important in understanding the chemistry near the bulb wall [19, 20]. In an earlier kinetic analysis of the transport processes in halogen-doped incandescent lamps, a mass transfer model was developed which is based on diffusion and counter-diffusion of gaseous species across the stagnant boundary layer surrounding the W filament with simultaneous reactions at the bulb wall and the filament. The model is applied to lamps with special geometry in which the initial gas phase consists of argon with small concentrations of oxygen and bromine. The tungsten flux from wall to filament and vice versa and, thus, the conditions of wall blackening are calculated as a function of the initial bromine and oxygen content, filament and wall temperatures, total pressure, and wall condensate, i.e., $W(s)$, $WO_2(s)$, and $WO_3(s)$ [13]. For an experimental study of the effect of the chemical reaction rates on the processes in W–Br lamps, see p. 99.

References:

[1] Langmuir, I. (J. Am. Chem. Soc. 37 [1915] 1139/67).

[2] Price, D. H. (J. Sci. Technol. [London] 39 [1972] 125/30).

[3] Zubler, E. G.; Mosby, F. A. (Illum. Eng. [N. Y.] 54 [1959] 734/40).

[4] Zubler, E. G.; Mosby, F. A. (U.S. 3160454 [1964] from [25]).

[5] Moore, J. A.; Jolly, C. M. (GEC J. Sci. Technol. 29 No. 2 [1962] 99/106).

[6] T'Jampens, G. (RGE Rev. Gen. Electr. 75 [1966] 990/6; Bull. Soc. R. Belge Electr. 83 No. 3 [1967] 274/56).

[7] T'Jampens, G. R.; van de Weijer, M. H. A. (Philips Tech. Rev. 27 [1966] 173/9; Philips Tech. Rundsch. 27 [1966] 165/71).

[8] Dettingmeijer, J. H.; Dittmer, G.; Klopfer, A.; et al. (Philips Tech. Rev. 35 [1975] 302/6; Philips Tech. Rundsch. 35 [1975] 324/7).

[9] Coaton, J. R.; Phillips, N. J. (Proc. Inst. Electr. Eng. 118 [1971] 871/4).

[10] Kirk-Othmer Encycl. Chem. Technol. 3rd Ed. 20 [1982] 811/2.

[11] Römpp Chemie Lexikon 9th Ed. **3** [1990] 1718.

[12] Neumann, G. M. (Thermochim. Acta **8** [1974] 369/79).

[13] Harvey, F. J. (Metall. Trans. A **7** [1976] 1167/76).

[14] Neumann, G. M. (Tech. Wiss. Abh. OSRAM-Ges. **11** [1973] 8/41).

[15] Ullmann's Encykl. Tech. Chem. 4th Ed. **2** [1972] 382.

[16] Geszti, T.; Vicsek, T. (J. Phys. D **9** [1976] 903/12).

[17] Dettingmeijer, J. H.; Meinders, B.; Nijland, L. M. (J. Less-Common Met. **35** [1974] 159/69).

[18] Neumann, G. M. (Z. Metallk. **64** [1973] 26/32).

[19] Gupta, S. K. (J. Electrochem. Soc. **125** [1978] 2064/70).

[20] Gupta, S. K. (Proc. Electrochem. Soc. **78**-1 [1978] 20/42; C.A. **89** [1978] No. 68444).

[21] Kopelman, B.; van Wormer, K. A. (Illum. Eng. [N.Y.] **63** [1968] 176/82).

[22] Kopelman, B.; van Wormer, K. A. (Illum. Eng. [N.Y.] **64** [1969] 230/5).

[23] Yannopoulos, L. N.; Pebler, A. (J. Appl. Phys. **42** [1971] 858/62).

[24] Yannopoulos, L. N.; Pebler, A. (J. Appl. Phys. **43** [1972] 2435/9).

[25] Rabenau, A. (Angew. Chem. **79** [1967] 43/9; Angew. Chem. Int. Ed. Engl. **6** [1967] 68/73).

[26] Schilling, W. (ETZ Elektrotech. Z. B **13** [1961] 485/7).

[27] Schilling, W. (Svetotekhnika **20** No. 12 [1968] 139A/142A).

[28] Strange, J. W.; Stewart, J. (Trans. Illum. Eng. Soc. [London] **28** [1963] 91/104; discussion pp. 104/9).

[29] Coaton, J. R.; Fitzpatrick, J. R. (IEE Proc. A **127** No. 3 [1980] 142/8).

[30] Dittmer, G.; Niemann, U. (Philips J. Res. **36** [1981] 87/111).

[31] Burgin, R.; Edwards, E. F. (Light. Res. Technol. **2** [1970] 95/104; discussion pp. 104/8).

[32] Ullmann's Encycl. Ind. Chem. 5th Ed. A **15** [1990] 129/31.

2.5.2 Tungsten–Fluorine Systems

W–F Containing Systems

Due to the high thermal stability of gaseous W fluorides, fluorine seems to offer the best prospects of establishing a true regenerative cycle in halogen incandescent lamps. Tungsten fluorides decompose at temperatures high enough to deposit tungsten preferentially at the hot spots of a filament burning at 3400 K. On the other hand, fluorine reacts with W already in the cold. Thus bulb blackening is prevented even at low wall temperatures and existing blackenings are removed. Unfortunately, the high reactivity of fluorine also leads to attack on the bulb walls and the cooler metal parts (leads, supports) of the lamps. This creates great difficulties in practical lamp design.

The actual existence of a regenerative cycle was proved by exploratory experiments [1, 2]. A tungsten bow, one half of which had been thinned by etching, was resistively heated to 3100°C (thin half) and 2800°C (thick half), respectively, in a 500 Torr Ar + 2 Torr WF_6 atmosphere. After 10 to 15 min, the bow showed equal thickness and temperature over its whole length. A tungsten coil which untreated exhibited a ragged temperature distribution with temperatures strongly decreasing towards its ends, acquired a homogeneous temperature (~ 3230 K) nearly up to the leads when heated for 5 min in 200 Torr Ar + 4 Torr NF_3. The bulb walls were protected by a thin layer of MgF_2 and CoF_2 in these experiments. Bulb blackening of a tungsten lamp filled with 500 Torr Ar after a burning period of 10 h (3000°C) could be removed by adding 2 Torr WF_6 and burning 15 min at 3100°C [1 to 3]; see also [4, 22]. The existence of a regenerative transport cycle was later confirmed by radiochemical tracer experiments. These proved that tungsten does transfer from a cold region (e.g., tungsten on the envelope) to a high temperature region (e.g., filament hot spot) in the presence of a suitable fluoride [23].

An analysis of the transport processes liable to occur in W–F lamps has been attempted by determining the local chemical equilibria. The temperature dependence of the tungsten "solubility" in the gas phase was measured by dynamic and stationary methods. An NF_3 flow in Ar of 40 cm/s with a corresponding fluorine content of 0.1 or 1% F_2 was directed on a tungsten filament at various temperatures between 1500 and 3500 K. The change of the electrical conductance of the filament was measured versus time as a relative measure of the tungsten transport from the wire surface. In the second experiments, the absolute value of dissolved tungsten n(W) was determined by measuring the weight loss of a W wire burned under stationary conditions in a known quantity of fluorine $n(F)_2$ initially introduced at temperatures (> 2500 K) where no tungsten transport to the colder parts of the lamp occurred. The tungsten "solubility", defined as the ratio $n(W)/n(F_2)$, relates to the partial pressures p_i of the components as follows:

$$\frac{n(W)}{n(F_2)} = \frac{\Sigma p(W)}{\Sigma p(F_2)} =$$

$$\frac{p(W) + p(WF) + p(WF_2) + p(WF_3) + p(WF_4) + p(WF_5) + p(WF_6)}{0.5\, p(F) + p(F_2) + 0.5\, p(WF) + p(WF_2) + 1.5\, p(WF_3) + 2\, p(WF_4) + 2.5\, p(WF_5) + 3\, p(WF_6)}$$

The solubility in the temperature range between 500 and 4500 K for $\Sigma p(F_2)$ equal to 10^{-2} and 10^{-3} atm is shown in **Fig. 23**. The characteristic maximum of the tungsten solubility at 2000 K corresponds to the species WF_4 in accordance with a solubility value of $\Sigma p(W)/\Sigma p(F_2) = 0.5$. At low temperatures the compounds WF_6 and WF_5 are stable. Lower valent fluorides occur only at high temperatures and when there is an excess of atomic F. Tungsten solubility measurements at temperatures of ~3400 K and fluorine partial pressures of 10^{-1} atm were realized by introducing SiF_4 of BF_3 which decompose at high temperatures. Calculation of gas phase compositions was difficult, since reliable thermodynamic data for the lower fluorides were not available. The chemical and thermodynamic properties of WF_6 and WF_5 are well known. With the restriction to constant pressures of introduced reactive gases, calculated partial pressures and measured W solubilities were approached iteratively by computation through variation of the reaction enthalpies of the lower fluorides. Thus the partial pressures of the W fluorides corresponding to the W solubility curve for $\Sigma p(F_2) = 10^{-2}$ atm in Fig. 23 were obtained and are shown in a figure in the paper [5]. A more recent representation of the gas phase composition under these conditions derived by the same authors from the W removal rate in flows of 10 mbar F_2 in 1 bar Ar [6] is given in **Fig. 24**. Further gas phase distributions over W(s)–F containing systems with different fluorine sources [6] are given on pp. 58/60. Studies in a closed system were performed [26]. A coiled tungsten wire was mounted in a glass bulb of about 300 cm³ volume, which was filled with NF_3 corresponding to 10 mbar F_2 (at 400 K) + 0.5 bar Kr. The weight changes of the heated coil were used to calculate the amount of W volatilized into the gaseous phase. The results were verified by quantitative analysis of the WF_6 IR absorption at 712 cm^{-1}. The summed (total effective) tungsten pressure obtained is given as a function of the W filament temperature. The gas phase composition as a function of the filament temperature is shown in **Fig. 25** for both the filament and bulb wall regions [26].

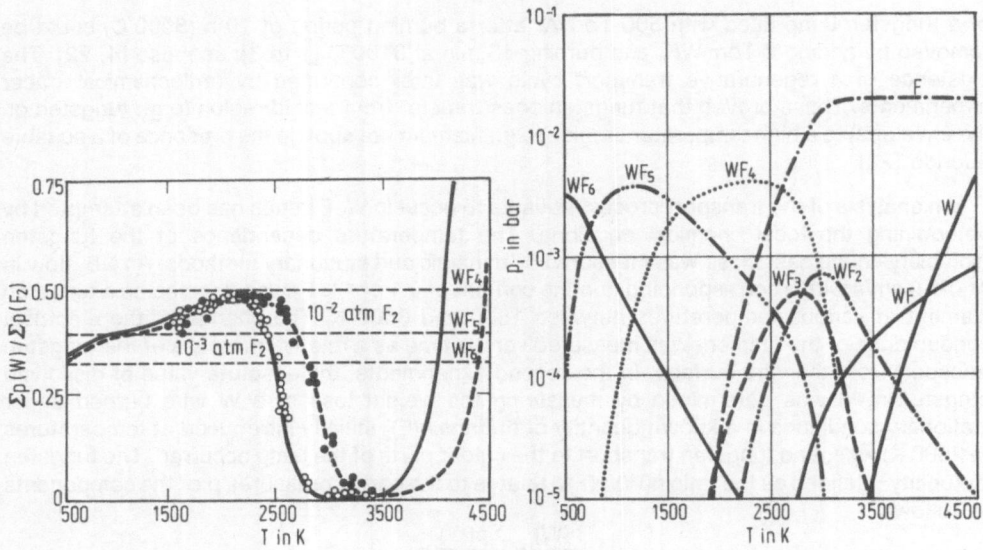

Fig. 23 (left).   Gas phase solubility of W in the W(s)–F(–Ar) system at initial NF$_3$ pressures corresponding to $\Sigma p(F_2)$ of 10^{-3} and 10^{-2} atm. For definitions, see text [5].

Fig. 24 (right).   Equilibrium gas phase composition (partial pressures p_i) of the W(s)–F$_2$(–Ar) system at an initial fluorine pressure $\Sigma p(F_2)$ of 10^{-2} bar [6].

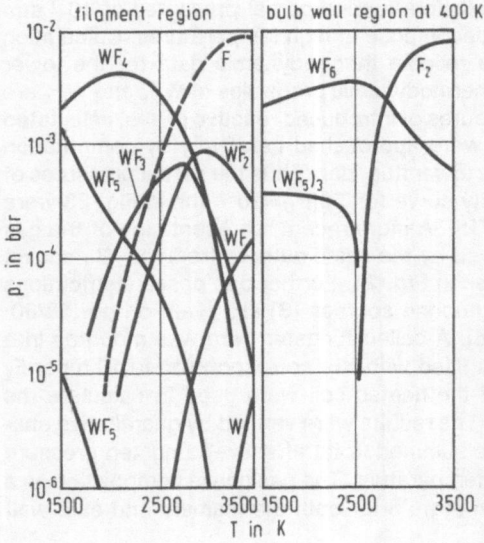

Fig. 25. Gas phase composition (partial pressures p_i) in a W–F(–Kr) lamp at the W filament and at the bulb wall as function of the filament temperature. Bulb wall at 400 K, NF$_3$ fill pressure corresponding to $p_{400K}^0(F_2) = 10^{-3}$ bar, 0.5 bar Kr [26].

These experimentally based data differ considerably from purely calculated ones [7 to 11] based on thermodynamic data for W chlorides taken from the JANAF Tables [12]. Results for an initial F_2 pressure of 10^{-2} atm [10] are given in **Fig. 26**; for comparison (data for $p^0(F_2)=10^{-3}$ atm), see Fig. 15, p. 36. As can be seen, the main difference from Fig. 24 consists of the absence of the WF_3 species and a considerable contribution and stability of the WF_2 species. According to [5] the data basis in [7 to 11] is questionable. This holds particularly for the derived reaction enthalphy of WF_2. Therefore, the partial pressures obtained are considered unrealistic; see also [26]. Earlier calculations of the gas phase composition [13] suffer from the same inadequacy [6, 8].

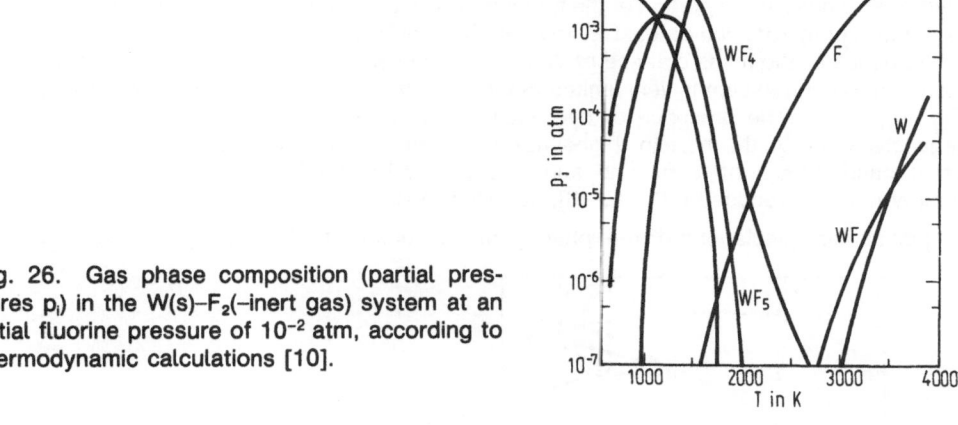

Fig. 26. Gas phase composition (partial pressures p_i) in the $W(s)–F_2(–inert gas)$ system at an initial fluorine pressure of 10^{-2} atm, according to thermodynamic calculations [10].

The trend of the curves obtained by [5, 6] and [8, 10, 11] for the temperature dependence of the solubility or the mass balance ratio (see p. 38) of W in the gas phase is similar. The curves show a maximum and a minimum, except possibly at high fluorine pressures. Accordingly, the tungsten transport cycle changes its direction distinctly with temperature and pressure. Generally, at low and at very high temperatures, the direction is from hot to cold, but it is reversed at medium temperatures. Within the latter range, corrosion of the cooler filament ends will occur in a W–F lamp. An increase in the initial fluorine pressure shifts the critical temperatures to higher values.

After a burning period of 80 h at 3200 to 3400 K, a W–F lamp filled with 1 to 10 Torr WF_6 and 3000 Torr Kr showed attack on the leads in the range of medium temperatures and needle-shaped deposits of W on the leads at lower temperatures. W transport from 2200 to 1000 K was observed [14].

In order to attain an even (total effective) tungsten pressure $\Sigma p(W)$ at all hot tungsten parts of a W–F lamp, additions of chlorine and SiF_4 to the lamp filling have been proposed. Chlorine raises $\Sigma p(W)$ at the lower temperatures, while SiF_4 dissociates above 2500 K into SiF_2 and 2F and thus increases the partial pressures of the lower-valent fluorides in this region with the same overall effect [15].

In contrast to other W–halogen systems, there exist lower as well as upper limits for the F_2 concentration below and above which bulb blackening has to be expected. At too low F_2 concentrations the W vaporizing from the incandescent filament is not entirely retained in the gas phase and deposits at the lamp walls. If the F_2 concentration is too high, fluoride formation (WF_2?) continues even at the temperatures of the incandescent W filament with the result that the mass balance will be higher than at the temperatures of the bulb wall, which leads to W transport from filament to bulb and blackening of the latter [10, 11, 22].

W–F–H Containing Systems

Hydrogen strongly influences the reactions in the W(s)–F system. Gas phase compositions at 500 to 3600 K were calculated from thermodynamic data assuming that thermodynamic equilibrium is achieved at the gas/solid interface and that the rate of reaction is not controlled by kinetic factors. The (initial) partial pressures were in the range $10^{-1} \leq p^0(F_2) \leq 10^{-6}$ atm and the $F_2 : H_2$ ratios varied from 10:1 to 1:4. Results for $p^0(F_2) = 10^{-2}$ atm and $F_2 : H_2 = 1:1$ are given in **Fig. 27**; for the gas phase composition at $p^0(F_2) = 10^{-3}$ atm and $F_2 : H_2 = 1:1$, see Fig. 17, p. 39. Hydrogen binds a large portion of the fluorine by formation of HF. The formation of W fluorides is thus severely reduced at least at low and medium high temperatures and the temperature dependence of the mass balance of W is affected in such a way that transport of W occurs from hot to cold, except in very limited ranges of temperature; see, for example, **Fig. 28** and Fig. 18, p. 39. The presence of hydrogen in the filling of W–F lamps thus impedes the regenerative cycle, the more the higher its concentration, and leads to lamp blackening. On the other hand, attack of leads and supports of the W filaments or coils at medium high temperatures is reduced [10, 11, 16]; see also [14].

For earlier calculations of gas phase compositions in the W–F–H system, see [13].

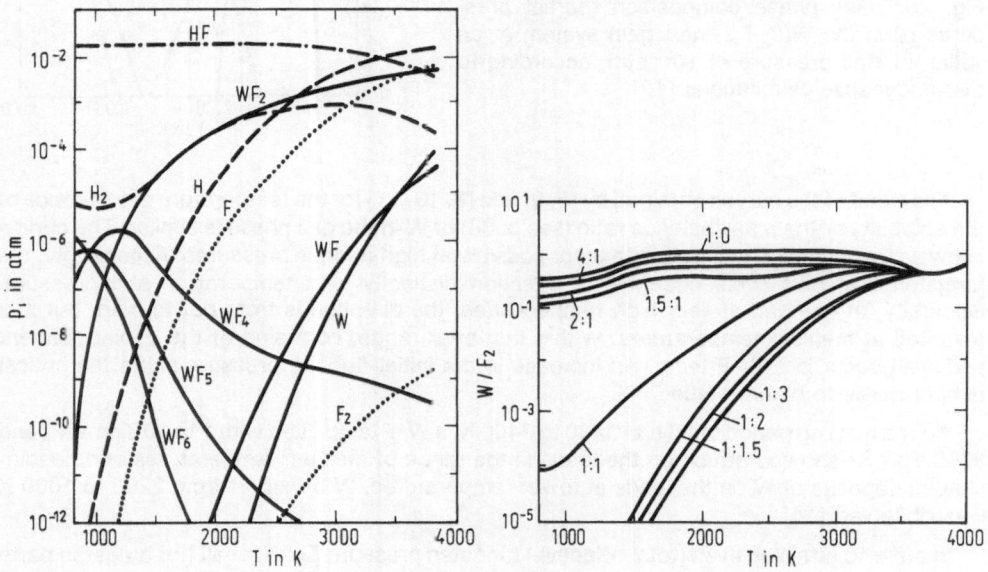

Fig. 27 (left).   Gas phase composition (partial pressures p_i) in the W(s)–F_2–H_2(–inert gas) system at an initial fluorine pressure $p^0(F_2) = 10^{-2}$ atm and $F_2 : H_2 = 1:1$ [16].

Fig. 28 (right).   Mass balance ratio W/F_2 in the gas phase of the W(s)–F_2–H_2(–inert gas) system at $p^0(F_2) = 10^{-3}$ atm and various $F_2 : H_2$ ratios given as parameter at the curves [11].

W–F–O Containing Systems

Equilibrium gas phase compositions in the W(s)–F–O containing system were calculated from the mass variation of a hot tungsten wire in a gas flow of 1 bar Ar with additions of 2×10^{-3} bar O_2 and $(2.2$ to $8.8) \times 10^{-3}$ bar NF_3 for fluorine supply (see p. 47). The corresponding $F_2 : O_2$ ratios established at the filament were 4:1, 3:1, 2:1, and 1:1. The required thermodynamic data for the fluorides were derived by the method described on p. 36. Data for the oxide fluorides were also obtained by interpolation of atomic formation enthalpy data for the ternary compounds estimating the change in the enthalpy by adding a further atom to a tungsten compound or by substituting F by O. The temperature dependence of the partial pressures of the gas phase components is shown in **Fig. 29** and **Fig. 30** for $F_2 : O_2$ ratios of 4:1 and 1:1, respectively. In Fig. 29, WOF_4 is seen to be the dominant species up to temperatures of about 2500 K. Fig. 30 indicates an enhanced formation of the WO_2F_2 species and the precipitation of solid WO_2 at rather high temperatures ($\leqq 1700$ K) [6]. Measurements of the weight changes of a W coil in a closed system, as described on p. 47, gave the partial pressures at the filament and at the bulb wall shown in **Fig. 33**. The glass bulb contained 20 mbar F_2 (at 400 K, added as NF_3), 5 mbar O_2 (400 K) and 0.5 bar Kr. As can be seen, tungsten transport in the bulb wall region is supported by WOF_4 and, to a minor extent, by WO_2F_2 [26].

Thermodynamic data given in the JANAF Tables [12] were used to calculate the phase equilibria in the W(s)–F_2–O_2 system for the ranges $10^{-1} \leqq p(F_2) \leqq 10^{-6}$ atm, $F_2 : O_2 = 10:1$ to 1:6, and $800 \leqq T \leqq 3600$ K, assuming that no kinetic factors influence the reactions. Gas phase compositions at $p(F_2) = 10^{-2}$ and 10^{-4} atm and $F_2 : O_2 = 1:1$ are represented in **Fig. 31** and **Fig. 32**, respectively (see also Fig. 19, p. 40, for data at $p(F_2) = 10^{-3}$ atm). The main component in the low and middle temperature ranges is the WO_2F_2 species; the WOF_4 species apparently does not exist under these conditions. The concentration of the binary W fluorides (particularly the higher-valent) is lowered by several orders of magnitude. With the beginning decay of WO_2F_2 at higher temperatures, WF_2 and W oxides (mainly monomeric) gain increasing importance and the W–F and W–O partial systems exist independent of one another. With increasing fluorine or oxygen contents in the gas phase, the phase equilibria approach those in the W–F and W–O systems, respectively. This is evident already at $F_2 : O_2 = 2:1$ and 1:2, respectively. Solid WO_2 is formed at low temperatures when the $F_2 : O_2$ ratio is below ~1:1 (see Fig. 29). The temperature dependence of the mass balance ratios which indicate the solubility of W in the gas phase is shown in **Fig. 34** and **Fig. 35** (see also Fig. 20, p. 42). A comparison with Fig. 16, p. 38 shows that the maximum found in the W–F system disappears with increasing oxygen content of the gas phase. This will have a beneficial effect in W–F lamps since, by the reduction of the W transport from hot to cold over an extended range of temperature, bulb blackening and attack of leads and cold ends of the incandescent filaments are greatly diminished. At very high temperatures, on the other hand, the transport from hot to cold is not influenced by the presence of oxygen. Thus W deposition at the hottest spots of the filament prolonging lamp life is ensured [11, 14, 17]; see also [20]. After a burning period of 80 h at 3200 to 3400 K, a practical W–F lamp filled with 1 to 10 Torr WF_6, 3000 Torr Kr, and an undefined amount of oxygen did not show any attack of the tungsten parts. There was neither a removal of W nor a deposition of W at the leads in the ranges of higher and lower temperatures, respectively [14]; see, however [24].

W–F–O–H Containing Systems

Usually the presence of oxygen in W–F lamps is linked with that of hydrogen. Both get into the lamp atmosphere in the form of H_2O as an impurity in the filling gas or adsorbate on the inner lamp parts, or else can be released from OH groups of the quartz bulb.

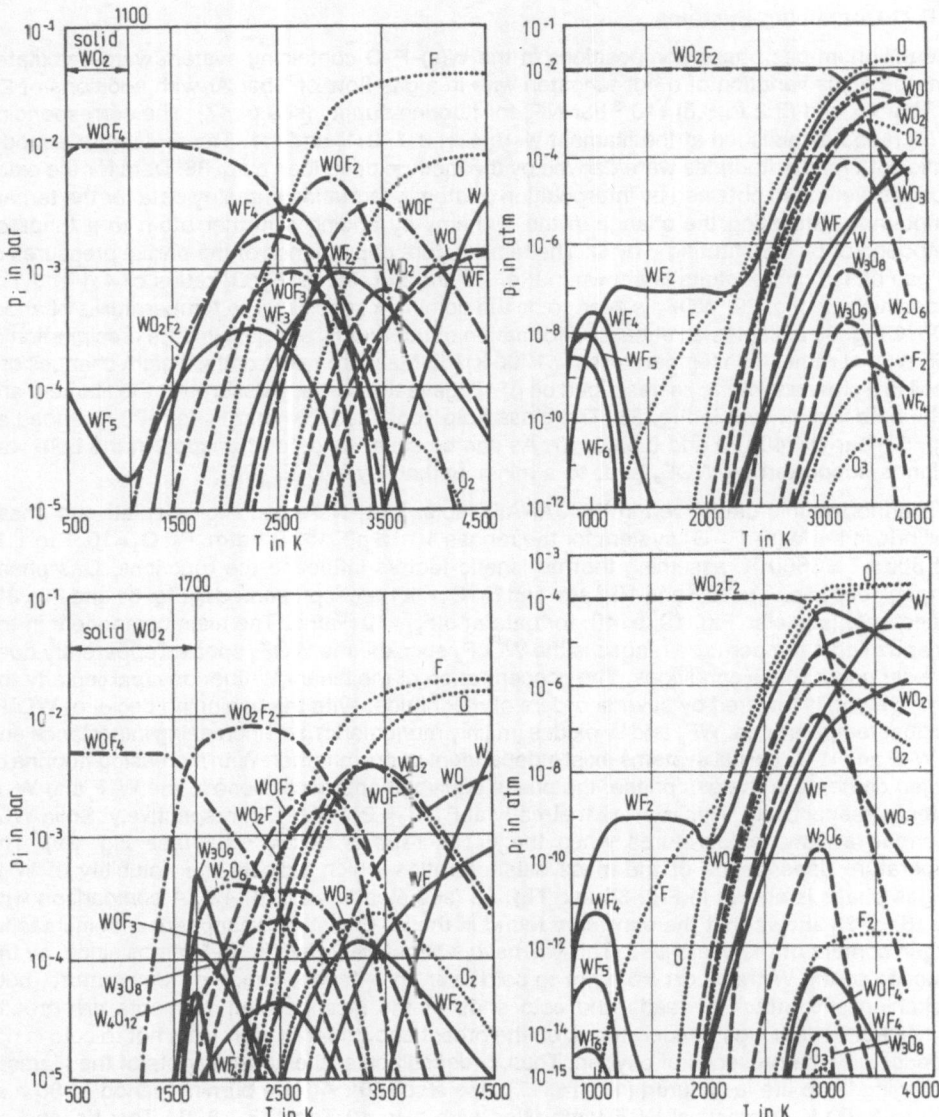

Fig. 29 (left above), Fig. 30 (left below).   Equilibrium composition (partial pressures p_i) of the gas phase in the W(s)–F–O(–Ar) containing system at corresponding ratios $F_2:O_2$ = 4:1 for $\Sigma p(WOF_4) = 10^{-2}$ bar (Fig. 29) and $F_2:O_2 = 1:1$ for $\Sigma p(WO_2F_2) = 10^{-2}$ bar (Fig. 30). The data are in part based on experimental determinations of the mass variation of a hot W wire in a flow of 1 bar Ar with additions of O_2 and NF_3 [6].

Fig. 31 (right above), Fig. 32 (right below).   Equilibrium composition (partial pressures p_i) of the gas phase in the W(s)–F_2–O_2(–inert gas) system at $F_2:O_2 = 1:1$ for initial fluorine pressures $p^0(F_2) = 10^{-2}$ atm (Fig. 31) and $p^0(F_2) = 10^{-4}$ atm (Fig. 32). For these calculations the thermodynamic data basis of the JANAF Tables [12] was used [17].

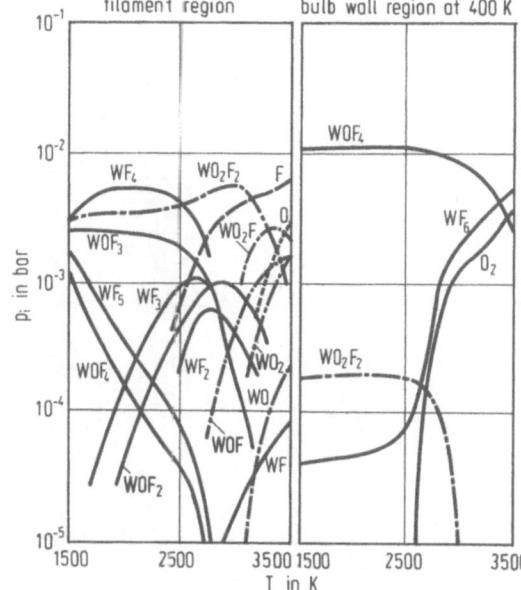

Fig. 33. Gas phase composition (partial pressures p_i) at the filament and at the bulb wall of a W–F–O(–Kr) containing lamp as a function of the filament temperature. Bulb wall at 400 K; fill pressures at 400 K: NF_3 corresponding to 20 mbar F_2, 5 mbar O_2, 0.5 bar Kr [26].

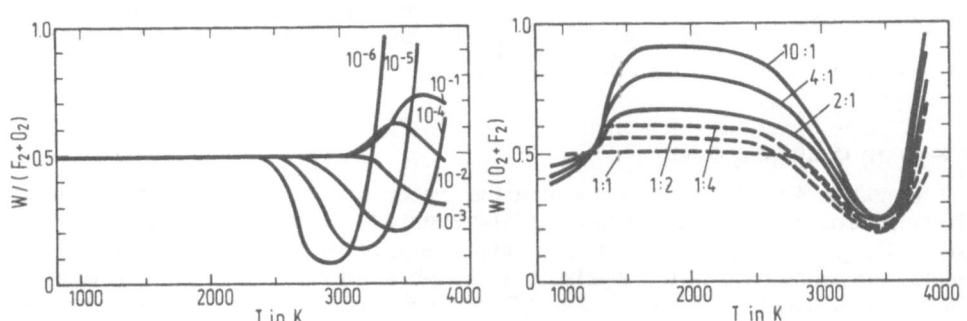

Fig. 34 (left). Mass balance ratio in the W(s)–F_2–O_2(–inert gas) system at $F_2:O_2$ = 1:1 and various initial fluorine pressures $p^0(F_2)$ given as parameter at the curves [14].

Fig. 35 (right). Mass balance ratios in the W(s)–F_2–O_2(–inert gas) system at $p^0(F_2)$ = 10^{-4} atm and various $F_2:O_2$ ratios given as parameter at the curves [14].

The gas phase compositions in the W–F–O–H system over solid W at 1000 to 3600 K were calculated from thermodynamic data for initial F_2 pressures $p^0(F_2)$ of 10^{-1} to 10^{-6} atm and different $O_2:F_2$ ratios. Results for $p^0(F_2)=10^{-3}$ atm and $F_2:O_2:H_2=1:1:1$ are given in **Fig. 36**; similar diagrams for $F_2:H_2:O_2=1:2:1$ and 1:1:0.5 are presented in the paper. These phase equilibria are controlled by the competition between the formation of WO_2F_2 on the one hand and HF and H_2O on the other. When oxygen and hydrogen are introduced into the W–F system in the form of H_2O, the influence of oxygen on the reactions is more pronounced than that of hydrogen. The influence of water vapor and hydrogen on the temperature dependence of the mass balance ratio $W/(F_2+O_2)$ is shown in figures in the paper; for the influence of oxygen on the mass balance ratio $W/(F_2+O_2)$ in the gas phase at $F_2:H_2=1:1$ ($p^0(F_2)=10^{-3}$), see Fig. 22, p. 43 [11, 18].

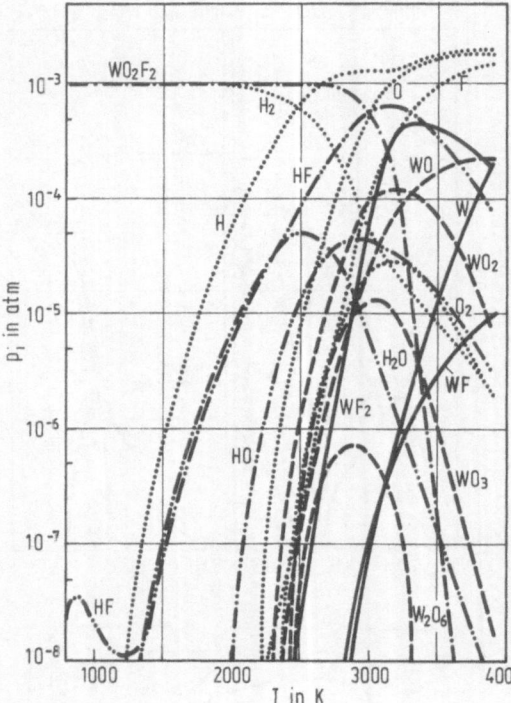

Fig. 36. Equilibrium composition (partial pressures p_i) of the gas phase in the $W(s)-F_2-O_2-H_2(-\text{inert gas})$ system at an initial fluorine pressure $p^0(F_2)=10^{-3}$ atm and a ratio $F_2:O_2:H_2=1:1:1$ [18].

W–F–C(–H) Containing Systems

Thermodynamic calculations show that the presence of carbon does not appreciably influence the gas phase compositions in the $W(s)-F$ and $W(s)-F-H$ systems. The compounds or radicals C_2H_2, CH, CH_2, CF, CF_2, etc. occur in the range of medium and higher temperatures, but their concentration is too low to affect the other phase equilibria; see **Fig. 37**. The absence of the rather stable compounds CF_4 and C_2F_6 in the presence of solid W is remarkable. Solid carbon is deposited at temperatures below 3000 K and can lead to the formation of W carbide resulting in an embrittlement of the incandescent filament in W–F lamps. The temperature dependence of the mass balance ratios W/F_2 under the conditions of Fig. 37 is represented in **Fig. 38**. A comparison with Fig. 18, p. 39, indicates that a noticeable influence of C on the chemical transport processes in the $W-F_2-H_2$ system is not to be expected [11, 19].

Calculated gas phase compositions and mass balance ratios for the case that carbon and fluorine are introduced in the form of 10^{-3} atm CF_4 are shown in **Fig. 39** and **Fig. 40**, respectively [21]; see also p. 43.

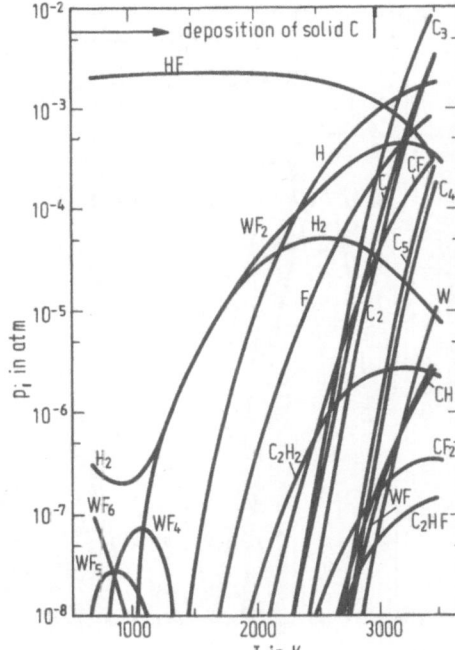

Fig. 37. Equilibrium composition (partial pressures p_i) of the gas phase in the W(s)–F_2–H_2–C(–inert gas) system at an initial fluorine pressure $p^0(F_2)=10^{-3}$ atm and a ratio $F_2:H_2:C=$ 1:1:1 [19].

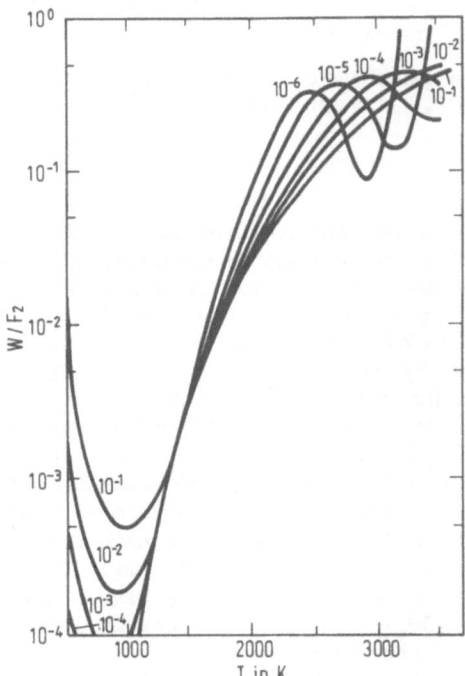

Fig. 38. Mass balance ratio W/F_2 in the gas phase of the W(s)–F_2–H_2–C(–inert gas) system at $F_2:H_2:C=1:1:1$ and various initial fluorine pressures $p^0(F_2)$ given as parameter at the curves [19].

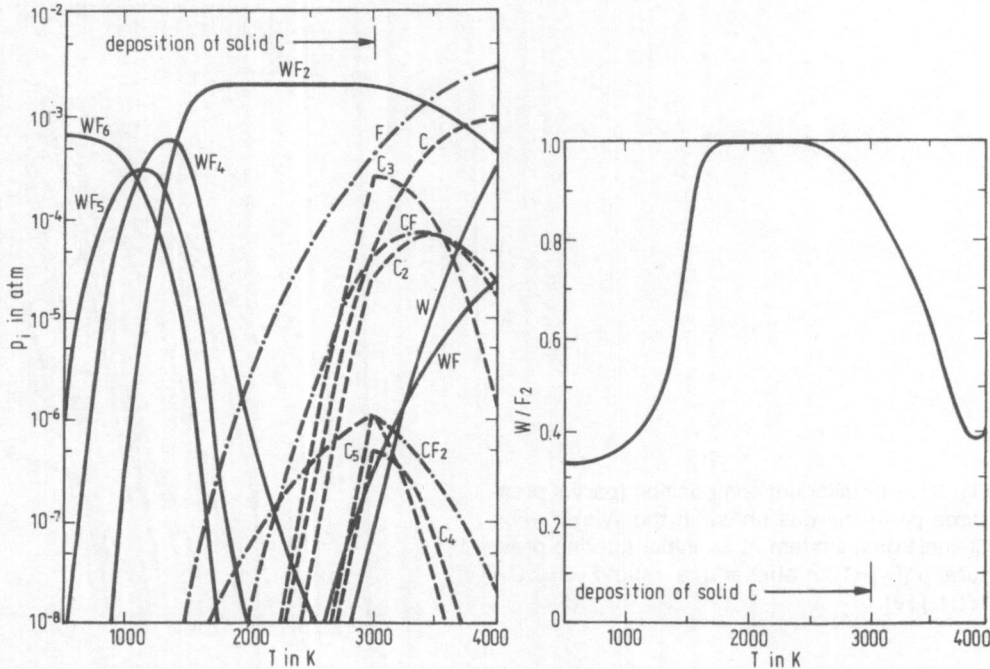

Fig. 39 (left).   Equilibrium composition (partial pressures p_i) of the gas phase in the W(s)–F–C(–inert gas) containing system for an initial CF_4 pressure of 10^{-3} atm [21]. Calculations are based on JANAF thermochemical data.

Fig. 40 (right).   Mass balance ratio W/F_2 in the gas phase of the W(s)–F–C(–inert gas) containing system at $p^0(CF_4) = 10^{-3}$ atm [21].

W–F–O–C Containing Systems

In this system a competition between the formation of CO or CO_2 on the one hand and that of W oxide fluorides on the other has to be expected. Thermodynamic calculations of the gas phase composition over solid W have been performed for the case that C and O are added in the form of CO_2 and CO. Results are presented in **Fig. 41** and **Fig. 42** for an initial F_2 pressure $p^0(F_2)$ of 10^{-3} atm. At temperatures above 2000 K the WO_2F_2 prevailing at lower temperatures is deprived of O and partially or completely replaced by CO which behaves as an inert gas with regard to the other phase relations in this range. In the case of CO addition (C : O_2 = 1 : 0.5), the formation of W oxides at high temperatures is completely suppressed. At lower temperatures added CO (and CO_2) acts as an O source promoting the formation of WO_2F_2. In this range of lower temperatures the deposition of solid carbon will be observed, while solid W oxides cannot exist. The capacity of CO as an O source can be used for a desired oxygen addition to the filling of a W–F lamp, since an overdose effect, as with elemental O_2 resulting in the undesired formation of W oxides at high temperatures, is not liable to occur due to the high stability of CO in this range. The temperature dependence of the mass balance ratios $W/(F_2 + O_2)$ is shown in **Fig. 43** and **Fig. 44** for $10^{-6} \leq p^0(F_2) \leq 10^{-1}$ atm. Depending on the amount of "effective" (not bound to C) oxygen, the shape of the curves resembles more those in the W–F system or more those in the W–F–O system. At low temperatures (<1000 K), the values of the mass balances are always higher than in the W–F system due to the formation of WO_2F_2. The general direction of the chemical transport processes is not influenced by the presence of carbon [11, 14, 19].

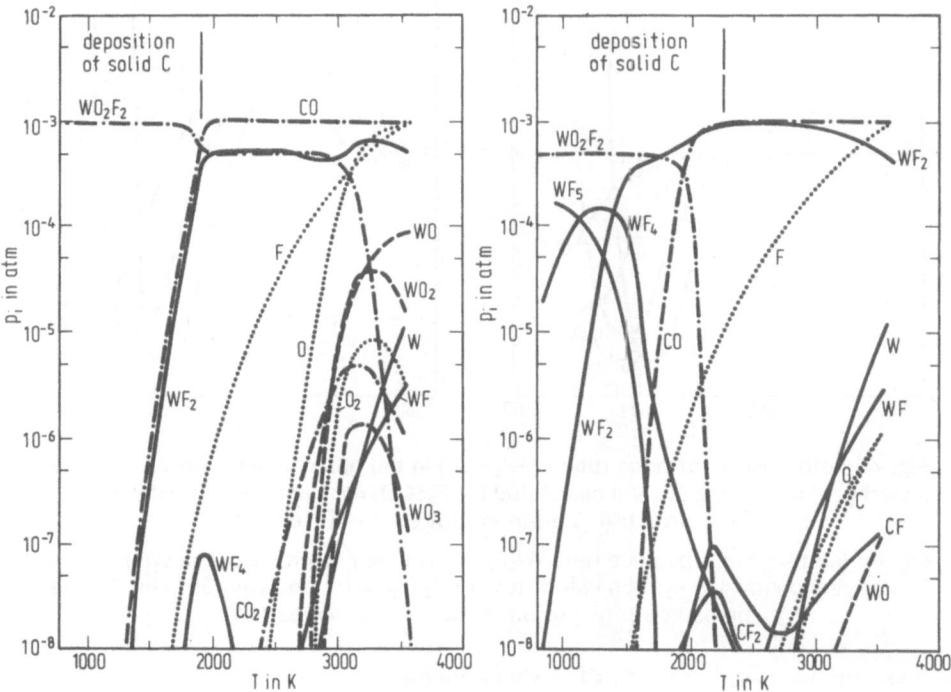

Fig. 41 (left). Equilibrium composition (partial pressures p_i) of the gas phase in the W(s)–F–O–C(–inert gas) containing system calculated for an initial fluorine pressure $p^0(F_2) = 10^{-3}$ atm and a ratio $F_2 : CO_2 = 1:1$ [14].

Fig. 42 (right). Equilibrium composition (partial pressures p_i) of the gas phase in the W(s)–F–O–C(–inert gas) containing system calculated for an initial fluorine pressure $p^0(F_2) = 10^{-3}$ atm and a ratio $F_2 : CO = 1:1$ [14].

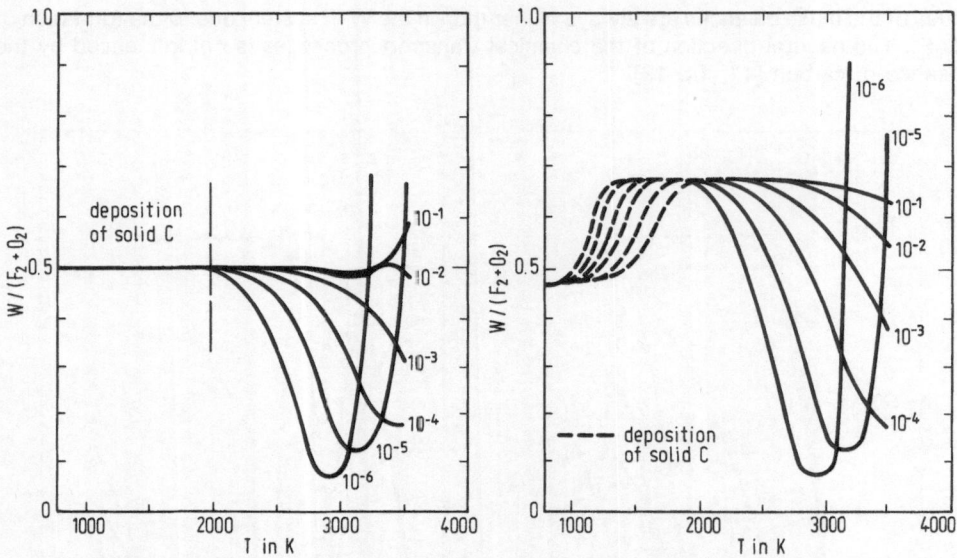

Fig. 43 (left).   Mass balance ratio $W/(F_2+O_2)$ in the gas phase of the W(s)–F–O–C (–inert gas) containing system calculated for $F_2:CO_2=1:1$ and various initial fluorine pressures $p^0(F_2)$ given in atm at the curves [14].

Fig. 44 (right).   Mass balance ratio $W/(F_2+O_2)$ in the gas phase of the W(s)–F–O–C (–inert gas) containing system calculated for $F_2:CO=1:1$ and various initial fluorine pressures $p^0(F_2)$ given in atm at the curves [14].

W–F Systems with Additional B, Si, P Components

Additions of easily dissociable NF_3 have been applied in a number of studies of the W–F systems at higher temperatures to avoid the handling problems and the material attack encountered with the use of the aggressive elemental fluorine, see e.g., pp. 12/4, 46. For the beneficial influence of $CBrF_3$ on the transport processes in halogen lamps, see [25].

The influence of the quartz glass of the bulbs on the gas phase composition and the transport processes in W–F lamps was studied [14]. **Fig. 45** and **Fig. 46** show the gas phase compositions thermodynamically calculated for solid SiO_2 kept at 1000 and 1500 K and an initial F_2 pressure of 10^{-3} atm. The main component of the gas phase up to rather high temperatures is the oxide fluoride WO_2F_2. Appreciable amounts of gaseous Si compounds will appear only at an SiO_2 temperature of 1500 K. An attack on SiO_2 without formation of volatile Si compounds is, however, to be expected also at lower temperatures. This should be connected with the formation of metallic Si. The temperature dependence of the mass balance ratios at $p^0(F_2)=10^{-6}$ to 10^{-1} atm corresponds to that of the $W–F_2–O_2$ system at an $F_2:O_2$ ratio of 1:1. The conclusions drawn from the above thermodynamic calculations could not be confirmed by practical lamp experiments. Deviating from the predictions, there were significant differences in the transport processes occurring at the tungsten parts in lamps with and without oxygen additions to the filling gas under otherwise identical conditions. This shows that kinetic factors play an important role. Evidently the reactions between quartz glass und fluorine or W fluorides are impeded and do not proceed fast enough to establish equilibrium conditions, even at the rather high temperatures of the bulb walls [14].

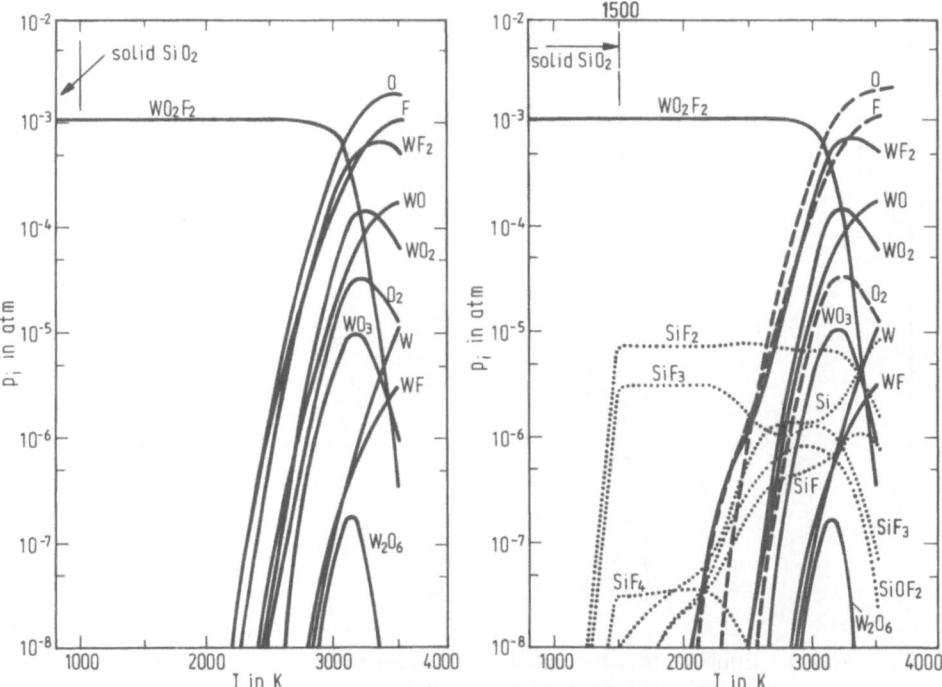

Fig. 45 (left). Equilibrium composition (partial pressures p_i) of the gas phase in the W(s)–F_2(–inert gas) system calculated for an initial fluorine pressure $p^0(F_2)=10^{-3}$ atm and for the presence of solid quartz glass kept at 1000 K [14].

Fig. 46 (right). Equilibrium composition of the gas phase in the W(s)–F_2(–inert gas) system calculated for an initial fluorine pressure $p^0(F_2)=10^{-3}$ atm and for the presence of solid quartz glass kept at 1500 K [14].

Equilibrium gas phase compositions in the W–F–B, W–F–Si, and W–F–P containing systems over solid W were calculated from the measured mass variation of a hot W wire in gas flows of 1 bar Ar with additions of 10^{-2} or 10^{-1} bar BF_3, SiF_4, and PF_3, respectively. Results are given in **Fig. 47**, **Fig. 48**, and **Fig. 49**, respectively. The partial pressure composition in the W–F–B system shows that no W fluorides other than WF_2 and WF are involved. In the W–F–Si and W–F–P systems, WF_4 and WF_2 are the main tungsten-fluorine species [6]. A mixture of F_2, Cl_2, and SiF_4 has been proposed to establish a chemical transport system in an incandescent filament lamp in which no spot is formed on the filament from which W is transported at a faster rate than elsewhere; see p. 78.

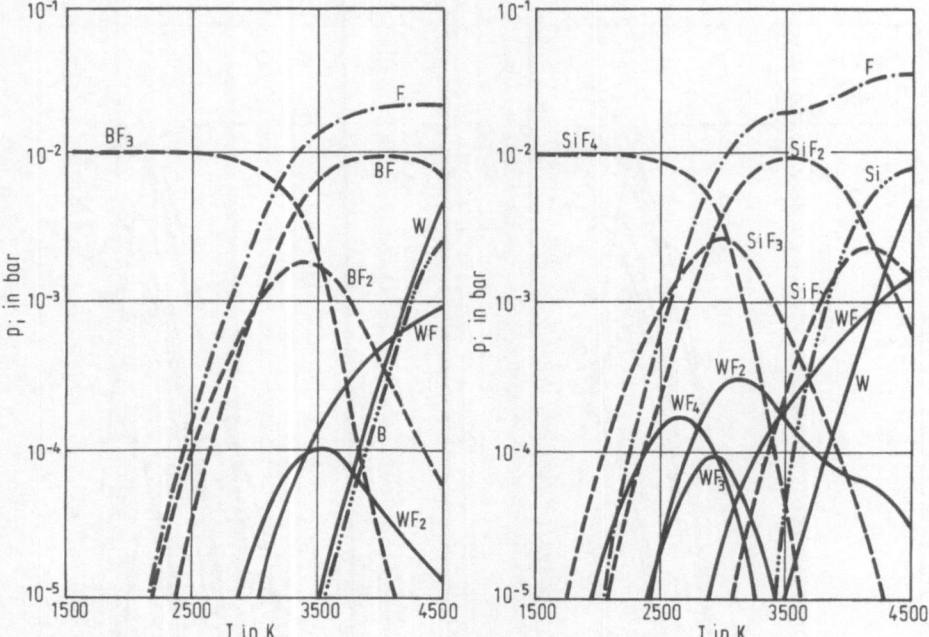

Fig. 47 (left).   Equilibrium composition (partial pressures p_i) of the gas phase in
the W(s)–BF$_3$(–Ar) system for $\Sigma p(BF_3) = 10^{-2}$ bar [6].

Fig. 48 (right).   Equilibrium composition (partial pressures p_i) of the gas phase in
the W(s)–SiF$_4$(–Ar) system for $\Sigma p(SiF_4) = 10^{-2}$ bar [6].

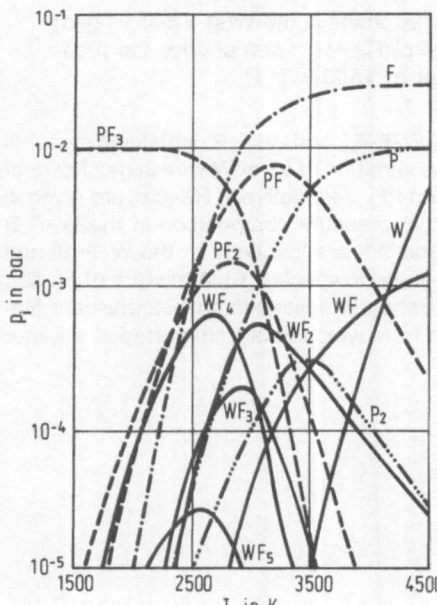

Fig. 49.   Equilibrium composition (partial pressures p_i) of the gas phase in the W(s)–PF$_3$(–Ar)
system for $\Sigma p(PF_3) = 10^{-2}$ bar [6].

References:

[1] Schröder, J. (Chem. Eng. News **42** No. 37 [1964] 77/8).
[2] Schröder, J. (Philips Tech. Rundsch. **25** [1963/64] 359/65).
[3] Schröder, J.; Grewe, F. J. (Chem. Ber. **103** [1970] 1536/46).
[4] Rabenau, A. (Angew. Chem. **79** [1967] 43/9; Angew. Chem. Int. Ed. Engl. **6** [1967] 68/73).
[5] Dittmer, G.; Klopfer, A.; Ross, D. S.; Schröder, J. (J. Chem. Soc. Chem. Commun. **1973** 846/7).
[6] Dittmer, G.; Klopfer, A.; Schröder, J. (Philips Res. Rep. **32** [1977] 341/64).
[7] Neumann, G. M.; Knatz, W. (Z. Naturforsch. **26a** [1971] 863/9).
[8] Neumann, G. M.; Gottschalk, G. (Z. Naturforsch. **26a** [1971] 870/81).
[9] Gottschalk, G.; Neumann, G. M. (Z. Metallk. **62** [1971] 910/5).
[10] Neumann, G. M. (J. Fluorine Chem. **1** [1971/72] 473/86).

[11] Neumann, G. M. (Tech. Wiss. Abh. OSRAM-Ges. **11** [1973] 8/41).
[12] JANAF Thermochemical Tables and Addenda I-III, Dow Chemical Company, Midland, Michigan, 1965/68.
[13] Kopelman, V.; van Wormer, K. A. (Illum. Eng. [N.Y.] **64** [1929] 230/5).
[14] Neumann, G. M. (J. Fluorine Chem. **3** [1973] 197/208).
[15] Dettingmeijer, J. H.; Dittmer, G.; Klopfer, A.; Schröder, J. (Philips Tech. Rev. **35** [1975] 302/6; Philips Tech. Rundsch. **35** [1975] 324/7).
[16] Neumann, G. M. (Z. Metallk. **64** [1973] 117/20).
[17] Neumann, G. M. (Z. Metallk. **64** [1973] 26/32).
[18] Neumann, G. M. (Z. Metallk. **64** [1973] 379/85).
[19] Neumann, G. M. (Z. Metallk. **64** [1973] 444/9).
[20] Neumann, G. M.; Knatz, W. (Z. Naturforsch. **26a** [1971] 1046/53).

[21] Neumann, G. M. (J. Fluorine Chem. **3** [1973] 209/25).
[22] Neumann, G. M.; Müller, U. (Tech. Wiss. Abh. OSRAM-Ges. **11** [1973] 42/54).
[23] Haigh, I. (Diss. Univ. Leicester 1973 from [24]).
[24] Coaton, J. R. (IEE Proc. A **127** No. 3 [1980] 142/8).
[25] Hill, J. C.; Dolenga, A. (J. Appl. Phys. **48** [1977] 3089/92).
[26] Dittmer, G.; Niemann, U. (Philips J. Res. **36** [1981] 87/111).

2.5.3 Tungsten–Chlorine Systems

W–Cl Containing Systems

According to thermodynamic calculations [1, 2], the binary chlorides are stable up to temperatures which are high enough to establish a regenerative cycle in W–Cl incandescent filament lamps. Equilibrium gas phase compositions over solid W at initial chlorine pressures $p^0(Cl_2)=10^{-1}$ and 1 atm, calculated from thermodynamic data in the JANAF Tables [3], are shown in **Fig. 50** and **Fig. 51**, respectively (for the gas phase composition at $p^0(Cl_2)=10^{-3}$ atm, see Fig. 15, p. 36; data for an open system at a total pressure of 10^{-2} atm are presented in a figure in [4]). The main W-containing gas phase components at lower $p^0(Cl_2)$ are WCl_4 and WCl_2. The higher-valent WCl_6 and WCl_5 are not formed in appreciable amounts. The endothermic WCl appears only at very high temperatures and even then plays only a minor role [1, 4 to 7]; see also earlier calculations of [8]. Tungsten compounds which were deposited on the bulb and on the current leads of incandescent lamps were identified by scanning electron microscopy and X-ray fluorescence. Primary condensates after a short-time operation (5 s) of small-volume (0.8 cm³) linear lamps filled with 800 mbar Cl_2 and up to 1 bar Kr, where bulb temperatures T_b were kept at 300 K, were WCl_6, WCl_5, and WCl_2 when filament temperatures T_f were varied from 1250, to 2100, and to 3100 K, respectively. At a fixed T_f value of 1250 K and for long-time operation with raising bulb temperatures to T_b=700 or 1050 K, the primary condensate WCl_6 was converted to WCl_4. Finally, the attack on quartz led to WO_2Cl_2 and $WOCl_4$ after 100 h of operation. In large-volume (300 cm³) spherical lamps which were filled with 10 mbar Cl_2 and up to 1 bar Kr the following primary condensates were observed: WCl_6 on the bulb for T_f=1500 K, T_b=300 K; traces of WCl_6 in the bulb region for T_f=3000 K, T_b=300 K. At bulb temperatures raised to 700 K lower-valent tungsten halides precipitated: WCl_4 and WCl_2 on the bulb and on current leads for T_f=1500 K; WCl_2 on the current leads (plus traces of WCl_6 in the bulb region) for T_f=3000 K [29].

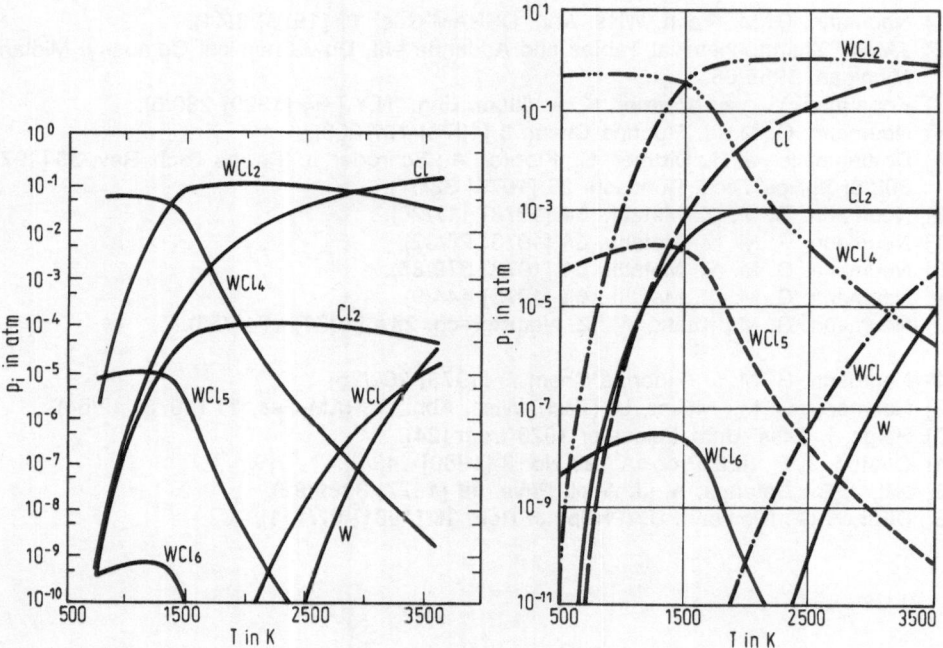

Fig. 50 (left).   Equilibrium composition (partial pressures p_i) of the gas phase in the W(s)–Cl_2(–inert gas) system at an initial chlorine pressure $p^0(Cl_2)$=0.1 atm [4].

Fig. 51 (right).   Equilibrium composition (partial pressures p_i) of the gas phase in the W(s)–Cl_2(–inert gas) system at an initial chlorine pressure $p^0(Cl_2)$=1 atm [6].

For the temperature dependence of the total effective tungsten pressure in a heterogeneous equilibrium system with 0.01 atm Cl_2, see p. 66.

The influence of inert gases on the distribution of the partial pressures of the gas phase components has been studied [4] and, as demonstrated in a figure for $p^0(Cl_2)$=10^{-1} atm, was found to be insignificant, except in rather limited ranges of temperature differing for the individual components [4].

The mass balance ratio W/Cl_2, which is indicative of the "solubility" of W in the gas phase, is represented in **Fig. 52** for $p^0(Cl_2) = 10^{-5}$ to 10 atm (see also Fig. 16, p. 38). The direction of the W transport which will occur in a W–Cl lamp can immediately be inferred from the slope of the curves. The point at which the transport reverses its direction at high temperatures is strongly dependent on the initial chlorine pressure. The characteristic minima of the curves lie at the following temperatures and W/Cl_2 ratios:

$p^0(Cl_2)$ in atm	10^{-2}	10^{-3}	10^{-4}	10^{-5}
T in K	3800	3400	3000	2700
W/Cl_2	3.86×10^{-2}	1.35×10^{-2}	4.21×10^{-3}	1.34×10^{-3}

The validity of the predictions could be confirmed by practical lamp experiments. W transport from 2500 to 3000 K was observed on a W coil at $p^0(Cl_2) = 5.3 \times 10^{-3}$ atm, and transport from 3400 to 3000 K at $p^0(Cl_2) = 5.3 \times 10^{-5}$ atm. The long lamps (220 V, 1000 W) used contained a coaxial W coil with several supports. They were filled with N_2 or Ar of ~1 atm with defined additions of CCl_4 for chlorine supply and burned for 10 h [4, 6, 9]; see also [23]. The chemical transport reactions in the high-temperature region of W–C containing systems were subsequently studied in greater detail in a quartz jacket containing two separate (bow-shaped) incandescent filaments (diameter ~250 µm) of W arranged opposite to each other. The gas filling of the jacket consisted of 1500 Torr Ar and varying amounts of CCl_4 to establish a desired chlorine pressure of 1×10^{-3} to 5×10^{-2} atm. The W filaments were resistively heated to different temperatures. The direction of W transport was judged from the growth of needle-like or dendritic deposits on one of the filaments. The observed directions of transport were in fair agreement with the above theoretical calculations of the reaction equilibria from the basic principles of thermodynamics. Macroscopic deposits formed in burning periods of several (e.g., 12) hours only when the temperature difference between the W filaments produced differences in the partial pressure of WCl_2 exceeding 6×10^{-3} atm. The latter compound was assumed to control the transport process in the studied range of temperatures [9, 20]. Preliminary measurements of tungsten transport in a chlorine atmosphere [10], by contrast, showed little agreement with the transport characteristics expected from estimations bases on the enthalpy values in the JANAF Tables [11]. The summed (total effective) tungsten pressure and the gas phase composition in the filament and bulb wall regions under operating condi-

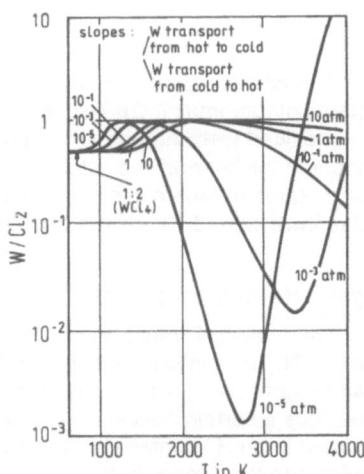

Fig. 52. Mass balance ratio W/Cl_2 in the gas phase of the W(s)–Cl_2 system at various initial chlorine pressures given as parameters at the curves [6].

tions of a W–Cl lamp are shown in **Fig. 53** a and b as functions of the tungsten filament temperature. The data are derived from more recent measurements of the weight change of a coiled tungsten wire mounted in a glass bulb containing 10 mbar Cl_2 (at 400 K) and 0.5 bar Kr. A new set of thermodynamic data for the tungsten-containing species was used. Fig. 53a indicates a considerable decrease of tungsten solubility in the gas phase above 1400 K. Filament temperatures below 1500 K give rise to blackening of the bulb because the tungsten solubility at the filament exceeds that at the bulb wall [26].

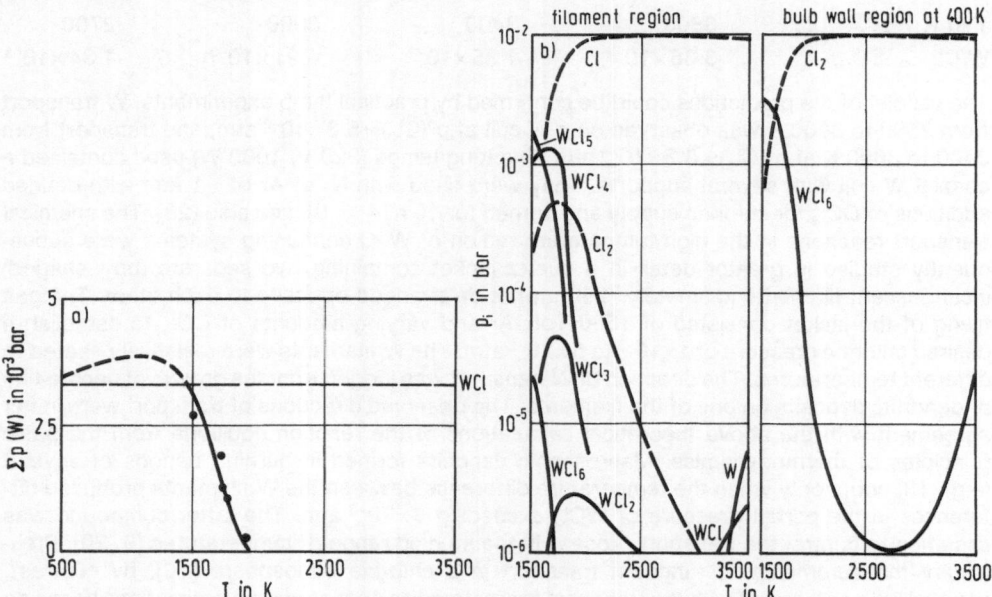

Fig. 53.   a) Summed (total effective) tungsten pressure $\Sigma p(W)$, and b) gas phase compositions (partial pressures p_i) at the filament region and the bulb wall region of a W–Cl_2(–Kr) lamp as function of the filament temperature. Bulb wall at 400 K, $p^0_{400\ K}(Cl_2) = 10^{-2}$ bar, 0.5 bar Kr [26].

The occurrence of a W transport from cold to hot was already established in early experiments of Langmuir [12]: Two W wires were simultaneously heated to different temperatures in a tube filled with chlorine at low pressure. At certain temperatures of the wires, the hotter wire became thicker and the colder wire thinner. A W deposit purposely produced in an incandescent filament lamp disappeared when it was exposed to the atomic chlorine formed by dissociation of added Cl_2 at the hot filament; see also [21].

W–Cl–H Containing Systems

The chemical transport reactions in a W–Cl lamp are substantially influenced by the presence of hydrogen in the filling gas. The transport from cold to hot preventing bulb blackening in lamps containing a pure Cl_2 (+inert gas) filling reverses its direction when the H_2:Cl_2 ratio exceeds a certain critical limit. In a series of experiments, different types of W–Cl lamps were burned in an H_2-containing atmosphere from which hydrogen diffused through the hot quartz bulbs into the interior of the lamps. Its amount could be calculated from the known permeability of quartz to H. The primary lamp filling consisted of 2000 Torr Ar and 2.7 to 5.5 Torr $CHCl_3$.

The lamps were preburned for 1 h to decompose the $CHCl_3$ and to establish the initial equilibrium conditions. A reversal of the transport direction associated with bulb blackening occurred when the $Cl_2:H_2$ ratio fell below 1.4:1 at a chlorine filling pressure of 10^{-3} atm. The experimental results were in fair agreement with simulated data based on chemical thermodynamics [7]. Equilibrium gas phase compositions in the $W–Cl_2–H_2$ system over solid W, calculated from thermodynamic data of the JANAF Tables [3], are shown in **Fig. 54**a and b for initial chlorine pressures $p^0(Cl_2)$ of 10^{-4} or 10^{-2} atm and a $Cl_2:H_2$ ratio of 1:1, and in Fig. 54c for $p^0(Cl_2)=10^{-4}$ atm and $Cl_2:H_2=2:1$ (see also Fig. 17, p. 39, for $p^0(Cl_2)=10^{-3}$ atm and $Cl_2:H_2=1:1$). Corresponding mass balances (total effective pressures, see p. 37) of W for various $Cl_2:H_2$ ratios at $p^0(Cl_2)=10^{-3}$ atm are presented in **Fig. 55**, and mass balance ratios W/Cl_2 for various $p^0(Cl_2)$ and $Cl_2:H_2=1:1$ in Fig. 18, p. 39. At lower temperatures (< 2000 K) and $Cl_2:H_2 \lesssim 1:1$, the formation of HCl is favored over that of the W chlorides (mainly WCl_4 and WCl_2), which leads to low values of the W mass balance in this range and can severely impede the desired back-transport of W from the bulb to the filament in W–Cl lamps. There may even be a reversal in the transport direction [5, 7, 13]. Total effective tungsten pressures calculated for H:Cl ratios in the range 0 to 1.2 and a total pressure of 0.01 atm are given in **Fig. 56**. As can be seen, considerable bulb blackening is to be expected when the H:Cl ratio is equal to or exceeds 1:1. Increasing transport from hot to cold (i.e., from the filament to the bulb wall in a W–Cl lamp) with increasing HCl (H:Cl=1) addition was also verified experimentally [14].

For earlier experiments with HCl-doped lamps, see [22, 23].

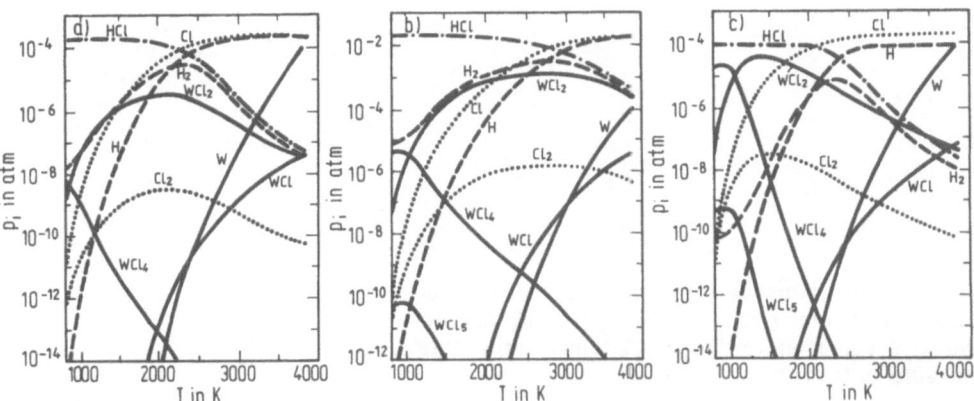

Fig. 54.   Gas phase compositions (partial pressures p_i) in W(s)–Cl_2–H_2(–inert gas) systems at various initial chlorine pressures $p^0(Cl_2)$ and $Cl_2:H_2$ ratios [13].

a) $p^0(Cl_2)=10^{-4}$ atm and $Cl_2:H_2=1:1$

b) $p^0(Cl_2)=10^{-2}$ atm and $Cl_2:H_2=1:1$

c) $p^0(Cl_2)=10^{-4}$ atm and $Cl_2:H_2=2:1$

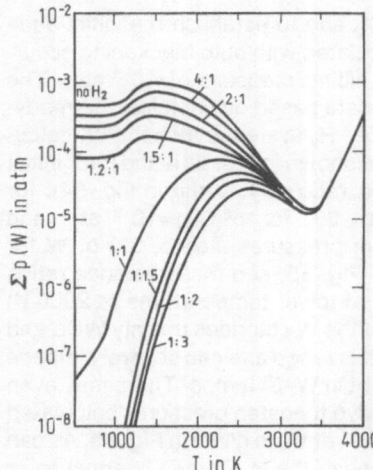

Fig. 55.   Mass balances $\Sigma p(W)$ of W in the gas phase of W(s)–Cl_2–H_2(–inert gas) systems at various Cl_2:H_2 ratios (given as parameters at the curves) and an initial chlorine pressure $p^0(Cl_2)=10^{-3}$ atm [7].

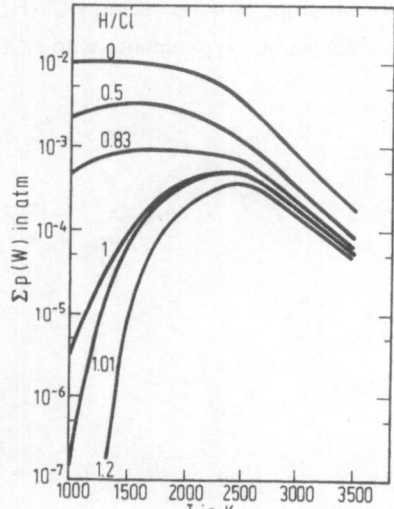

Fig. 56.   Total effective tungsten pressure (mass balance) $\Sigma p(W)$ in the gas phase of W(s)–Cl–H(–Ar) containing systems at various H:Cl ratios and a total pressure of 0.01 atm [14].

W–Cl–O Containing Systems

According to thermodynamic calculations [2], addition of oxygen to the filling of W–Cl lamps will result in the formation of thermally rather stable oxide chlorides, which should play a prominent part in the transport processes taking place and enhance the W transport.

Calculations of the partial pressures of the gas phase components in the W(s)–Cl_2–O_2 system in the ranges $500 < T < 3600$ K, $10^{-6} < p^0(Cl_2) < 10^{-1}$ atm, and Cl_2:O_2 = 10:1 to 1:10 showed that WO_2Cl_2 is the main gas phase species in the low and middle temperature ranges, while $WOCl_4$ plays only a minor role. Gaseous W oxides appear at higher temperatures. Solid WO_2 is formed at lower temperatures when the oxygen content of the gas phase exceeds ~40%; see Fig. 21, p. 42. With increasing chlorine or oxygen content in the gas phase, the phase equilibria approach those in the W–Cl_2 and W–O_2 systems, respectively. This is evident already at

$Cl_2:O_2=2:1$ and $1:2$, respectively. Gas phase compositions at $p^0(Cl_2)=10^{-2}$ and 10^{-4} atm and $Cl_2:O_2=1:1$ are represented in **Fig. 57** a and b, respectively, and results for $p^0(Cl_2)=10^{-3}$ atm and the same $Cl_2:O_2$ ratio in Fig. 19, p. 40 (for gas phase compositions at $p^0(Cl_2)=10^{-4}$ atm and $Cl_2:O_2=2:1$ and $1:2$, see figures in [15]). The mass balance ratios $W/(Cl_2+O_2)$ for various $Cl_2:O_2$ ratios and $p^0(Cl_2)=10^{-4}$ atm, and for $Cl_2:O_2=1:1$ and $p^0(Cl_2)$ varying between 10^{-6} and 10^{-1} atm, are shown in Fig. 20, p. 42 (see [15] for additional data). The addition of oxygen increases the solubility of W in the gas phase in the lower temperature range. The low temperature solubility maximum found in the $W–Cl_2$ system (see p. 63) disappears with increasing oxygen concentration [5, 15].

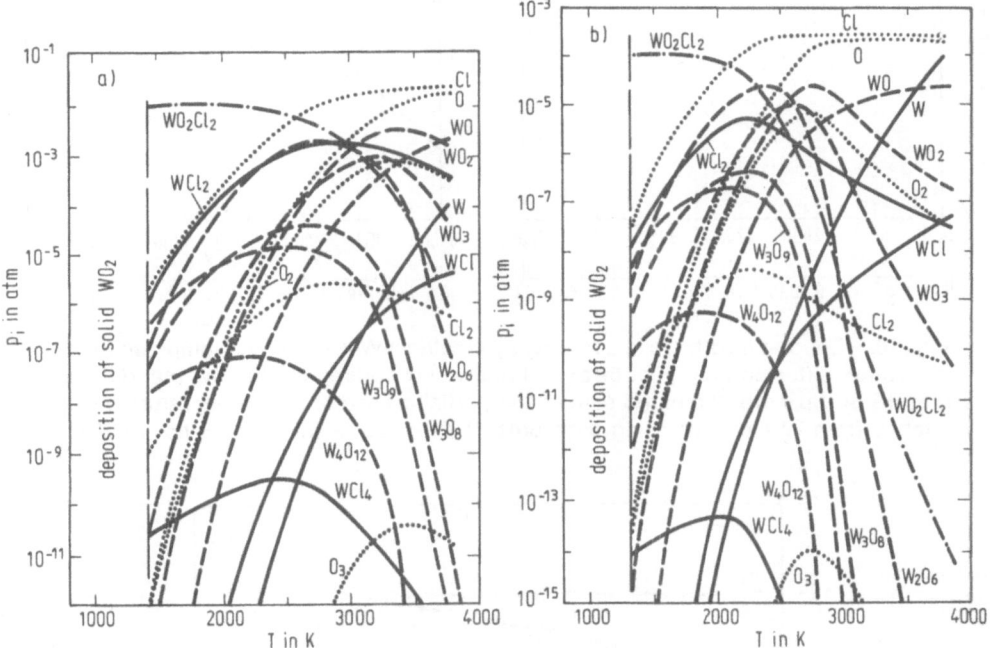

Fig. 57.   Gas phase composition (partial pressures p_i) in the $W(s)–Cl_2–O_2(–$inert gas) system at $Cl_2:O_2=1:1$ and initial chlorine pressures $p^0(Cl_2)=10^{-2}$ atm a) and 10^{-4} atm b) [15].

The importance of thermal diffusion for the radial tungsten transport in W–Cl–O lamps is shown by comparative calculations [27]. **Fig. 58** a shows the separation of the reactive gases in a radially symmetrical lamp due to concentration diffusion but neglecting thermal diffusion. As can be seen, the stoichiometric sum of the partial pressures Σp_i of the constituents varies remarkably with the temperature, i.e., the radial distance from the filament. The filling pressure of the cold lamp was composed of 2 mbar Cl_2, 2 mbar O_2, and 5 bar Ar. The Σp_i distribution in a lamp with the same filling when thermal diffusion effects are additionally considered are shown in Fig. 58b. The significantly smaller amount of volatilized tungsten at the incandescent wire in this case will affect the axial mass transport. **Fig. 59** shows the partial pressure distribution corresponding to Fig. 58b. Again, it can be seen that the tungsten solubility at the bulb wall is maintained only by WO_2Cl_2. Even in the region near the filament, tungsten is volatilized mainly in the form of O-containing high-temperature compounds [27].

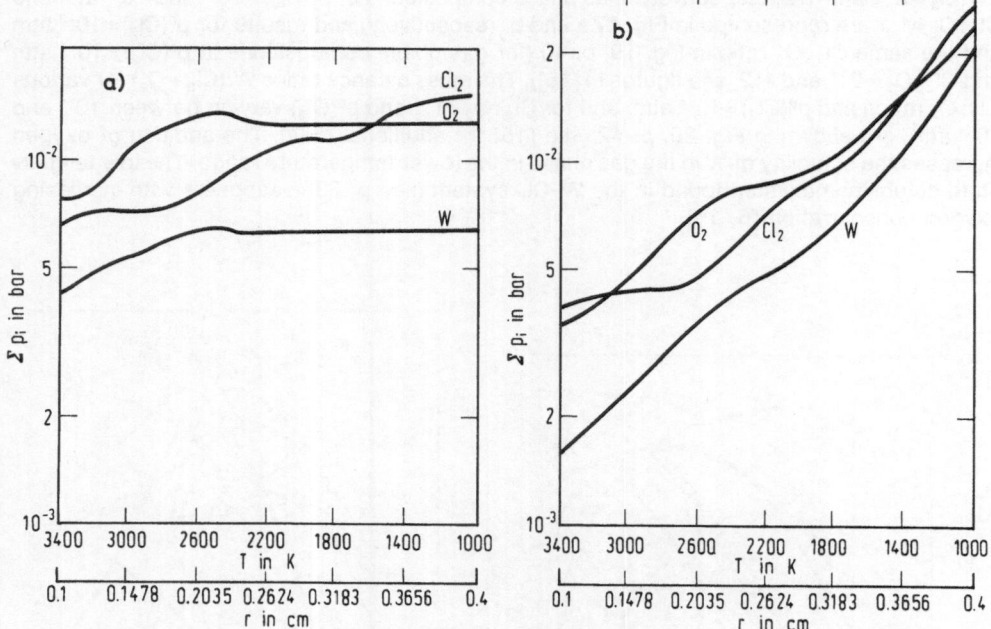

Fig. 58.   Separation effects in a radially symmetrical W–Cl$_2$–O$_2$(–Ar) lamp due to concentration diffusion only (Fig. 58a) and due to concentration plus thermal diffusion
(Fig. 58b); Σp_i = stoichiometric sum of the partial pressures, T = temperature, r = distance from W filament; filling pressures: 2 mbar Cl$_2$, 2 mbar O$_2$, 5 bar Ar [27].

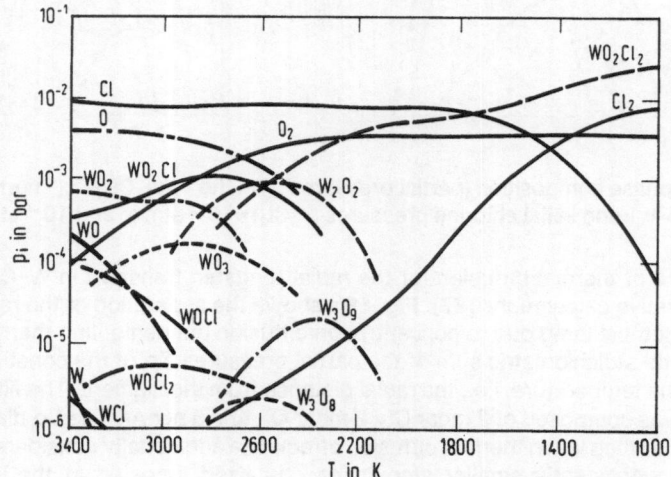

Fig. 59   Partial pressures in a radially symmetrical W–Cl$_2$–
O$_2$(–Ar) lamp at conditions corresponding to Fig. 58b [27].

The primary crystalline condensate observed after short-time operation (5 s) of a small-volume linear incandescent lamp filled with 400 mbar Cl_2, 200 mbar O_2, and up to 1 bar Kr was identified as a mixture of WO_2Cl_2 and $WOCl_4$ when the filament temperature T_f was 2100 K and the bulb temperature T_b was 300 K. Operation of the lamp for 5 min with T_b raised to 700 K caused partial decomposition of the primary condensate into $WOCl_2$. Complete reaction of primary WO_2Cl_2 crystals into $WOCl_4$ occurred in a residual Cl_2 atmosphere at temperatures around 530 K. In a large-volume spherical lamp filled with 10 mbar Cl_2, 5 mbar O_2, and up to 1 bar Kr a mixture of WO_2Cl_2 and $WOCl_4$ condensed on the bulb when T_f and T_b were 1500 and 300 K, respectively. An increase in T_f to 3000 K led to small amounts of WO_2Cl_2 deposited on the bulb. At longer lamp operation with T_b raised to 700 K, decomposition of the primary condensate (for $T_f = 1500$ K) led to $WOCl_3$, $WOCl_2$, $WOCl$, plus residual WO_2Cl_2 on the bulb and to $WOCl_2$ and $WOCl$ on the current leads. At $T_f = 3000$ K and $T_b = 700$ K only volatile tungsten oxichlorides were observed [29].

W–Cl–O–H Containing Systems

Small amounts of water vapor (and oxygen) are present even in well-processed halogen lamps. The W oxide chlorides which form in the low-temperature gas phase region of a W–Cl lamp have been analyzed by a sampling method using diffusion-limited mass transport in the temperature gradient of a small attached capillary and forced precipitation of the respective compounds. The lamp was filled with 3×10^{-2} bar HCl and 0.5 bar Ar; the filament temperature was 3150 K. X-ray diffraction diagrams of the precipitates in the capillary indicated the presence of WO_2Cl_2 and of minor traces of $W_{20}O_{58}$. Compounds $WOCl_3$ and $WOCl_4$ were absent. These results were confirmed by IR absorption measurements. The experiments supported thermodynamic equilibrium calculations assuming a water vapor pressure of the order of 10^{-4} bar in practical lamps and WO_2Cl_2 as the dominate gaseous W species in the wall region [28].

The gas phase compositions in the $W–Cl_2–O_2–H_2$ system over solid W at 1000 to 3600 K were calculated from thermodynamic data in the JANAF Tables [3] for initial Cl_2 pressures $p^0(Cl_2)$ of 10^{-1} to 10^{-6} atm and different $O_2 : Cl_2$ ratios. Results for $p^0(Cl_2) = 10^{-3}$ atm and $Cl_2 : O_2 : H_2 = 1 : 0.5 : 1$ are given in **Fig. 60**. The phase equilibria are controlled by the competition between the formation of WO_2Cl_2 on one hand and HCl and H_2O on the other. Below 1500 K, deposition of solid WO_2 occurs when the proportion of oxygen is comparable to that of chlorine. The gas phase is thus gradually depleted of oxygen and the transport processes occurring become similar to those in the oxygen-free system. When oxygen and hydrogen are introduced into the W–Cl system in the form of H_2O, the influence of oxygen on the reactions is more pronounced than that of hydrogen [5, 16]. Gas phase compositions calculated for ratios $H : Cl = 1 : 1$, $Cl : O = 100 : 1$ and a total pressure of 0.01 atm are shown in **Fig. 61**. Besides the components indicated, the presence of 18 other species of minor importance was included in the calculations. The relative importance of the dichloride WCl_2 under the above conditions is striking, which seems to be responsible for the observed blackening of the envelope of practical lamps filled with HCl, even when traces of oxygen are present [14].

A partial impairment of the W transport from cold to hot in W–Cl–O–H containing systems, particularly in the range of medium temperatures, can be concluded from the temperature dependence of the mass balance ratios $W/(Cl_2 + O_2)$. The influence of water vapor and hydrogen on the temperature dependence of $W/(Cl_2 + O_2)$ is shown in figures in the paper; for the influence of oxygen on $W/(Cl_2 + O_2)$ in the gas phase at $Cl_2 : H_2 = 1 : 1$ ($p^0(Cl_2) = 10^{-3}$ atm), see Fig. 22, p. 43 [5, 16]. The total effective W pressure (for definition, see p. 37) for $Cl : O$ ratios of 10 to 10^4, $H : Cl = 1 : 1$, and a total pressure of 0.01 atm is depicted in **Fig. 62** [14].

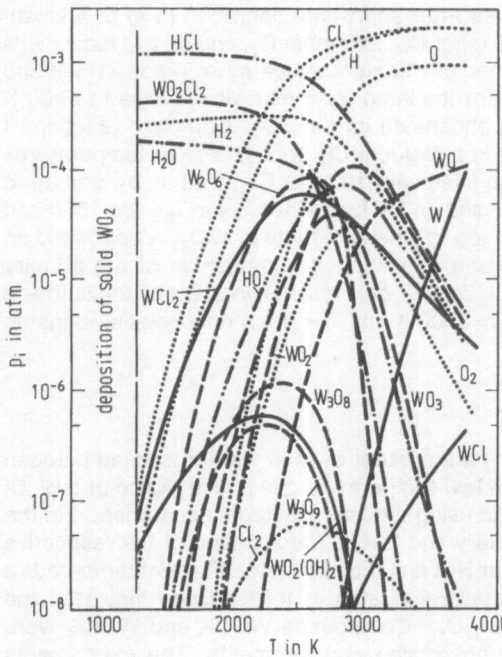

Fig. 60. Equilibrium composition (partial pressures p_i) of the gas phase in the W(s)–Cl$_2$–O$_2$–H$_2$(–inert gas) system at an initial chlorine pressure $p^0(Cl_2)=10^{-3}$ atm and a ratio $Cl_2:O_2:H_2=1:0.5:1$ [16].

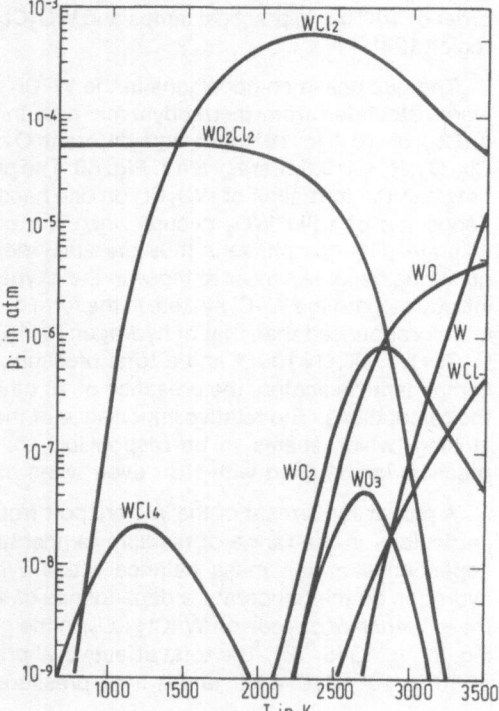

Fig. 61. Gas phase composition calculated for a W(s)–Cl–O–H containing system at ratios H:Cl=1, Cl:O=100, and a total pressure of 0.01 atm. Only the partial pressures p_i of the W-containing components are given [14].

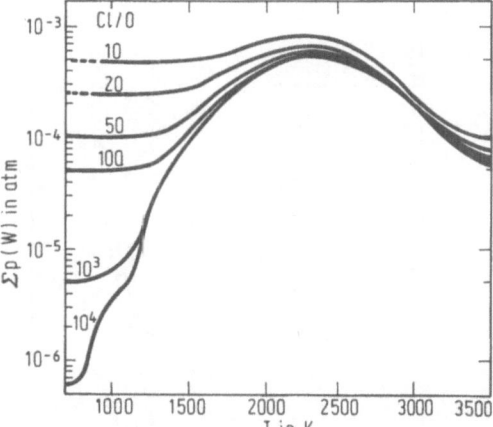

Fig. 62.   Total effective tungsten pressure $\Sigma p(W)$ in the gas phase calculated for W(s)–Cl–O–H containing systems at a ratio H : Cl =1, a total pressure of 0.01 atm, and various Cl : O ratios given as parameters at the curves [14].

W–Cl–N Containing Systems

The potential use of chlorophosphonitriles as chlorine sources in tungsten-halogen lamps has been discussed [24, 25].

W–Cl–C(–H) Containing Systems

The presence of carbon does not appreciably influence the reactions between tungsten and chlorine or between chlorine and hydrogen. Compound formation between carbon and chlorine and carbon and hydrogen is negligible under the conditions prevailing in W–Cl incandescent lamps. Carbon is often introduced into these lamps in the form of the compounds CCl_4, $CHCl_3$, CH_2Cl_2, or CH_3Cl, which dissociate at higher temperatures [5, 9, 17, 22, 23]. Transport directions of W observed in lamps with CCl_4 and $CHCl_3$ additions are described on pp. 63/5. Bulb blackening is observed in lamps containing CH_2Cl_2 or CH_3Cl additions, while bulbs containing $CHCl_3$ remain clear. This shows that only the $H_2 : Cl_2$ ratio decides whether or not a regenerative cycle is established. In the presence of oxygen, lamps with CH_2Cl_2 additions do not show bulb blackening [9].

Thermodynamically calculated compositions of the gas phase over solid W at an initial chlorine pressure of 10^{-3} atm and a ratio $Cl_2 : H_2 : C = 1:1:1$ are shown in **Fig. 63**. Solid carbon is deposited at temperatures below 3000 K and can lead to the formation of W carbide resulting in an embrittlement of the incandescent filament in W–Cl lamps. The temperature dependence of the mass balance ratios W/Cl_2 under the conditions of Fig. 63, but with various inital chlorine pressures, is represented in **Fig. 64**. A comparison with Fig. 18, p. 39, indicates that a noticeable influence of C on the chemical transport processes in the W–Cl–H system is not to be expected [5, 17].

W–Cl–O–C Containing Systems

The influence of carbon on the chemical reactions and the transport processes in W–Cl–O systems is qualitatively similar to that in W–F–O systems; see p. 56. The gas phase compositions thermodynamically calculated for CO_2 and CO additions at an initial Cl_2 pressure of 10^{-3} atm are shown in **Fig. 65** and **Fig. 66**, respectively. Corresponding mass balance ratios $W/(Cl_2 + O_2)$ are depicted in **Fig. 67** and **Fig. 68**, respectively [5, 17].

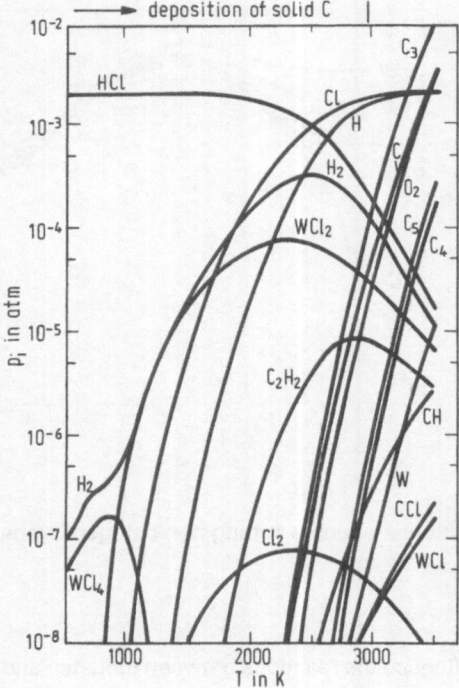

Fig. 63. Equilibrium composition (partial pressures p_i) of the gas phase in a W(s)–Cl–H–C(–inert gas) containing system calculated for an initial chlorine pressure $p^0(Cl_2)=10^{-3}$ atm and a ratio $Cl_2:H_2:C=1:1:1$ [17].

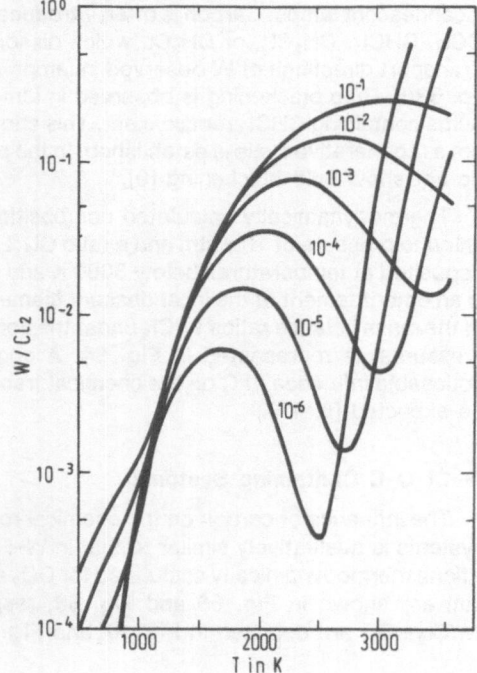

Fig. 64. Mass balance ratio W/Cl_2 in the gas phase of the W(s)–Cl–H–C(–inert gas) containing system calculated for a ratio $Cl_2:H_2:C=1:1:1$ and various initial chlorine pressures $p^0(Cl_2)$ given as parameter at the curves [17].

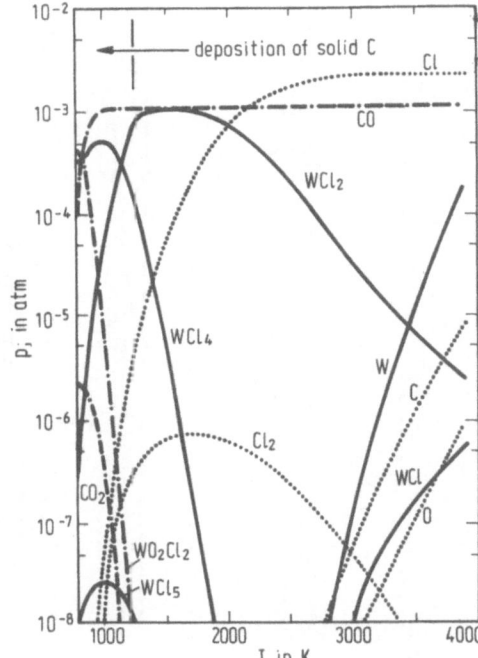

Fig. 65. Equilibrium composition (partial pressures p_i) of the gas phase in the W(s)–Cl$_2$–CO$_2$(–inert gas) system at an initial chlorine pressure $p^0(Cl_2) = 10^{-3}$ atm and a ratio Cl$_2$:CO$_2$ =1:1 [17].

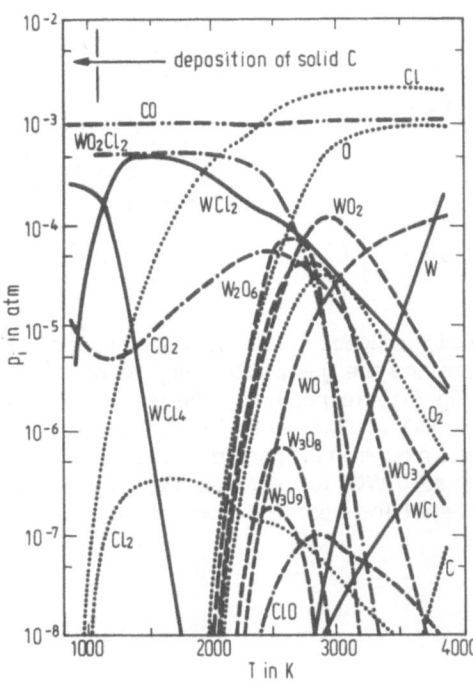

Fig. 66. Equilibrium composition (partial pressures p_i) of the gas phase in the W(s)–Cl$_2$–CO(–inert gas) system at an initial chlorine pressure $p^0(Cl_2) = 10^{-3}$ atm and a ratio Cl$_2$:CO =1:1 [17].

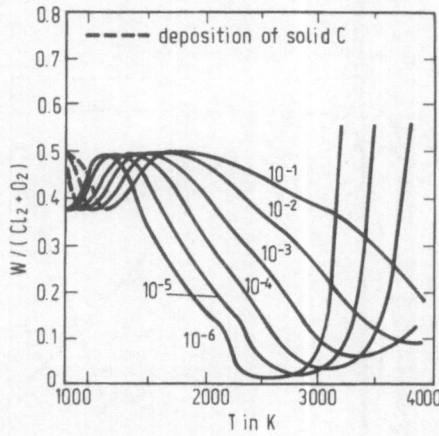

Fig. 67. Mass balance ratio $W/(Cl_2 + O_2)$ in the gas phase of the $W(s)$–Cl_2–CO_2(–inert gas) system at $Cl_2 : CO_2 = 1:1$ and various initial chlorine pressures $p^0(Cl_2)$ (in atm) given as parameter at the curves [17].

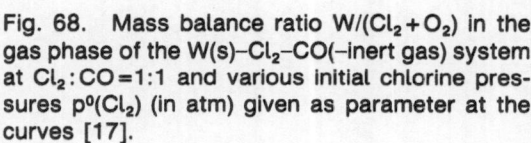

Fig. 68. Mass balance ratio $W/(Cl_2 + O_2)$ in the gas phase of the $W(s)$–Cl_2–CO(–inert gas) system at $Cl_2 : CO = 1:1$ and various initial chlorine pressures $p^0(Cl_2)$ (in atm) given as parameter at the curves [17].

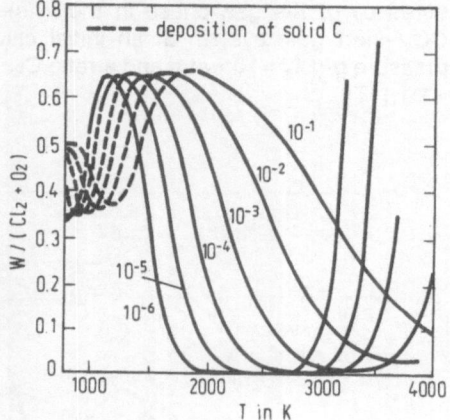

Measurements of the weight changes of a W coil in a closed system gave the dependence of the summed (total effective) tungsten pressure and the gas phase composition on the W filament temperature shown in **Fig. 69**a and Fig. 69b, respectively. The glass bulb contained (at 700 K) 22 mbar Cl_2, 5.5 mbar O_2, 11 mbar CO, and 0.5 bar Kr. The CO added to the W–Cl–O system acts as an efficient getter for excess oxygen in the parts of the lamp, thus preventing W oxide chlorides from decomposing into solid WO_3 and chlorine. Fig. 69b demonstrates the dominating role of the WO_2Cl_2 species for the transport process [26].

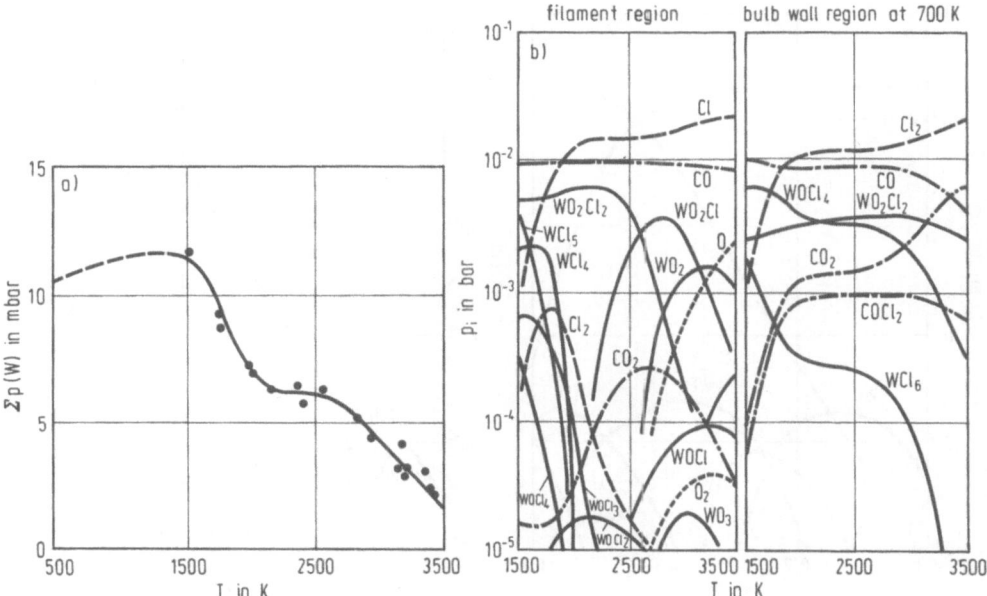

Fig. 69. Summed (total effective) tungsten pressure $\Sigma p(W)$ (Fig. 69a) and gas phase composition (partial pressures p_i) (Fig. 69b) at the W filament and bulb wall of a W–Cl$_2$–CO(–Kr) lamp as a function of the filament temperature. Bulb wall at 700 K; fill pressures at 700 K: 22 mbar Cl$_2$, 5.5 mbar O$_2$, 11 mbar CO, 0.5 bar Kr [26].

W–Cl–O–Si Containing Systems

The influence of the bulb silica on the distribution of partial pressures in a radially symmetrical W–Cl$_2$ lamp was demonstrated. The lamp filling was assumed to consist of 5 mbar Cl$_2$ and 5 bar Kr. Concentration as well as thermal diffusion was presumed. It can be seen from **Fig. 70**a and Fig. 70b that the tungsten solubility cannot be significantly increased by the reaction of tungsten chlorides with the silica of the bulb. Especially in the high-temperature region near the filament, no oxygen-containing tungsten compounds should appear which would increase the W transport along the filament of a real lamp [27]. Deposits of WO$_2$Cl$_2$ and WOCl$_4$ crystals in a W–Cl$_2$ incandescent lamp resulted from the attack on quartz during long-time operation of the lamp [29].

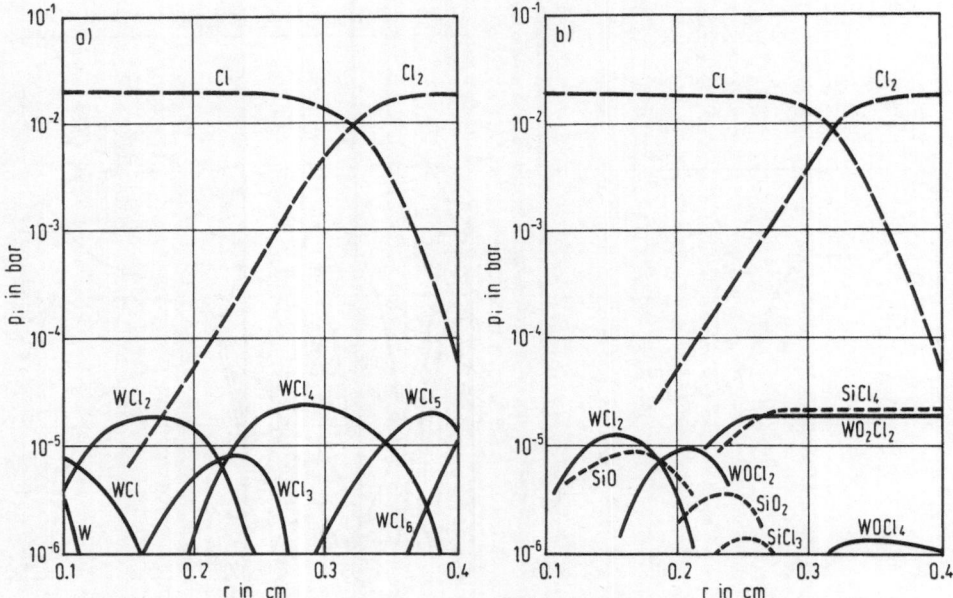

Fig. 70.   Calculated partial pressures in a radially symmetrical W–Cl$_2$(–Kr) lamp as
function of the distance from the W filament for 5 mbar Cl$_2$ and 5 bar Kr filling
pressures: a) reactions with the bulb silica neglected, b) reactions with the bulb
silica included [27].

W–Cl–F–C and W–Cl–F–Si Containing Systems

The gas phase composition of the W(s)–Cl–F–C containing system has been calculated
assuming that thermodynamic equilibrium is attained at the gas/solid interface with tungsten
and that the rate of reaction is not kinetically controlled. The calculations were performed
using data for the equilibrium constants taken from the JANAF Tables [11] or derived from
empirically estimated values for the enthalpies and entropies of reaction. Halogens and carbon
were assumed to be introduced by addition of CCl$_3$F, CCl$_2$F$_2$, CClF$_3$, or CF$_4$ to the input gas,
producing corresponding (initial) halogen pressures of 2×10^{-6} to 2×10^{-1} atm. Results for
$p^0(\text{Cl}_2 + \text{F}_2) = 2 \times 10^{-3}$ atm are presented in **Fig. 71**a to Fig. 71c. While compounds of the type
C$_n$X$_m$ (X = halogen) are quite stable in the C–Cl–F systems, they are relatively unstable in the
presence of solid tungsten where the simple binary W chlorides and fluorides are formed. The
main components of the gas phase at low temperatures are WF$_6$ and WCl$_4$, while at higher
temperatures the dihalides WF$_2$ and WCl$_2$ predominate, in addition to atomic F and Cl. At very
high temperatures, radicals CF$_2$, CF, and CCl are also formed. Solid carbon may appear at
temperatures up to 3000 K giving rise to W carbide formation. The curves representing the
temperature dependence of the mass balance ratios W/(Cl$_2$ + F$_2$) at various $p^0(\text{Cl}_2 + \text{F}_2)$ have
characteristic maxima in the range of medium temperatures and minima at high temperatures.
This has the meaning that the W transport in incandescent filament lamps will proceed down
the temperature gradient (from hot to cold) at lower temperatures, reverse its direction at
moderate temperatures and proceed down the temperature gradient once more at high
temperatures. The influence of an increasing chlorine content in the gas phase on the mass
balance ratios is demonstrated in **Fig. 72**. As can be seen, the gas phase solubility of W in the
range of medium temperatures is substantially decreased. This leads to an increased gradient

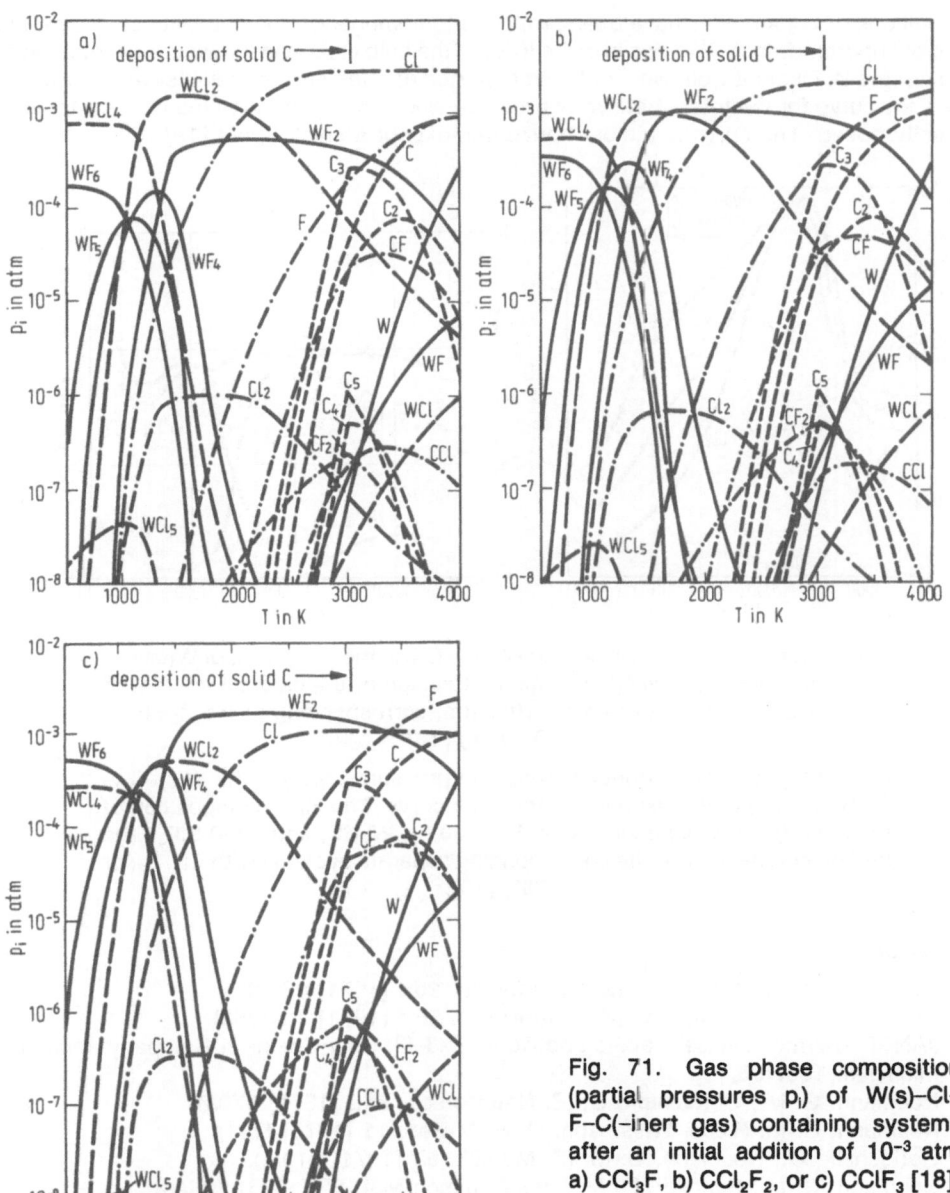

Fig. 71. Gas phase composition (partial pressures p_i) of W(s)–Cl–F–C(–inert gas) containing systems after an initial addition of 10^{-3} atm a) CCl_3F, b) CCl_2F_2, or c) $CClF_3$ [18].

of the W solubility in this range, which will result in an intensified attack on the leads in practical W–halogen lamps. Experiments with 12 V/100 W lamps, which were burned at 3400 K and contained 3000 Torr Kr and 1 to 10 Torr of various CCl_mF_{4-m} compounds, confirmed these predictions [18].

Additions of Cl_2 and SiF_4 have been proposed to homogenize the solubility of W over the range of importance in W–F lamps in order to keep the bulb clear and to avoid the formation of spots on the W filament from which W is transported at a faster rate than elsewhere. **Fig. 73** shows the curve for the total effective W pressure when the additions to the W–F system are optimally dosed. The curve is almost horizontal, except a small ripple [19].

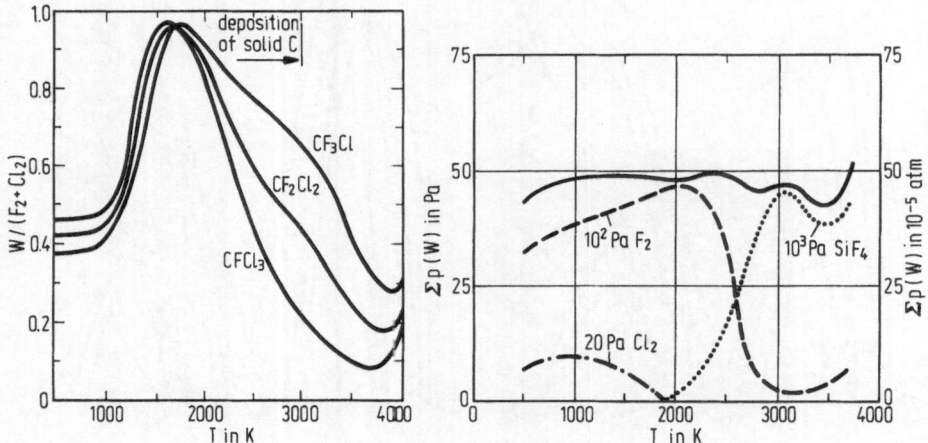

Fig. 72 (left).   Mass balance ratio $W/(F_2 + Cl_2)$ in the gas phase of W(s)–Cl–F–C(–inert gas) containing systems as function of the chlorine content of added CCl_mF_{4-m} (initial pressure 10^{-3} atm, corresponding to $p^0(Cl_2 + F_2) = 2 \times 10^{-3}$ atm) [18].

Fig. 73 (right).   Total effective tungsten pressure $\Sigma p(W)$ in the W(s)–Cl_2–F_2–SiF_4 system, represented by the solid curve. The initial composition of the input gas corresponded to (in Pa) 100 F_2, 20 Cl_2, and 1000 SiF_4. The broken curves refer to the corresponding separate systems with Cl_2, F_2, or SiF_4 [19].

References:

[1] Neumann, G. M.; Knatz, W. (Z. Naturforsch. **26**a [1971] 863/9).
[2] Neumann, G. M.; Knatz, W. (Z. Naturforsch. **26**a [1971] 1046/53).
[3] JANAF Thermochemical Tables and Addenda I–III, Dow Chemical Company, Midland, Michigan, 1965/68.
[4] Neumann, G. M.; Gottschalk, G. (Z. Naturforsch. **26**a [1971] 870/81).
[5] Neumann, G. M. (Tech. Wiss. Abh. OSRAM-Ges. **11** [1973] 8/41).
[6] Gottschalk, G.; Neumann, G. M. (Z. Metallk. **62** [1971] 910/5).
[7] Neumann, G. M.; Schmidt, D. (J. Less-Common Met. **33** [1973] 209/18).
[8] Kopelman, B.; van Wormer, K. A. (Illum. Eng. [N.Y.] **64** [1969] 230/5).
[9] Neumann, G. M.; Müller, U. (Tech. Wiss. Abh. OSRAM-Ges. **11** [1973] 42/54).
[10] Dittmer, G.; Klopfer, A.; Ross, D. S.; Schröder, J. (J. Chem. Soc. Chem. Commun. **1973** 846/7).

[11] JANAF Thermochemical Tables, 2nd Ed., Dow Chemical Company, Midland, Michigan, 1971.
[12] Langmuir, I. (J. Am. Chem. Soc. **37** [1915] 1139/67).

[13] Neumann, G. M. (Z. Metallk. **64** [1973] 117/20).
[14] Dettingmeijer, J. H.; Meinders, B.; Nijland, L. M. (J. Less-Common Met. **35** [1974] 159/69).
[15] Neumann, G. M. (Z. Metallk. **64** [1973] 26/32).
[16] Neumann, G. M. (Z. Metallk. **64** [1973] 379/85).
[17] Neumann, G. M. (Z. Metallk. **64** [1973] 444/9).
[18] Neumann, G. M. (J. Fluorine Chem. **3** [1973] 209/25).
[19] Dettingmeijer, J. H.; Dittmer, G.; Klopfer, A.; Schröder, J. (Philips Tech. Rev. **35** [1975] 302/6; Philips Tech. Rundsch. **35** [1975] 324/7).
[20] Neumann, G. M.; Müller, U. (J. Less-Common Met. **26** [1972] 391/7).

[21] Rabenau, A. (Angew. Chem. **79** [1967] 43/9; Angew. Chem. Int. Ed. Engl. **6** [1967] 68/73).
[22] T'Jampens, G. (RGE Rev. Gen. Electr. **75** [1966] 990/6; Bull. Soc. R. Belge Electr. **83** No. 3 [1967] 247/56).
[23] T'Jampens, G. R.; van de Weijer, M. H. A. (Philips Tech. Rev. **27** [1966] 173/9; Philips Tech. Rundsch. **27** [1966] 165/71).
[24] Rees, J. M. (Light. Res. Tech. Rev. **2** [1970] 257/60).
[25] Coaton, J. R.; Philips, N. J. (Proc. Inst. Electr. Eng. **118** [1971] 871/4).
[26] Dittmer, G.; Niemann, U. (Philips J. Res. **36** [1981] 87/111).
[27] Schnedler, E. (Philips J. Res. **38** [1983] 236/47).
[28] Eckerlin, P.; Garbe, S. (Philips J. Res. **35** [1980] 320/5).
[29] Dittmer, G.; Niemann, U. (Philips J. Res. **42** [1987] 41/57).

2.5.4 Tungsten–Bromine Systems

General

In accordance with thermodynamic calculations, the first lamp experiments with bromine as the transport gas revealed that an effective cycle could be realized, but at the usual bromine dosage of a few Torr a serious filament end attack occurred. This attack could be suppressed by lowering the initial bromine pressure, but then, in the course of the lifetime of the lamp, bulb blackening appeared [27, 28] which was assumed to be due to an irreversible reaction of bromine with lamp impurities [3]. The solution was to introduce bromine in the form of HBr. While there is a buffer quantity of bromine present in the lamp (as HBr), the partial pressure of elemental bromine at low temperatures is small. For practical reasons, bromomethanes are often used instead of HBr. Normally, the compound used depends on the type of lamp being considered. Thus, a short-life lamp, with its high filament temperature and evaporation rate, may require excess bromine over hydrogen, as provited by $CHBr_3$, while for a long-life lamp, susceptible to tail (lead) erosion and loss of hydrogen by diffusion through the silica envelope, excess hydrogen, as in CH_3Br would be more appropriate. Although the use of these bromine compounds facilitates lamp processing, carbon (and hydrogen) are also introduced into the lamp atmosphere together with bromine. A further complication comes from oxidizing impurities, i.e., oxygen as well as mixtures such as water vapor-hydrogen or carbon dioxide-carbon monoxide. Therefore, the control of these impurities during lamp manufacture is very important for long-life lamps.

Hydrogen excess (beyond H : Br = 1:1) promotes bulb blackening, though as mentioned above, part of the hydrogen disappears from the lamp atmosphere by diffusion through the bulb wall during longer operation times of the lamps. The sensitivity of the cycle to hydrogen excess strongly increases if carbon is present. This suggests that perhaps oxygen, bound by carbon as rather stable CO, plays a greater role in bromine lamps than is usually assumed. It is fairly well established that the presence of oxygen increases the rate of the tungsten-bromine cycle, thus requiring less halogen to maintain bulb clarity. This is apparently due to operation of the cycle through the kinetically and thermodynamically favored W oxide bromides rather than the simple bromides. In the presence of even a little oxygen, the compound WO_2Br_2 is more important than the bromide WBr_4 at low temperature. A small quantity of oxygen appears to be necessary for the cycle to function if $H:Br=1:1$; a large quantity of oxygen would give filament end attack problems, notably trough the deleterious "water-vapor cycle". In this case, the finely divided carbon liberated by the thermal decomposition of alkyl bromides proves to be a good buffer or getter by CO formation. Deliberate CO addition to the filling gas has been proposed to reliably establish a desired oxygen level in the lamp atmosphere. Also, the addition of phosphonitrilic bromides to lamps has been advocated, which, besides supplying Br, releases phosphorus to act as a getter for oxygen.

Comparative data for the summed tungsten pressure (definition on p. 37) in incandescent lamps were evaluated for W–Br, W–Br–O, and W–Br–O–H containing transport systems (approximated by assumed lamp fillings of 10^{-6} bar WBr_6, or WO_2Br_2, or 10^{-6} bar $WO_2Br_2 + 10^{-3}$ bar HBr, respectively). The poor thermal stability of binary tungsten bromides causes a gap of the summed tungsten pressure in the W–Br system for filament temperatures between 1000 and 3000 K. Admittance of hydrogen to the W–Br–O system leads to a drastic reduction of the summed tungsten pressure in the 1500 to 2500 K range because the transport controlling compounds are reduced [39].

For discussions of the problems associated with the manufacturing, processing, and operation of W–Br lamps, see, e.g. [3, 22, 28, 31, 33].

W–Br Containing Systems

According to thermodynamic calculations [1, 2], the bromides in the binary system W–Br are stable up to temperatures high enough to establish a regenerative cycle in tungsten-bromine lamps, although the main species predicted in the gas phase, WBr_4 and WBr_2, should have a lower thermal stability than the binary W fluorides and chlorides; see also [28].

Equilibrium gas phase compositions over solid W at initial bromine pressures $p^0(Br_2)$ of 1 to 10^{-6} atm and temperatures of 500 to 3600 K were calculated from thermodynamic data in the JANAF Tables [4]. Results obtained for $p^0(Br_2)=10^{-1}$ atm are shown in **Fig. 74** (for the gas phase composition at $p^0(Br_2)=10^{-3}$ atm, see Fig. 15, p. 36). The main W-containing gas phase components at lower $p^0(Br_2)$ are WBr_4 and WBr_2; higher valent bromides play a minor role [5, 6, 21]. For gaseous bromides formed by reaction of solid W with bromine, see also p. 24.

The temperature dependence of the mass balance ratio W/Br_2 is depicted in Fig. 16, p. 38, for various $p^0(Br_2)$. As in the $W–F_2$ and $W–Cl_2$ systems, several of these curves exhibit maxima and/or minima and at certain pressures the slopes of the curves indicate W transport from hot to cold at low temperatures, a reversal of the transport direction in the range of medium temperatures, and again transport from hot to cold at very high temperatures. The point at which this latter change in the transport direction occurs strongly depends on the initial bromine pressure. It shifts with increasing $p^0(Br_2)$ from 2800 K at $\sim 2\times10^{-5}$ Torr to 3700 K at $\sim 5\times10^{-3}$ Torr [5, 6, 21]. The position of the minima of the W/Br_2 vs. T curves is given as follows [7]:

$p^0(Br_2)$ in atm	10^{-2}	10^{-3}	10^{-4}	10^{-5}
T in K	3500	3100	2800	2500
W/Br_2	7.57×10^{-3}	1.57×10^{-3}	3.21×10^{-4}	6.23×10^{-5}

For earlier calculations of the gas phase composition in the $W(s)–Br_2$ system under the conditions prevailing in incandescent filament lamps, see [8, 9].

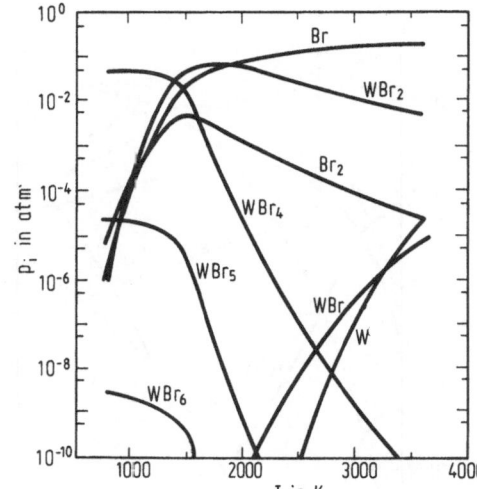

Fig. 74.   Equilibrium gas phase composition
(partial pressures p_i) of the W(s)–Br_2(–inert
gas) system for an initial bromine pressure
$p^0(Br_2)=10^{-1}$ atm [5].

Crystalline tungsten bromides which were deposited on the bulb of incandescent lamps
were identified by scanning electron microscopy and X-ray fluorescence. In a small volume
(0.8 cm³) linear lamp filled with 800 mbar Br_2 and up to 1 bar Kr, WBr_6 and WBr_5 condensed at
filament temperatures T_f of 1250 and 2100 K, respectively. During short-time operation (5 sec)
of the lamp the bulb temperature T_b was kept at 300 K. A condensate of WBr_6 was also
observed when a large-volume (300 cm³) spherical lamp filled with 10 mbar Br_2 and up to 1 bar
Kr was shortly operated at low filament temperatures ($T_f=1500$ K, $T_b=300$ K). At T_b increased
to 700 K only volatile WBr_5 was observed which was thought to arise from unstable WBr_4 and
residual Br_2 [39].

W–Br–H Containing Systems

Thermodynamic analysis show that hydrogen strongly affects the reactions and equilibria in
the W–Br system. At low temperatures, the bromine is bound as HBr, and, by the lowered concentration of free bromine, bromide formation is diminished. Formation of higher bromides is
almost completely suppressed. The degree of influence of hydrogen on the W–Br system is,
however, smaller than that on the W–F and W–Cl systems due to the lower stability of the
hydrogen halide. Equilibrium compositions of the gas phase at $Br_2:H_2$ ratios of 1:1 and initial
bromine pressures $p^0(Br_2)=10^{-2}$ and 10^{-4} atm, as calculated from thermodynamic data in the
JANAF Tables [4], are shown in **Fig. 75** and **Fig. 76**, respectively (for the gas phase composition at $p^0(Br_2)=10^{-3}$ atm and $Br_2:H_2=1:1$, see Fig. 17, p. 39). Mass balance ratios W/Br_2 for
$Br_2:H_2=1:1$ and $10^{-6}\leq p^0(Br_2)\leq10^{-1}$ atm (computed from data in the JANAF Tables [11]
or estimated relations for the reaction enthalpies and entropies) are depicted in **Fig. 77**
(for results based on earlier thermodynamic values [4], see Fig. 18, p. 39). As can be seen, the
transport from cold to hot at low temperatures, required for prevention of bulb blackening in
incandescent lamps, will be severely impaired and partially suppressed by the presence of H_2.
The detrimental influence of hydrogen increases with decreasing $Br_2:H_2$ ratio, as **Fig. 78**
demonstrates [6, 10, 21].

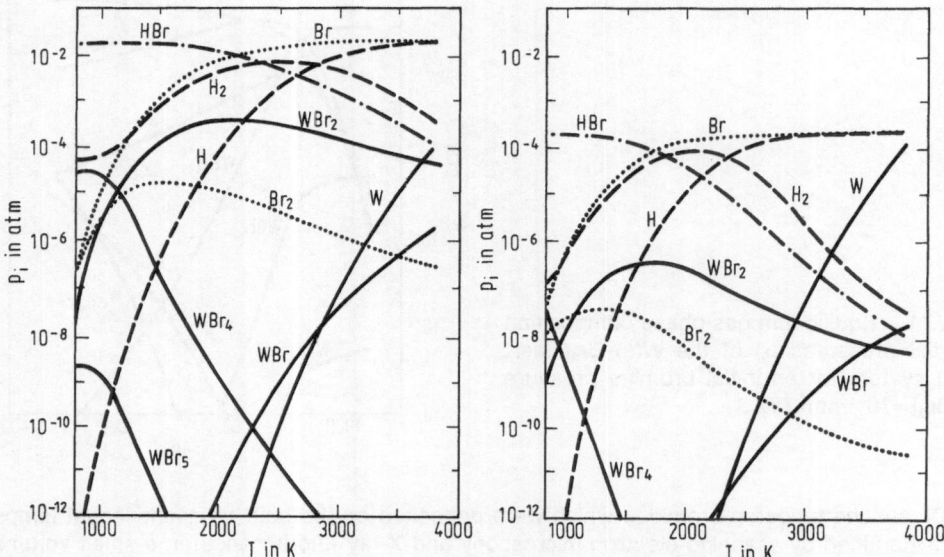

Figs. 75, 76. Equilibrium gas phase composition (partial pressures p_i) of the W(s)–Br_2–H_2(–inert gas) system for Br_2:H_2=1:1 and initial bromine pressures $p^0(Br_2)$=10^{-2} atm (Fig. 75, left) and 10^{-4} atm (Fig. 76, right) [10].

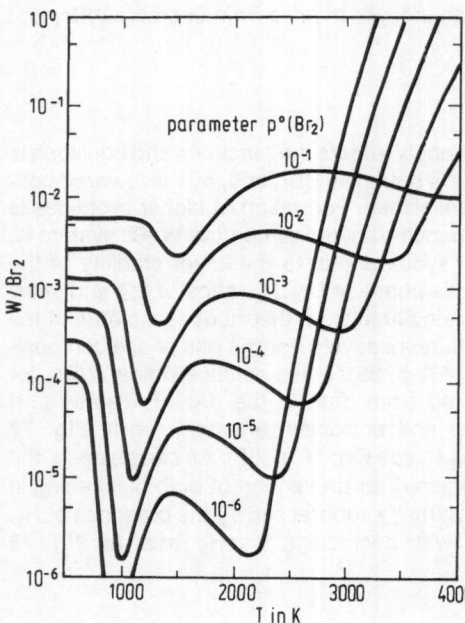

Fig. 77. Mass balance ratio W/Br_2 in the gas phase of the W(s)–Br_2–H_2(–inert gas) system for Br_2:H_2=1:1 and various initial bromine pressures $p^0(Br_2)$ (in atm) given as parameters at the curves [21].

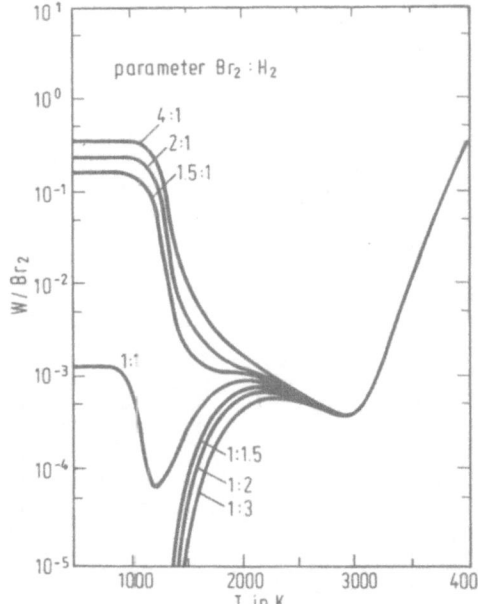

Fig. 78.   Mass balance ratio W/Br$_2$ in the gas phase of the W(s)–Br$_2$–H$_2$(–inert gas) system for an initial bromine pressure of 10^{-3} atm and various Br$_2$:H$_2$ ratios given as parameters at the curves [21].

More recent calculations [3], based on thermochemical data from the JANAF Tables [11] and the authors' own measurements and estimations of reaction enthalpies and entropies, essentially confirmed the above conclusions. The gas phase in the W(s)–Br–H containing system at H:Br=1:1 and lower temperatures was found to consist mainly of HBr. Thus the free bromine pressure in this temperature range is low, which will lead to less filament end attack in W–Br lamps. A strong reduction of the WBr$_4$ pressure is predicted, which is proportional to $p^2(Br_2)$. On the other hand, the partial pressure of WBr$_2$, which becomes important at higher temperatures, is only proportional to $p(Br_2)$. Therefore, $p(WBr_2)$ is less sensitive to hydrogen addition, even more so because HBr begins to dissociate in this temperature region. The influence of the H:Br ratio on the effective total W pressure $\Sigma p(W)$ for some temperatures is demonstrated in **Fig. 79** for a total pressure of 0.01 atm. In **Fig. 80**, $\Sigma p(W)$ is shown as a function of the temperature at fixed H:Br ratios. Fig. 80 reveals an increasing risk of bulb blackening in practical W–Br lamps with increasing H:Br ratio. At H:Br=1, one would expect a small transport of W from filament to bulb wall. In fact, it is known from lamp experiments that a large excess of hydrogen will lead to blackening problems, but the practical situation appears to be less critical than theory predicts. At a ratio H:Br=1, for example, bromine lamps of good quality can be made without great difficulty [3]. The calculated radial separation of bromine and hydrogen in a radially symmetrical W–Br–H lamp due to concentration and thermal diffusion is shown in **Fig. 81**. The lamp filling is assumed to consist of 1.5 mbar HBr and Kr. It can be seen that H$_2$ is enriched in the bulb region, which may lead to an increased reduction of W bromides. The corresponding partial pressure distribution of the gas phase components is given in **Fig. 82**. The partial pressures of the tungsten bromides are below 10^{-5} bar at the bulb [36].

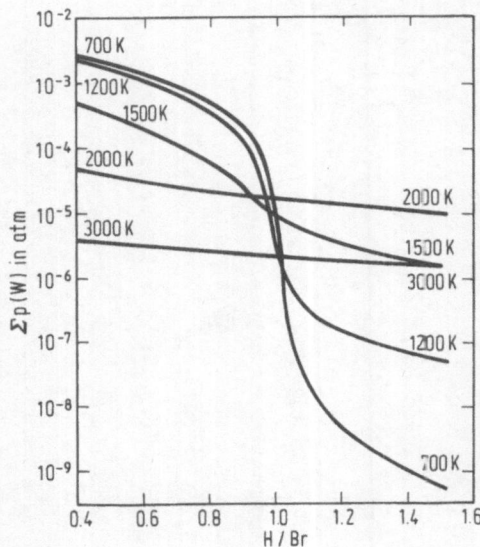

Fig. 79.   Effect of the H : Br ratio on the effective total W pressure $\Sigma p(W)$ in the gas phase of W(s)–Br–H containing systems at various constant temperatures and a total pressure of 0.01 atm [3].

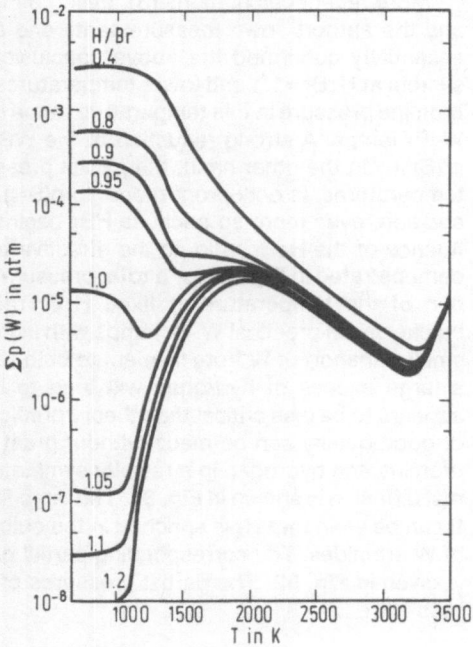

Fig. 80.   Temperature dependence of the effective total W pressure $\Sigma p(W)$ in the gas phase of W(s)–Br–H containing systems at various fixed H : Br ratios (given as parameters at the curves) and a total pressure of 0.01 atm [3].

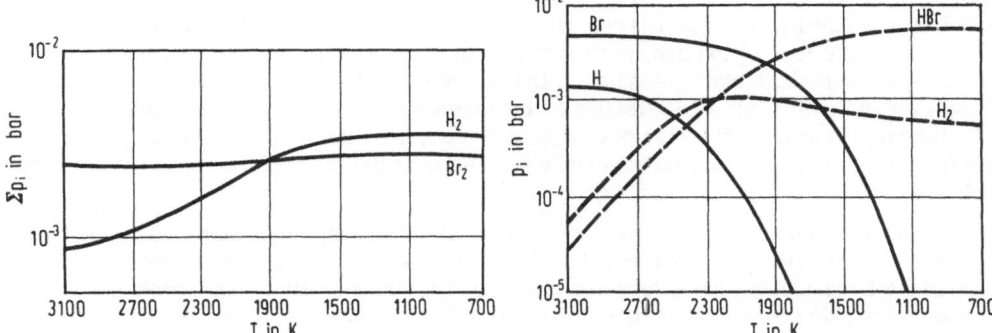

Fig. 81 (left).   Separation effects due to concentration and thermal diffusion calculated for a W–HBr(–Kr) containing lamp; Σp_i = stoichiometric sum of partial pressures, assumed HBr filling pressure 1.5 mbar [36].

Fig. 82 (right).   Equilibrium gas phase composition (partial pressures p_i) in a radially symmetrical W–HBr(–Kr) lamp with an assumed HBr filling pressure of 1.5 mbar [36].

The direction of the chemical transport reactions in the high-temperature region of the W–Br–H system was experimentally studied in a quartz jacket containing two separate (bow-shaped) incandescent filaments (diameter ~ 250 µm) of W arranged opposite to each other. The gas filling of the jacket consisted of 1500 Torr Ar and varying amounts of $CHBr_3$ to establish a desired bromine pressure of 10^{-4} to 10^{-2} atm. The W filaments were resistively heated to different temperatures. The direction of W transport was judged from the transition of ^{185}W from the one filament activated by neutron irradiation to the other (originally inactive) filament. Qualitative results are presented in the following table:

No. of test	initial Br_2 pressure in atm	temperature in K T_1	T_2	mass balance ratio W/Br_2 (calculated) T_1	T_2	transport direction observed
1	10^{-2}	1100	1400	0.33×10^{-2}	0.38×10^{-2}	$T_2 \rightarrow T_1$
2	10^{-2}	1100	1600	0.33×10^{-2}	0.28×10^{-2}	$T_1 \rightarrow T_2$
3	10^{-3}	1100	2500	0.32×10^{-3}	0.01×10^{-3}	$T_1 \rightarrow T_2$
4	10^{-4}	1100	2500	0.23×10^{-4}	0.01×10^{-4}	$T_1 \rightarrow T_2$
5	10^{-4}	2800	3300	0.01×10^{-4}	0.02×10^{-4}	$T_2 \rightarrow T_1$

Attempts at quantitative determinations were discontinued because of strongly scattered results. The observed directions of transport were in fair agreement with the above theoretical calculations [6, 10] of the reaction equilibria based on thermodynamic principles [20].

For additional information on the W(s)–Br–H system, particularly for the case where bromine and hydrogen are introduced as HBr, see "Tungsten" Suppl. Vol. A 7, 1987, pp. 74/7.

W–Br–O Containing Systems

Although the stability of the binary bromides in the W–Br system should be sufficient to support a regenerative cycle in bromine incandescent lamps, addition of oxygen greatly enhances the transport processes. According to thermodynamic calculations [2], addition of oxygen to the filling gas will result in the formation of thermally rather stable oxide bromides, which should play a prominent part in the processes taking place and enhance the W transport. The beneficial effect of oxygen, particularly in the presence of carbon, in certain bromine lamps to eliminate bulb blackening has been analyzed by [3]; see the discussion in "Tungsten" Suppl. Vol. A 7, 1987, pp. 77/9. Results on the kinetics of the reaction of oxygen and bromine with tungsten indicate that traces of oxygen will in any case be required in W–Br lamps, since the formation of WO_2Br_2 is also kinetically favored over the formation of binary W bromides [16, 17].

Equilibrium compositions of the gas phase in the $W(s)–Br_2–O_2$ system, calculated from thermochemical data in the JANAF Tables [4], are shown in **Fig. 83** and **Fig. 84** for a $Br_2:O_2$ ratio of 1:1 and initial bromine pressures $p^0(Br_2)$ of 10^{-2} and 10^{-4} atm, respectively (for the gas phase composition at $p^0(Br_2)$ of 10^{-3} atm and $Br_2:O_2=1:1$, see Fig. 19, p. 40). The main constituent at lower temperatures is, besides atomic Br, the oxide bromide WO_2Br_2, while $WOBr_4$ does not occur in significant amounts. In the range of lower temperatures, solid WO_2 will also be deposited at widely varying $Cl_2:O_2$ ratios. Its deposition temperature depends strongly on $p^0(Br_2)$ and the bromine content of the gas phase; see Fig. 21, p. 42. Gaseous W oxides appear at higher temperatures. In this range, the W–Br und W–O systems exist independently of one another. Mass balance ratios $W/(Br_2+O_2)$ were derived for a ratio $Br_2:O_2=1$ and various $p^0(Br_2)$ between 10^{-1} and 10^{-6} atm as well as for $Br_2:O_2$ varying between 1:0 and 0:1 and a fixed $p^0(Br_2)$ of 10^{-4} atm; see Fig. 20, p. 42. The addition of oxygen increases the solubility of W in the gas phase in the range of lower temperatures and conserves this increased solubility up to higher temperatures, due to the higher thermal stability of the oxide bromides formed [1, 12, 21]. Independent calculations of the equilibria and vapor pressures of

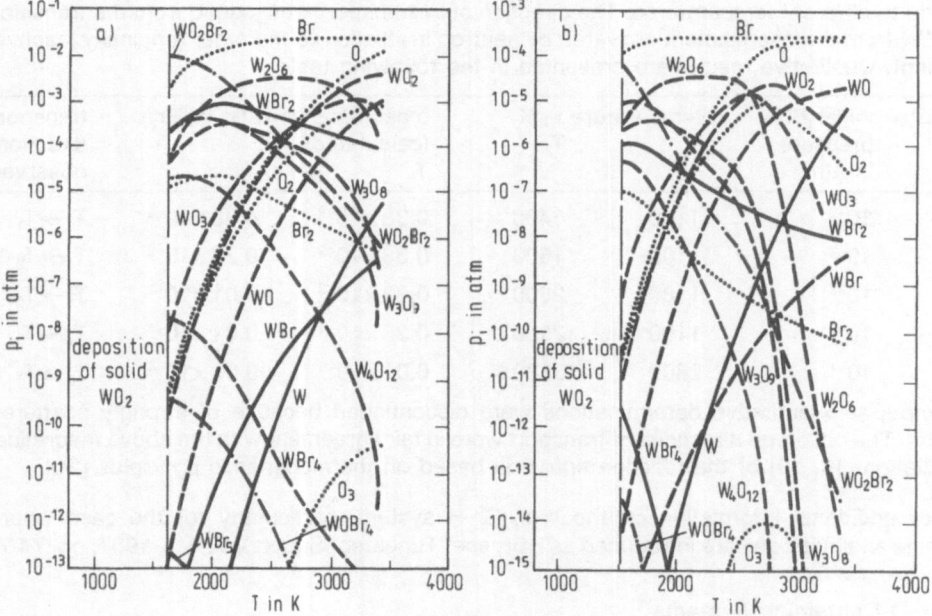

Figs. 83, 84.　Equilibrium gas phase composition (partial pressures p_i) of the $W(s)–Br_2–O_2(–inert\ gas)$ system at $Br_2:O_2=1:1$ and initial bromine pressures $p^0(Br_2)=10^{-2}$ atm (Fig. 83, left) and 10^{-4} atm (Fig. 84, right) [12].

the W-transporting species in the W–Br_2–O_2 system were carried out for the following initial and operating conditions: $p^0(Br_2) \sim 10^{-3}$ to 10^{-2} atm, $p^0(O_2) = 10^{-6}$ to 10^{-2} atm, Ar fill pressure 1 atm, operating pressure 4 atm, temperature range 500 to 3500 K [13, 14].

An extended mass-transfer model of halogen-doped incandescent lamps is formulated including both thermochemical considerations, and kinetic analyses. In this model, the transport processes occurring in these lamps are represented by diffusion and counter diffusion of gaseous species across the stagnant boundary layer surrounding the filament with simultaneous reactions at the bulb wall and the filament. The model is applied to a specific type of 500 W/120 V quartz lamp in which the initial gas phase is argon with small concentrations of bromine and oxygen. The tungsten flux from wall to filament and vice versa and, thus, the conditions of wall blackening are calculated as a function of the initial bromine and oxygen mole fractions, the filament and wall temperature, total pressure, and wall condensate, i.e., W(s), WO_2(s), and WO_3(s) [15].

WO_2Br_2 and a mixture of $WOBr_2$, WO_3, and WO_2 were deposited on the walls of small volume (0.8 cm³) incandescent lamps which were filled with 400 mbar Br_2, 200 mbar O_2, and up to 1 bar Kr and were operated at filament temperatures of 1250 and 2100 K, respectively. During short-time operation the bulb temperature was kept at 300 K. In large volume (300 cm³) incandescent lamps which were filled with 10 mbar Br_2, 5 mbar O_2 and up to 1 bar Kr, a mixture of $WOBr_4$ and WO_2Br_2 condensed on the bulb when the lamp was shortly operated at high and low filament temperatures, $T_f = 1500$ and 3000 K, $T_b = 300$ K. An increase of the bulb temperature T_b to 700 K led to $WOBr_3$ and $WOBr_2$ deposits at the current leads for $T_f = 1500$ K. At high filament and bulb temperatures ($T_f = 3000$ K, $T_b = 700$ K) deposits of $WOBr_2$, WO_2, and WO_3 were observed [39].

W–Br–O–H Containing Systems

The sensitivity of the transport cycle in tungsten-bromine lamps to hydrogen excess strongly increases when carbon is present; see p. 90. This suggests that perhaps oxygen plays a greater role in bromine lamps than is usually assumed. The use of oxygen in certain bromine lamps to eliminate blackening can perhaps be seen in this light [3]. Usually the presence of oxygen in W–Br lamps is linked with that of hydrogen. Both get into the lamp atmosphere in the form of H_2O as an impurity in the filling gas or adsorbate on the inner lamp parts, or else can be released from OH groups of the quartz bulb [6, 37].

In order to gain insight into the role of oxygen in bromine lamps, calculations on the W–Br–O–H system have been made. A total of 25 gaseous species was taken into account. The results obtained for a total pressure of 0.01 atm and ratios Br : H = 1 : 1 and Br : O = 100 : 1 are shown in **Fig. 85** and support the view that in the presence of even very small amounts of oxygen, the compound WO_2Br_2 is more important than the bromide WBr_4 at low temperatures. The influence of oxygen on the solubility (total effective pressure) of W in the gas phase under these conditions is depicted in "Tungsten" Suppl. Vol. A 7, 1987, p. 77, Fig. 11). A small quantity of oxygen is necessary for the transport cycle to function at Br : H = 1 : 1; a large quantity of oxygen would give problems with respect to the attack on cooler tungsten parts (filament ends). If the Br : H ratio falls below unity, the presence of oxygen becomes as indispensible for the regenerative cycle as it is in the W–I system [3].

Equilibrium gas phase compositions calculated for initial bromine pressures $p^0(Br_2) = 10^{-6}$ to 10^{-1} atm and different $Br_2 : O_2 : H_2$ ratios at 1000 to 3600 K have been published [18]. Results obtained for $p^0(Br_2) = 10^{-3}$ atm and $Br_2 : O_2 : H_2 = 1 : 0.5 : 1$ are given in **Fig. 86**. The phase equilibria are controlled by the competition between the formation of WO_2Br_2 on one hand and HBr and H_2O on the other. Below 1600 K, deposition of solid WO_2, occurs when the proportion of oxygen is comparable to that of bromine. The gas phase is thus gradually depleted of O and the transport processes occurring become similar to those in the O-free system. When oxygen and hydrogen are introduced into the W–Br system in the form of H_2O, the influence of oxygen on the reactions is more pronounced than that of hydrogen. The influence of water vapor and hydrogen on the temperature dependence of the mass balance ratio $W/(Br_2 + O_2)$ is shown in figures in the paper; for the influence of oxygen on $W/(Br_2 + O_2)$ in the gas phase at $Br_2 : H_2 = 1:1$ ($p^0(Br_2) = 10^{-3}$ atm), see Fig. 22, p. 43 [6, 18].

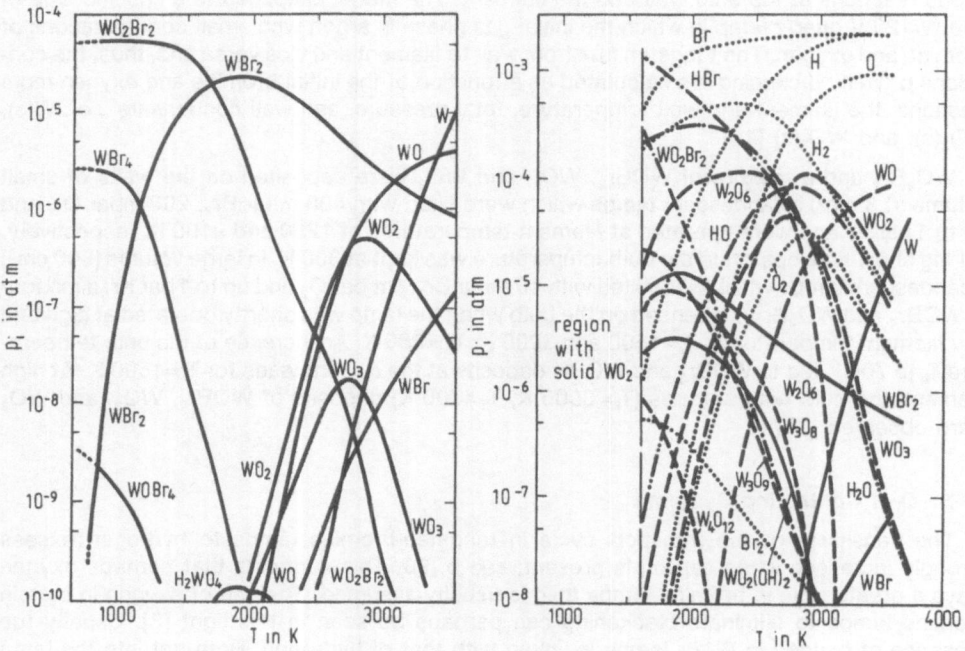

Fig. 85 (left). Equilibrium composition (partial pressures p_i) of the gas phase in the W(s)–Br–O–H containing system at H:Br=1, Br:O=100, and a total pressure of 0.01 atm. Only the partial pressures p_i of the W-containing components are given [3].

Fig. 86 (right). Equilibrium composition (partial pressures p_i) of the gas phase in the W(s)–Br_2–O_2–H_2(–inert gas) system at an initial bromine pressure $p^0(Br_2) = 10^{-3}$ atm and a ratio $Br_2 : O_2 : H_2 = 1 : 0.5 : 1$ [18].

The calculated partial pressure distribution for a radially symmetrical lamp containing 1.5 mbar HBr, 10^{-2} mbar O_2, and Kr is shown in **Fig. 87**. Concentration as well as thermal diffusion is assumed. A comparison with Fig. 82, p. 85, demonstrates that very small amounts of oxygen have a major effect on the tungsten solubility. The addition of 10^{-2} mbar O_2 increases the partial pressures of the tungsten compounds by about a factor 7 in the bulb region. A higher W transport rate will result along the axis of the filament of a real lamp [36].

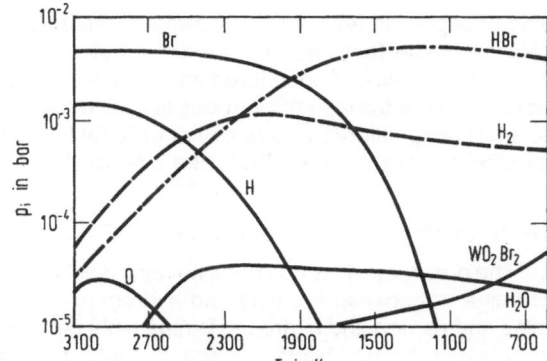

Fig. 87. Equilibrium gas phase composition (partial pressures p_i) in a radially symmetrical W–HBr–O$_2$(–Kr) lamp. Assumed filling pressures: 1.5 mbar HBr, 10^{-2} mbar O$_2$ [36].

The W oxide bromides which form in the low-temperature gas phase region of a W–Br lamp have been analysed by a sampling method using diffusion-limited mass transport in the temperature gradient of a small attached capillary and forced precipitation of the respective compounds. The lamp was filled 3×10^{-2} bar HBr and 0.5 bar Ar; the filament temperature was 3150 K. X-ray diffraction diagrams of the precipitates in the capillary indicated the presence of WO$_2$Br$_2$, WOBr$_2$, and W$_6$Br$_{18}$ in the intensity ratio 400:10:1. The amount of WO$_2$Br$_2$ corresponded to a pressure of (2 to 5)$\times 10^{-3}$ bar in the bulb wall region. These results were qualitatively confirmed by IR absorption measurements. Deposits of W oxide bromides which are sometimes formed in the low-temperature regions of the bulbs of tubular lamps were likewise indentified to be composed of WO$_2$Br$_2$. These deposits are formed only at filling pressures of HBr (or CH$_2$Br$_2$) above 10^{-2} bar. The experiments support thermodynamic equilibrium calculations assuming a water pressure of the order of 10^{-4} bar in practical lamps and WO$_2$Br$_2$ as the dominate species in the bulb wall region [37].

For earlier calculations of the equilibrium gas phase composition at 500 to 3000 K, p^0(Br$_2$)$=$ 10^{-2}, p^0(O$_2$)$=10^{-3}$, p^0(H$_2$)$=10^{-3}$, or 10^{-2} atm, and an Ar fill pressure of 1 atm (total operating pressure 4 atm), see [13]. Total effective W pressures were calculated for temperatures between 500 and 3500 K, an assumed initial bromine pressure of 2×10^{-3} atm, and various oxygen to bromine as well as hydrogen to bromine ratios [14].

W–Br–N Containing Systems

The use of halophosphonitriles (phosphorus nitride chlorides) as halogen sources in incandescent lamps has been proposed [33]. These compounds offer a novel and reliable method for the manufacture of such lamps. In the case of bromine, the cyclic polymers (PNBr$_2$)$_3$ and (PNBr$_2$)$_4$ can be dissolved in a volatile nonpolar solvent such as benzene or ether, and an accurate volume dispensed rapidly and repeatably into a lamp. The solvent can then be readily removed either by evacuation or by flushing with an inert gas. This leaves the bromophosphonitrile as a crystalline deposit on the bulb wall or filament, which can then be thermally decomposed by the heat of the filament when the lamp is aged:

$$(PNBr_2)_3 \longleftrightarrow (PN)_3 + Br_2$$
$$\updownarrow$$
$$P_3N_5$$
$$\updownarrow$$
$$PNBr_n + PBr_3 \xleftarrow{\ Br_2\ } \begin{cases} PN + N_2 \\ \updownarrow \\ P_2 + N_2 \end{cases}$$

The phosphorus and the PN radicals formed can recombine with the bromine in the colder regions of the lamp and thus exert a buffering action in the same way as hydrogen from hydrogen bromide. Phosphorus has the considerable advantage over hydrogen in that it cannot diffuse out through the vitreous silica envelope to leave an excess of bromine, which would result in lamp failure by corrosion of the cold tungsten. It acts as a getter for impurities such as oxygen and water vapor [33], see also [34].

W–Br–C(–H) Containing Systems

The presence of carbon does not appreciably influence the reactions between tungsten and bromine or between bromine and hydrogen. Compound formation between carbon and bromine and carbon and hydrogen is negligible under the conditions prevailing in W–Br incandescent lamps. Carbon is often introduced into these lamps in the form of the compounds CBr_4, $CHBr_3$, CH_2Br_2, or CH_3Br, which dissociate at higher temperatures [6, 19, 21, 27, 28, 31]. In the case of a $CHBr_3$ addition to the fill gas, the transport reactions are very similar to those in the W–Br(–inert gas) system, due to the high excess of Br (for the transport directions experimentally observed, see p. 85) [20].

The effect of the different H : Br ratios established in the lamp atmosphere by the addition of $CHBr_3$ or CH_2Br_2 to the filling gas (3.2 atm Ar) on the performance (particularly, the useful lifetime) of W–Br lamps has been experimentally studied [31]. The filament temperatures were in the region of 3000 K, the bulb temperatures ranged between ~470 and ~620 K. The internal lamp volume was ~0.85 cm³. At too low levels of the additives (<1.2 µg of $CHBr_3$, 2 µg of CH_2Br_2), bulb blackening of the lamps occurred in both cases. Above ~25 µg, a drastic reduction of lamp life by filament tail attack was observed with $CHBr_3$, while CH_2Br_2 additions up to 110 µg did not lead to any tail attack. The likely reason for the tail protection in the latter case is the availability of additional hydrogen which combines with free bromine to form HBr in the cooler regions of the lamps, the HBr thus formed being less reactive towards the cool tungsten. Thus the use in long-life lamps of compounds having low H : B ratios, such as bromoform, requires careful control of the dose added to the lamp to avoid bulb blackening at low doses and tail attack at high doses. The lower availability of bromine in lamps containing compounds with a high H : Br ratio, such as dibromomethane, demands larger doses to prevent bulb blackening [31].

Thermodynamically calculated compositions of the gas phase over solid W at an initial bromine pressure of 10^{-3} atm and a ratio $Br_2 : H_2 : C = 1:1:1$ are shown in **Fig. 88**. Solid carbon is deposited at temperatures below 3000 K and can lead to the formation of W carbide resulting in an embrittlement of the incandescent filament in W–Br lamps. The temperature dependence of the mass balance ratios W/Br_2 under the conditions of Fig. 88. but with various initial bromine pressures, is represented in **Fig. 89**. A comparison with Fig. 77, p. 82, indicates that a noticeable influence of C on the chemical transport processes in the W–Br–H system is not to be expected [6, 19].

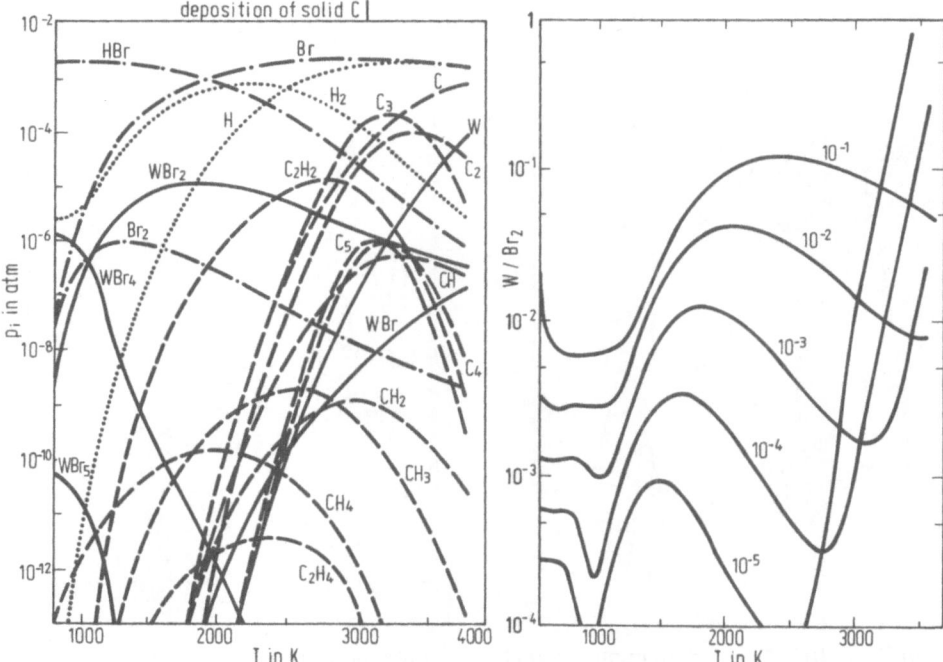

Fig. 88 (left). Equilibrium composition of the gas phase (partial pressures p_i) in the W(s)–Br_2–H_2–C(–inert gas) system at an initial bromine pressure $p^0(Br_2)=10^{-3}$ atm and a ratio $Br_2:H_2:C=1:1:1$ [19].

Fig. 89 (right). Mass balance ratio W/Br_2 in the gas phase of the W(s)–Br_2–H_2–C (–inert gas) system at $Br_2:H_2:C=1:1:1$ and various initial bromine pressures $p^0(Br_2)$ (in atm) given as parameter at the curves [19].

W–Br–O–C Containing Systems

The influence of carbon on the chemical reactions and the transport processes in the W–Br–O system is qualitatively similar to that in the W–F–O system; see p. 51. Deviating from the W–F–O system, solid WO_2 is deposited at temperatures below 1000 K when larger amounts of oxygen are present. The gas phase compositions thermodynamically calculated for CO_2 and CO additions at an initial Br_2 pressure of 10^{-3} atm are shown in **Fig. 90** and **Fig. 91**, respectively. Corresponding mass balance ratios W/(Br+O_2) are depicted in **Fig. 92** and **Fig. 93**, respectively [5, 19, 21].

Calculations of the equilibrium partial pressures in the W(s)–Br–O–C containing system had earlier been performed for initial fill and operating conditions compatible with those in W–Br lamps. A figure presented in the paper shows the results obtained for $p^0(Br_2)=2.8\times10^{-4}$, $p^0(O_2)=10.0\times10^{-4}$ atm, a C amount of 3.2×10^{-8} g-atom, and an Ar fill pressure of 1 atm (which corresponds to ~4 atm at operating conditions) [13]. In agreement with other findings [5, 19], it is concluded that carbon in amounts equal to or larger than the residual (free) oxygen ties up the latter to form CO, which is very stable at intermediate and high temperatures. It is further assumed that it should be thermodynamically possible for carbon to react with tungsten to form tungsten carbide over the entire temperature range of 500 to 3500 K sudied. Because of

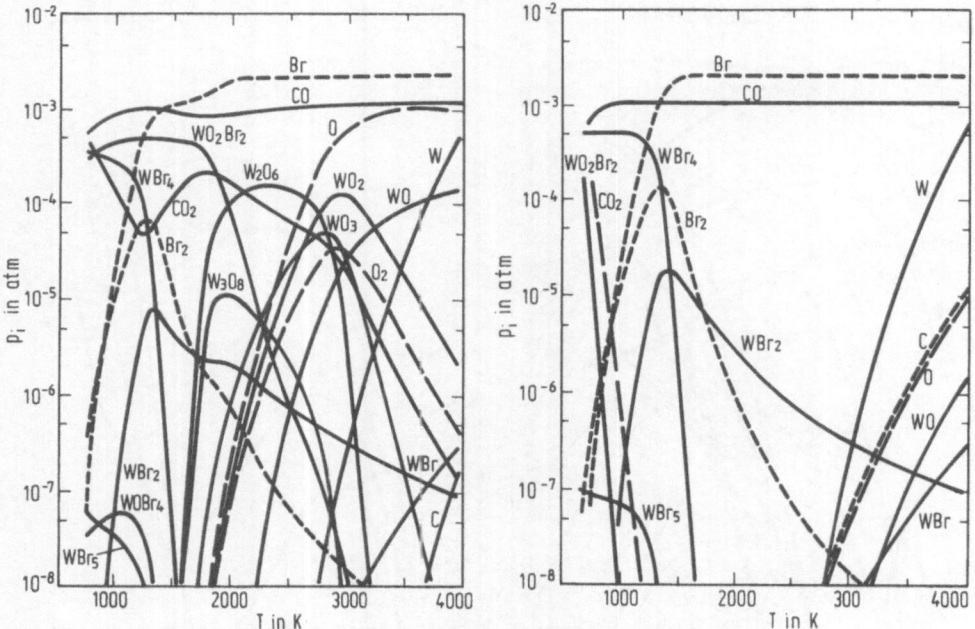

Fig. 90 (left).   Equilibrium composition (partial pressures p_i) of the gas phase in the
W(s)–Br_2–CO_2(–inert gas) system at an initial bromine pressure $p^0(Br_2) = 10^{-3}$ atm and
a ratio $Br_2 : CO_2 = 1:1$ [21].

Fig. 91 (right).   Equilibrium composition (partial pressures p_i) of the gas phase in the
W(s)–Br_2–CO(–inert gas) system at an initial bromine pressure $p^0(Br_2) = 10^{-3}$ atm and
a ratio $Br_2 : CO = 1:1$ [21].

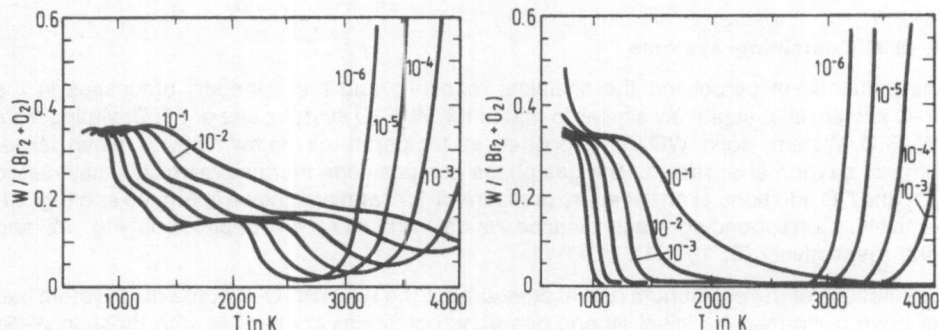

Fig. 92 (left).   Mass balance ratio $W/(Br_2 + O_2)$ in the gas phase of the W(s)–Br_2–
CO_2(–inert gas) system at $Br_2 : CO_2 = 1:1$ and various initial bromine pressures
$p^0(Br_2)$ (in atm) given as parameter at the curves [21].

Fig. 93 (right).   Mass balance ratio $W/(Br_2 + O_2)$ in the gas phase of the W(s)–Br_2–
CO(–inert gas) system at $Br_2 : CO = 1:1$ and various initial bromine pressures $p^0(Br_2)$
(in atm) given as parameter at the curves [21].

the high thermal stability of CO, the volatilization of W via tungsten oxides should be practically eliminated and gaseous WO_2Br_2 should be stable to lower temperatures. The stability of $WOBr_4(g)$ and $WO_2Br_2(g)$ should not greatly be affected in the presence of C at low temperatures, at which $WOBr_4(g)$ is assumed to prevail over $WO_2Br_2(g)$, at least under the conditions of the graph mentioned. The respective pressures of the oxide bromides should be sufficiently high to prevent precipitation of WO_2 at the wall [13]. The temperature dependence of the total effective W pressure $\Sigma p(W)$ in the gas phase is demonstrated in Fig. 97 for various ratios oxygen:bromine and oxygen:carbon. Whenever the amount of oxygen is less or equal to that of carbon, an upgradient transport of W is predicted between 500 and approximately 2100 K. In the neighborhood of 2100 K, the $\Sigma p(W)$ values reach a minimum and above this temperature, the W transport is downgradient. If the amount of oxygen exceeds that of carbon, the transport resembles that in the W–Br–O system for an oxygen:bromine ratio corresponding to the excess amount of oxygen after fixation of carbon as CO [14].

Measurements of the weight changes of a W coil in a closed system gave the dependence of the summed (total effective) tungsten pressure and the gas phase composition on the W filament temperature shown in **Fig. 94** and **Fig. 95**, respectively. The glass bulb contained 20 mbar Br_2, 5 mbar O_2, 10 mbar CO, and 0.5 bar Kr. The CO added to the W–Br–O system acts as an efficient getter for excess oxygen in the colder parts of the lamp, thus preventing W oxide bromides from decomposing into solid WO_3 and bromine. Fig. 95 demonstrates the dominating role of the WO_2Br_2 species for the transport process [35].

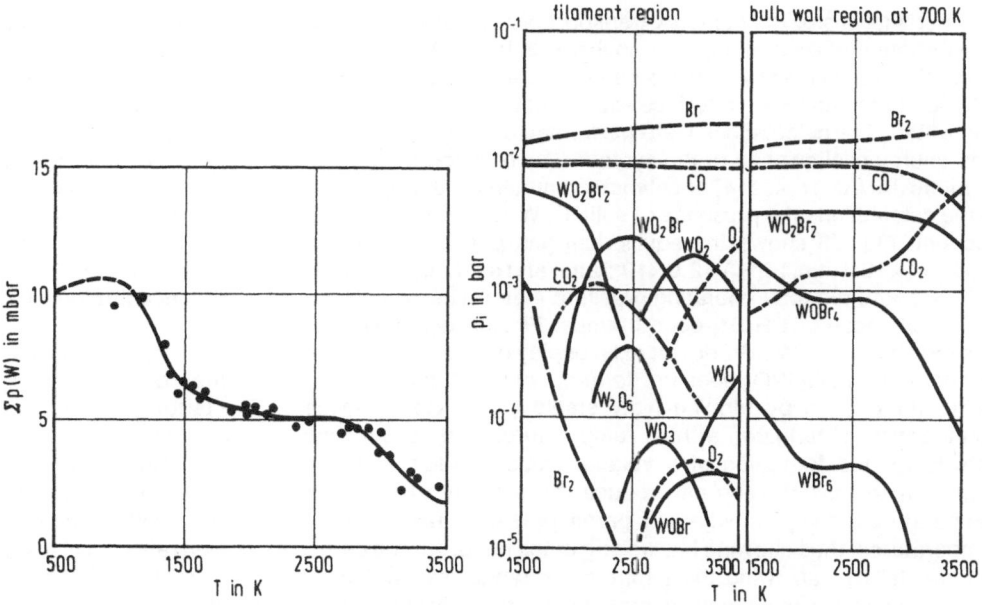

Figs. 94, 95. Summed (total effective) tungsten pressure $\Sigma p(W)$ (Fig. 94, left) and gas phase composition (partial pressures p_i) at the filament and the bulb wall of a W–Br_2–O_2–CO(–Kr) lamp (Fig. 95, right) as a function of the filament temperature. Bulb wall at 700 K; fill pressures at 700 K: 20 mbar Br_2, 5 mbar O_2, 10 mbar CO [35].

W–Br–O–C–H Containing Systems

Oxygen, hydrogen, and carbon are usual additions or impurities in practical W–Br lamps, see also p. 79. Usually the presence of oxygen in W–Br lamps is linked with that of hydrogen. Both get into the lamp atmosphere in the form of H_2O as an impurity in the filling gas or adsorbate on the inner lamp parts, or else can be released from OH groups of the quartz bulb; see the discussions in [3, 6, 22 to 24, 31].

The sensitivity of the transport cycle in bromine lamps to hydrogen excess (see p. 84) strongly increases if carbon is present. This suggests that perhaps oxygen plays a greater role in bromine lamps than is usually assumed. When H and C are introduced in the form of CH_2Br_2, the removal of oxygen by carbon as CO must be expected. Calculations on the W–Br–O–C–H containing system show that in this case more oxygen is needed for an effective cycle. This can certainly be considered a drawback, but, however, a greater tolerance for oxygen results, since the larger part reacts with carbon to form the stable compound CO; see "Tungsten" Suppl. Vol. A 7, 1987, pp. 77/8, Fig. 12. Lamp blackening can be expected if the wall temperature is of the order of 1100 K [3]. On the other hand, the minimum amount of CH_2Br_2 (e.g., 0.6%) required for maintaining the halogen cycle and preventing bulb blackening can be reduced to less than a tenth, if a few Torr of CO are added to the lamp filling. The lifetime of the lamps remains unchanged. Only at CH_2Br_2 concentrations $\geq 0.2\%$ and CO partial pressures ≥ 10 Torr, do tungsten crystals form on the incandescent filament, and the lifetime of the lamps is thus reduced. The growth rate of the crystals increases with increasing proportions of CH_2Br_2 and CO and the lifetime of the lamps decreases drastically [26].

First analysis of the complex transport processes occurring in practical W–Br lamps have been attempted on a purely thermochemical basis. The gas phase composition of the W(s)–Br–O–C–H containing system has been calculated assuming that thermodynamic equilibrium is attained at the gas/solid interface with tungsten and that the rate of reaction is not kinetically controlled. The calculations were performed employing a procedure which was based on the free-energy minimization of the system. The required thermodynamic data were either taken from the JANAF Tables [4] or estimated enthalpies, entropies, and heat capacities were used. In the calculations the presence of solid C, W_2C, and WC in addition to solid W was taken into account. **Fig. 96** shows the equilibrium gas phase composition for $p^0(Br_2) = 2.8 \times 10^{-4}$ atm, $p^0(O_2) = 1 \times 10^{-4}$ atm, $p^0(H_2) = 2.8 \times 10^{-4}$ atm, and a C amount of 3.2×10^{-8} g-atom; the Ar fill pressure is 1 atm, the total operating pressure 4 atm. The initial Br_2, H_2, and C concentrations are typical of those in CH_2Br_2-doped lamps. Fig. 96 demonstrates that volatile W oxides and gaseous WBr_5 and $WOBr_5$ do not exist or play only an insignificant role under these conditions. Furthermore, solid WO_2 does not form at lamp wall temperatures. The importance of C as a buffer for oxygen by forming higly stable CO is clearly evident. The latter prevents the volatilization of tungsten at high temperatures and yet provides a sufficiently high oxygen activity for the formation of a volatile oxide bromide at bulb wall temperatures. A large hydrogen to oxygen ratio can reduce the stability of the oxide bromide at the bulb wall temperature to the point where its partial pressures falls below that of W at practical filament temperatures, thus allowing blackening of the lamp. A large carbon to oxygen ratio can also be detrimental by narrowing the temperature range over which WO_2Br_2 is stable [13]. These thermochemical calculations are stated to be in conflict with the available thermodynamic data on tungsten oxide bromides. Furthermore, the existence of the condensed WO_2 phase is considered essential in the analysis of the W–Br–O–C and W–Br–O–C–H containing systems [32].

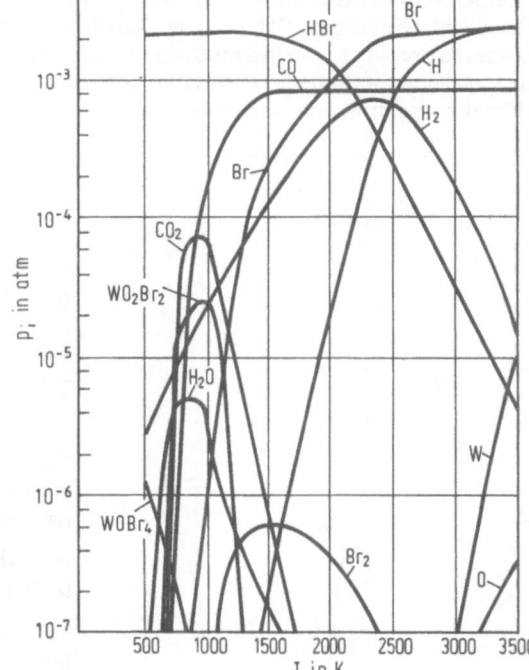

Fig. 96. Equilibrium gas phase composition (partial pressures p_i) of the W(s)–
Br_2–O_2–C–H_2(–Ar) system for $p^0(Br_2)$=
2.8×10^{-4} atm, $p^0(O_2) = 1 \times 10^{-4}$ atm, $p^0(H_2)$
$= 2.8 \times 10^{-4}$ atm, and a C amount of
3.2×10^{-8} g-atom. The Ar fill pressure is
1 atm, the total operating pressure 4 atm
[13].

The total effective W pressure in the gas phase of the W–Br_2–O_2–C and W–Br_2–O_2–H_2–C
systems calculated for various conditions is depicted in **Fig. 97**. As can be seen, the pressure
of hydrogen in a carbon-containing W–Br lamp, at a level comparable to that of bromine
(i.e., H:Br=1), affects the W transport significantly only at lamp wall temperatures, as long as
O:Br<1 and O:C<1. From the trend of the curves in the high-temperature region at O:C<1, a
shortened lifetime of the W filaments could be expected. However, experience has shown that
no accumulation of W in the colder parts of the filament occurs if the bromine is introduced in
the form of an alkyl bromide (e.g., CH_2Br_2). It may be that most of the carbon is deposited on
the low-temperature parts of the lamp upon the initial pyrolysis of the alkyl bromide and thus
cannot participate in the high-temperature reactions of the system. Presumably, this would
result in O:C>1 at filament temperatures, and if so, the W transport behavior should become
similar to that which is predicted for the W–Br–O–H and W–Br–O containing systems where the
minimum is closer to practical filament temperatures of tungsten halogen lamps [14]. Similar
calculations of $\Sigma p(W)$ were performed for various other combinations of bromine, oxygen,
hydrogen, and carbon. These were assumed to be established by adding appropriate amounts
of CBr_4, CH_2Br_2, O_2, H_2, H_2O, or C to the lamp filling. The ratios of the constituents varied
between 0 and 3.0 for H:Br, 0.1 and 1.0 for O:Br, and 0.2 to 2.0 for O:C. The $\Sigma p(W)$ vs. T
curves all showed minima between 2000 and 2200 K and (except in the absence of hydrogen)
maxima at ~950 K. The results were compared with bulb blackening experiments on lamps
with additions of 0.02 to 0.1% bromomethanes or 0.2 HBr to the fill gas. The basic inert fill gas
consisted of an Ar–10% N_2 mixture; the oxygen concentration of the inert gas was between 1
and 10 ppm O_2. Discrepancies between calculations and experiments under certain conditions
were explained by additional release of carbon and water vapor from the filament and the silica
bulb, respectively, in amounts comparable to that of bromine (i.e., ~10^{-3} atm) [22]. Some of
these discrepancies were enumerated [23, 24]. More blackening occurred in the $CHBr_3$ lamps

compared to CH_2Br_2 lamps, even thought the latter contained the far higher H : Br ratio. Lamps made with HBr, $CBr_4 + CH_4$ (1:1), and CH_2Br_2 behaved differently inspite of same, i.e., 1:1 H : Br ratios. According to the thermodynamic calculations, lamps made with CH_3Br and low levels of oxygen, such as 0.01%, should darken if the controlling wall temperature is below 800 K. However, such lamps have been found to stay clear.

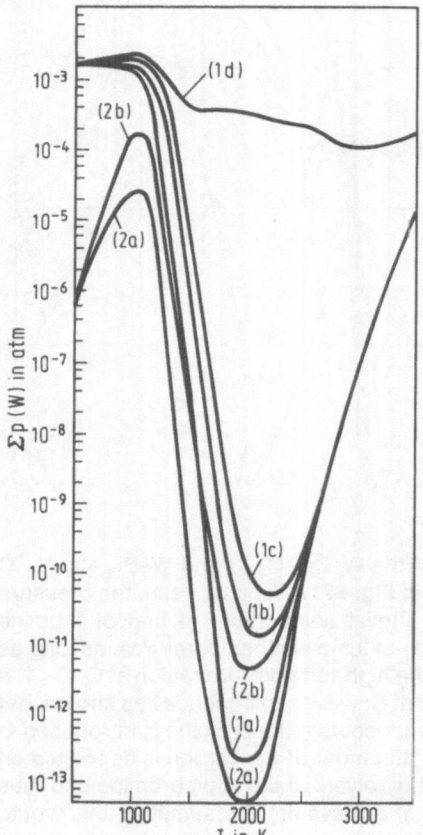

Fig. 97.   Total effective tungsten pressure in the gas phase of the $W-Br_2-O_2-C$ (curves 1a to 1d) and $W-Br_2-O_2-C-H_2$ (curves 2a, 2b) systems [14].

curve:	1a	1b	1c	1d
$O_2 : Br_2$	0.01	0.10	0.25	0.35
$O_2 : C$	0.04	0.4	1.0	1.4
$H_2 : Br_2$	—	—	—	—
$p^0(C)$*)	5×10^{-4}	5×10^{-4}	5×10^{-4}	5×10^{-4}
$p^0(Br_2)$	2×10^{-3}	2×10^{-3}	2×10^{-3}	2×10^{-3}
$p^0(O_2)$	2×10^{-5}	2×10^{-4}	5×10^{-4}	7×10^{-4}

curve:	2a	2b
$O_2 : Br_2$	0.01	0.10
$O_2 : C$	0.02	0.20
$H_2 : Br_2$	1.0	1.0
$p^0(C)$*)	1×10^{-3}	1×10^{-3}
$p^0(Br_2)$	2×10^{-3}	2×10^{-3}
$p^0(O_2)$	2×10^{-5}	2×10^{-4}
$p^0(H_2)$	2×10^{-3}	2×10^{-3}

*) in equivalent atm

The effect of water on the performance of W–Br lamps was also studied with the $CHBr_3$- and CH_2Br_2-containing lamps described on p. 90. Two different bromomethane doses of ~3.5 and ~33 µg were used. To each of these doses water was added at four different concentrations (0.9 to 3.0 µg) als well as a "dry" level of a few ppm. A few lamps showed bulb blackening, but its occurrence could not be definitely related to the water additions. Three types of filament failure were observed: (1) normal failure at a point near the centre of the filament coil where it is hottest, probably due to normal evaporation of tungsten; (2) tail failure at the base of the tail where it enters the pinch seal, associated with attack by free bromine in high-dosed bromoform lamps with little or no added water; (3) water failure either by thinning and melting of the penultimate turn of the filament or by crystal growth on the end turn and upper region of the filament tail, finally extending across the first two turns. The destructive action of the water-vapor cycle was severely reduced by large additions of $CHBr_3$ of CH_2Br_2. Bromoform appears to have a particularly strong nullifying effect on the water cycle. This is ascribed to its lack of sufficient hydrogen for binding all the available Br as HBr. The free bromine suppresses the water cycle either by removal of hydrogen by HBr formation or possibly also by removal of oxygen by formation of W oxide bromides [31].

Equilibrium calculation results for CH_3Br-doped lamps are presented in **Fig. 98**. Since WO_2Br_2 is believed to be the major W-bearing species near the bulb wall, only its partial pressures are given at different levels of CH_3Br and oxygen additions to the lamps. The important influence of the solid phases assumed (W, WO_2, and W_2C or W and W_2C only) on the

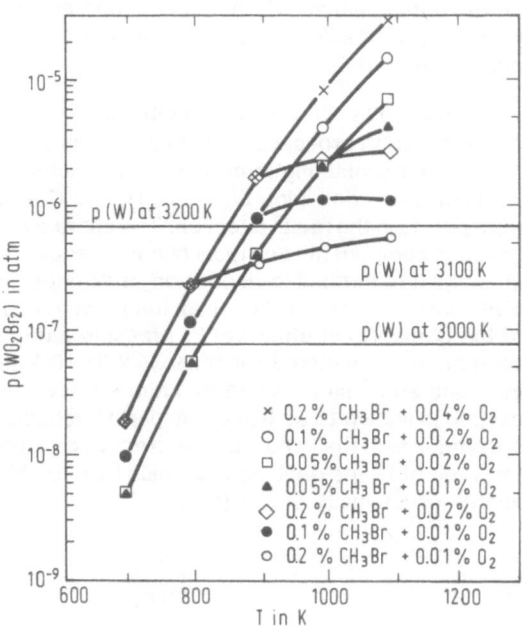

Fig. 98. Equilibrium partial pressures $p(WO_2Br_2)$ for different lamp dosing levels with CH_3Br computed for the W(s)–Br–O–C–H containing system with W_2C as the solid carbon-containing phase. The branched curves for each CH_3Br level represent calculations with W and W_2C as condensed phases in contrast to the three solid phases W, WO_2, and W_2C for the main curves [24].

calculated WO_2Br_2 pressures is clearly evident from the figure. The effect of using carbon as graphite or W_2C condensed phase on $p(WO_2Br_2)$ in calculations for CH_3Br and CH_2Br_2 lamp systems is disclosed by the following tables [24]:

0.2% CH_3Br Lamps (3000 Torr fill pressure); $p(WO_2Br_2)$ in atm.

wall temperature:	700 K		900 K		1100 K	
O_2 level	C	W_2C	C	W_2C	C	W_2C
0.01%	1.93×10^{-8}	1.94×10^{-8}	2.72×10^{-9}	3.28×10^{-7}	8.22×10^{-12}	5.23×10^{-7}
0.02%	1.93×10^{-8}	1.94×10^{-8}	9.23×10^{-9}	1.58×10^{-6}	2.86×10^{-11}	2.45×10^{-6}

0.1% CH_2Br_2 Lamps (3000 Torr fill pressure); $p(WO_2Br_2)$ in atm.

wall temperature:	700 K		900 K		1100 K	
O_2 level	C	W_2C	C	W_2C	C	W_2C
0.01%	2.24×10^{-5}	2.12×10^{-5}	4.90×10^{-6}	2.17×10^{-4}	3.42×10^{-9}	7.70×10^{-4}
0.02%	2.24×10^{-5}	2.12×10^{-5}	1.45×10^{-5}	2.12×10^{-4}	1.68×10^{-8}	8.64×10^{-4}

Most likely elemental carbon exists early in lamp life, but gets converted gradually to W_2C. As can be seen from the table, the use of either the graphite or carbide phase effects dramatically the $p(WO_2Br_2)$ values at the wall temperatures. The temperature region of importance to the bulb wall chemistry was selected because of its application in understanding wall blackening phenomena. The total effective W pressure at filament temperatures, by contrast, is hardly affected by the nature of the solid C-containing phase, since gaseous W is still the major species with WO(g) orders of magnitude lower for a typical lamp system. Proper inclusion of the solid WO_2 phase in the calculations for the CH_2Br_2 lamp reduces $p(WO_2Br_2)$ by one or two orders of magnitude, which does not appreciably alter the conclusions regarding blackening. However, misleading conclusions would follow for systems with $H:Br > 1$, such as CH_3Br-doped lamps [24].

In order to avoid some of the difficulties connected with the usual thermodynamic analysis of the transport processes in halogen incandescent lamps, a different calculation scheme was worked out simulating more closely the actual situation in a stationary (steady state), well-working lamp containing W, Br, O, H, and C. In this lamp, neither W nor C is assumed to be transported in the radial direction. The approximations of a common diffusion coefficient for all gaseous components and local chemical equilibrium are used. Non-blackening is connected to an inequality forbidding deposition of W onto the bulb wall. Chemical attack or growth of a given part of the W coil, i.e., axial tungsten transport, is characterized by the second derivative of the global (total effective) W pressure along the coil. Partial pressures of the gas phase components calculated for $p^0(Br_2) = 2.8 \times 10^{-4}$ atm, $p^0(O_2) = 1.0 \times 10^{-4}$ atm, $p^0(H_2) = 2.8 \times 10^{-4}$ atm, and an Ar fill pressure of 1 atm are given in **Fig. 99**. The gettering effect of carbon on oxygen is shown to be limited by partial deposition of C as well as the formation of CO_2 that decomposes at the W coil releasing active oxygen. Carbon deposited on the bulb wall acts as a reservoir of C, the depletion of which by a too high amount of oxygen causes a sharply rising degree of chemical attack [25].

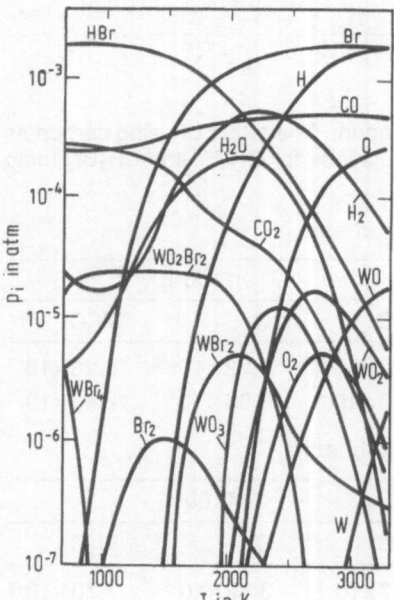

Fig. 99.   Equilibrium gas phase composition (partial pressures p_i) of the $W-Br_2-O_2-C-H_2$ system at $p^0(Br_2)$ $= p^0(H_2) = 2.8 \times 10^{-4}$ atm and $p^0(O_2) = 1.0 \times 10^{-4}$ atm. The Ar filling pressure is 1 atm, the total operating pressure 4 atm. A carbon deposit on the bulb wall at 700 K is assumed; otherwise the conditions correspond to those prevailing in Fig. 96, p. 95 [25].

As mentioned above, the thermochemical approach failed in explaining some of the observed phenomena in W–Br (and other W–halogen) lamps. This may be due to a number of reasons: First, the blackening or W transport criterion needs to be modified for the lamp, which because of its design geometry develops not only temperature gradients, but also stagnant pockets. Second, the thermodynamic treatment may be incomplete and/or incorrect by exclusion of important yet unknown species and inclusion of nonexistent phases or else suffer from the lack of reliable basis data. Third, the lamp filament impurities may play a significant role in altering the regenerative activity. Finally, the basic assumption of thermodynamic equilibria under steady-state operating conditions may not be valid due to some kinetically slow reactions, especially at the bulb wall and involving atomic H and CO. This last factor invited the exploration of a kinetic model which is based on the rates of various processes and the mass flow within the lamp. Calculations of the W mass flow, based on a simple diffusion model, have demonstrated the importance of gas phase and wall reactions. The role of atomic species has been considered since these are produced at the filament. Hydrogen atom recombination rates have strong dependence on the fill pressure; these are also found to be slower than their diffusion rates to the wall. The doubtful validity of the equilibrium assumption at the walls is clearly demonstrated by these calculations. Atomic hydrogen concentrations at the wall, however, are expected to be extremely low when H:Br<1. Application of the kinetic model to lamp chemistry is limited due to lack of appropriate kinetic data on wall reactions [23, 24].

The effect of the chemical reaction rates on the processes in bromine-containing lamps was studied by activation analysis of the amount and composition of the solid deposit formed on the bulb wall after burning the lamps for various times and cooling down after switching-off. The results showed that a stationary state is established much sooner for the main process maintaining the operation of the halogen lamp, i.e., the W transport, than for the side-processes induced by the impurities in the filament. In the latter processes, the diffusion of the impurity atoms to the surface of the incandescent filament is the slowest step, determining the rate of the reaction. The spatial distribution of the W content of the deposit along the bulb wall (i.e., its longitudinal axis) and its variation with time indicated that a part of the W halides in the gas phase condenses on the colder regions of the lamp, but another part decreases the vaporization of the W filament via its effect on the main reaction, and then stops further vaporization completely [29]. The spectral emission of lamps with CH_2Br_2 and Br_2 filling and different W source material, is a superposition of the continuous thermal radiation from the tungsten coil and a line spectrum originating from different impurity and doping elements in the W. For details, see the paper [30].

W–Br–F–C Containing Systems

The equilibrium gas phase composition of the W(s)–Br–F–C system has been thermodynamically calculated using the same procedure and assumptions as described for the W–Cl–F–C system on p. 76. Again, halogens and carbon were assumed to be introduced by addition of CBr_xF_{4-x} (x = 0 to 3) compounds to the input gas, producing (initial) halogen pressures of 2×10^{-6} to 2×10^{-1} atm. Results for $p^0(Br_2 + F_2) = 2 \times 10^{-3}$ atm are presented in **Fig. 100**. The phases occurring are generally analogous to those in the W–Cl–F–C system. The same holds for the curves representing the temperature dependence of the mass balance ratios $W/(Br_2 + F_2)$ at various $p^0(Br_2 + F_2)$, except that the curves for W–Br–F–C show a more complex behavior in the temperature range around 1500 K. This results from the superposition of the decay of W–Br compounds with the formation of WF_2. The influence of an increasing bromine content in the gas phase on the mass balance ratios is demonstrated in **Fig. 101**. As can be seen, an increasing Br content decreases the gas phase solubility of W at medium temperatures and thus should reduce the attack of the leads in practical W–halogen lamps.

Experiments with 12 V/100 W lamps containing 3000 Torr Kr and 1 to 10 Torr of different CBr_xF_{4-x}, which were burned at 3400 K, confirmed these predictions. Although a certain thinning of the leads and attack of the W filament were observed, the lifetime of such lamps was on the order of that attainable by CH_2Br_2 additions [38].

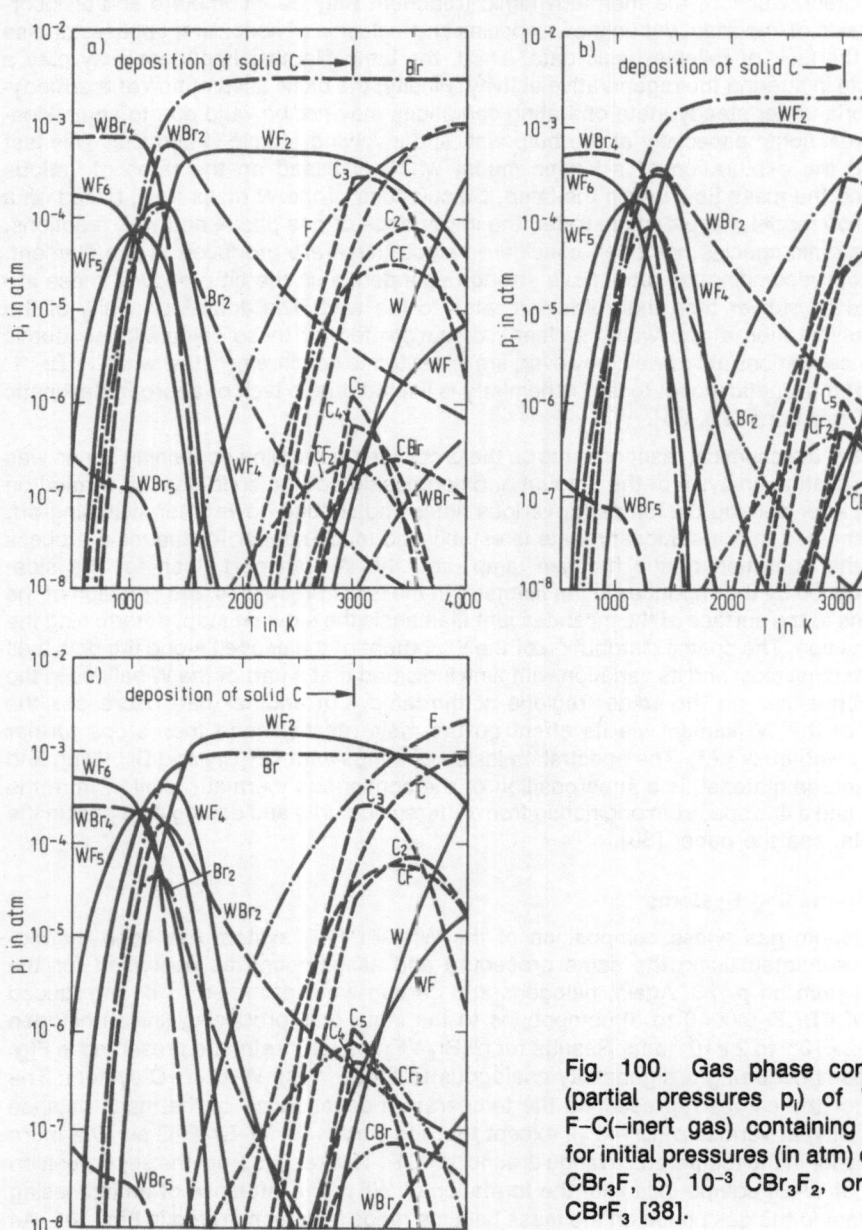

Fig. 100. Gas phase composition (partial pressures p_i) of W(s)–Br–F–C(–inert gas) containing systems for initial pressures (in atm) of a) 10^{-3} CBr_3F, b) 10^{-3} CBr_3F_2, or c) 10^{-3} $CBrF_3$ [38].

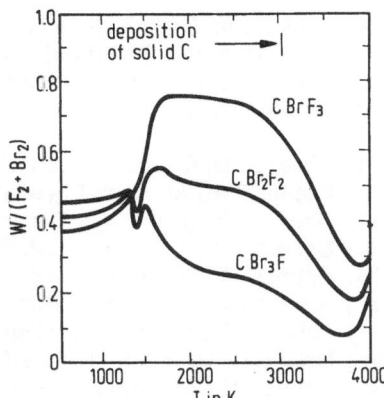

Fig. 101. Mass balance ratio $W/(Br_2 + F_2)$ in the gas phase of W(s)–Br–F–C(–inert gas) containing systems as a function of the bromine content of added CBr_xF_{4-x} (initial pressure 10^{-3} atm, corresponding to $p^0(Br_2 + F_2)$ $= 2 \times 10^{-3}$ atm) [38].

References:

[1] Neumann, G. M.; Knatz, W. (Z. Naturforsch. **26a** [1971] 863/9).
[2] Neumann, G. M.; Knatz, W. (Z. Naturforsch. **26a** [1971] 1046/53).
[3] Dettingmeijer, J. H.; Meinders, B.; Nijland, L. M. (J. Less-Common Met. **35** [1974] 159/69).
[4] JANAF Thermochemical Tables and Addenda I–III, Dow Chemical Company, Midland, Michigan, 1965/68.
[5] Neumann, G. M.; Gottschalk, G. (Z. Naturforsch. **26a** [1971] 870/81).
[6] Neumann, G. M. (Tech. Wiss. Abh. OSRAM-Ges. **11** [1973] 8/41).
[7] Gottschalk, G.; Neumann, G. M. (Z. Metallk. **62** [1971] 910/5).
[8] Kopelman, B.; van Wormer, K. A. (Illum. Eng. [N.Y.] **63** [1968] 176/82).
[9] Kopelman, B.; van Wormer, K. A. (Illum. Eng. [N.Y.] **64** [1969] 230/5).
[10] Neumann, G. M. (Z. Metallk. **64** [1973] 117/20).

[11] JANAF Thermochemical Tables, 2nd Ed., Dow Chemical Company, Midland, Michigan, 1971.
[12] Neumann, G. M. (Z. Metallk. **64** [1973] 26/32).
[13] Yannopoulos, L. N.; Pebler, A. (J. Appl. Phys. **42** [1971] 858/62).
[14] Yannopoulos, L. N.; Pebler, A. (J. Appl. Phys. **43** [1972] 2435/9).
[15] Harvey, F. J. (Metall. Trans. A 7 [1976] 1167/76).
[16] Zubler, E. G. (J. Phys. Chem. **74** [1970] 2479/84).
[17] Zubler, E. G. (J. Phys. Chem. **76** [1972] 320/2).
[18] Neumann, G. M. (Z. Metallk. **64** [1973] 379/85).
[19] Neumann, G. M. (Z. Metallk. **64** [1973] 444/9).
[20] Neumann, G. M.; Müller, U. (Tech. Wiss. Abh. OSRAM-Ges. **11** [1973] 42/54).

[21] Neumann, G. M. (Thermochim. Acta **8** [1974] 369/79).
[22] Yannopoulos, L. N.; Pebler, A. (J. Illum. Eng. Soc. **1** [1971] 21/4).
[23] Gupta, S. K. (J. Electrochem. Soc. **125** [1978] 2064/70).
[24] Gupta, S. K. (Proc. Electrochem. Soc. **78**-1 [1978] 20/42; C.A. **89** [1978] No. 68444).
[25] Geszti, T.; Vicsek, T. (J. Phys. D **9** [1976] 903/12).
[26] Maier, G. (Tech. Wiss. Abh. OSRAM-Ges. **11** [1973] 55/9).
[27] T'Jampens, G. (RGE Rev. Gen. Electr. **75** [1966] 990/6; Bull. Soc. R. Belge Electr. **83** No. 3 [1967] 247/56).

[28] T'Jampens, G. R.; van de Weijer, M. H. A. (Philips Tech. Rev. **27** [1966] 173/9; Philips Tech. Rundsch. **27** [1966] 165/71).
[29] Hangos, I.; Salamon, A.; Bartha, L. (Acta Tech. Acad. Sci. Hung. **78** [1974] 417/26).
[30] van den Hoek, W. J.; Berns, E. G. (Light. Res. Technol. **7** No. 2 [1975] 143/6).

[31] Price, D. H. (J. Sci. & Technol. [London] **39** [1972] 125/30).
[32] Gupta, S. K. (J. Appl. Phys. **42** [1971] 5855/6).
[33] Rees, J. M. (Light. Res. Tech. Rev. **2** [1970] 257/60).
[34] Coaton, J. R.; Phillips, N. J. (Proc. Inst. Electr. Eng. **118** [1971] 871/4).
[35] Dittmer, G.; Niemann, U. (Philips J. Res. **36** [1981] 87/111).
[36] Schnedler, E. (Philips J. Res. **38** [1983] 236/47).
[37] Eckerlin, P.; Garbe, S. (Philips J. Res. **35** [1980] 320/5).
[38] Neumann, G. M. (J. Fluorine Chem. **3** [1973] 209/25).
[39] Dittmer, G.; Niemann, U. (Philips J. Res. **42** [1987] 41/57).

2.5.5 Tungsten–Iodine Systems

General

The first tungsten-halogen incandescent filament lamps to be marketed contained elemental iodine as the agent for maintaining bulb clarity [1, 2]. Iodine was chosen as an additive because it is the least reactive of the halogens. The use of iodine brought problems, however, especially in its introduction into the lamps since it is a solid room temperature. Furthermore, its optical absorption lowers the color temperature of the lamps and brings about a color distortion. The high molecular weight of I_2 leads to thermal segregation in single coil, long life lamps. These lamps must be burned within ± 4 degrees of the horizontal position or else blackening of their upper ends by W deposition occurs due to a decrease in the iodine concentration below the level required to sustain an effective halogen cycle.

Most of these shortcomings of the standard iodine lamp can be eliminated if the concentration of molecular iodine I_2 is kept as low as possible by promoting its dissociation into atomic iodine I. This can be achieved by raising the lamp temperature (reducing the diameter of the lamp envelope) or by lowering the iodine (as I_2) dosing concentration. The first approach is limited by practical difficulties. Iodine additions usually are limited to at most a few μmol per cm³ of lamp volume. Theoretically, optimum lamp performance would be expected for an iodine level just sufficient to keep the lamp wall clean. This level is in the range of 0.07 Torr I_2. Practical iodine concentrations must be significantly higher (in excess of about 1.5 Torr) to assure that residual trace impurities do not remove too much iodine from the halogen cycle by gettering. One possible way of keeping the iodine activity in a lamp to a desired low value, and at the same time providing additional iodine for any that is lost due to impurity gettering, is by dosing with a suitable (colorless) iodine compound (e.g., SnI_4, see p. 109).

The low thermal stability of tungsten-iodine compounds compared with that of other tungsten-halogen compounds favors decomposition in zones with temperatures below the normal filament temperatures. Thus tungsten is not deposited preferentially onto the hottest spots of the incandescent coils and no improvement in lamp life attributable to a regenerative cycle is observed. Experimental proof of the axial transport of tungsten from hot to cold parts of the filament coil was found with a tracer technique [3, 4]. In order to establish the points of tungsten deposition, a test lamp containing three identical coils was used, of which the middle one had been activated in a nuclear reactor. The neutron irradiation produced (in addition to rapidly decaying [187]W) the isotope [185]W, a β-emitter. After a burning period of 220 h, [185]W had preferentially deposited at the colder ends of the nonactivated coils.

In view of the problems encountered with the iodine lamp, there has been a tendency to replace iodine by other halogens. Successful lamps have, for example, been made with bromine, but iodine proved to be much more tolerant of variations in impurities that may be left in the lamp during manufacturing. In particular, O_2 and water vapor do not have to be as carefully controlled in iodine lamps. The other halogens are more aggressive to leads, supports, and cool parts of the W coils, and none of the other halogens seems capable of providing the reliability obtained with the iodine lamp.

For a discussion of the above aspects, see, e.g. [5 to 9, 31 to 33, 36].

Reaction Mechanisms. Thermodynamics

Originally it was assumed that tungsten transport in iodine lamps took place by means of the compound WI_2 [1, 3, 4, 10, 31], for further literature, see [11, 32, 36]. This also appeared plausible from thermodynamic calculations [12, pp. 178/9], which, however, were founded on a, in part questionable, data basis [13]. It was noted that small traces of oxygen or moisture were indispensable to the cyclic process in the iodine lamp [2, 3]. This is in accordance with experimental evidence (see p. 104) and more recent thermodynamic calculations [14 to 16], according to which pure W does not react with iodine under the conditions prevailing in the lamps, and gaseous W iodides are thermodynamically unstable, at least at temperatures above 800 K and (initial) I_2 pressures below 1 atm. Iodine alone therefore cannot produce the desired cyclic process, see also [17, p. 16]. Free energies of formation for tungsten iodides between 500 and 2500 K and at 1 atm pressure which are presented in **Fig. 102** underline the instability of compounds in the W–I system [14]. Maximal attainable gas phase concentrations c_i (in mol%) of the species W, I_2, I, and WI_2 at respective temperatures (in K, given in parentheses) as calculated for different initial I_2 pressures, are [16]:

compo- nent	initial I_2 pressure in atm						
	10	1	10^{-1}	10^{-2}	10^{-3}	10^{-4}	10^{-5}
W	—	—	0.2 (4000)	1.8 (4000)	15.3 (4000)	64.4 (4000)	94.8 (4000)
I_2	100 (700)	100 (600)	100 (600)	100 (500)	100 (500)	99.9 (500)	99.8 (500)
I	97.3 (4000)	99.7 (4000)	100 (3600)	100 (3400)	100 (3200)	100 (3000)	100 (2800)
WI_2	2.5 (3600)	0.3 (3600)	—	—	—	—	—

For the temperature dependence of the equilibrium constants of iodide formation, see also [9, 32].

Several studies showed that WO_2I_2 is a stable compound with sufficient volatility enabling the cyclic process [18 to 22]. The cycle stopped as soon as the system was deprived of oxygen by the addition of carbon [23].

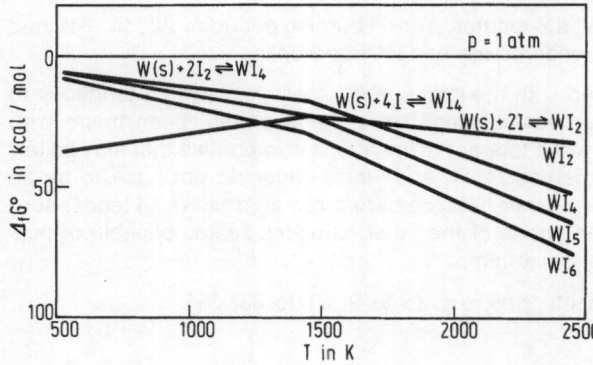

Fig. 102. Free energies of formation $\Delta_f G°$ for tungsten iodides as a function of temperature [14].

W–I–H Containing Systems

The use of gaseous hydrogen iodide in tungsten halogen lamps was claimed to have certain advantages over that of elemental iodine [34]. HI can be premixed with Ar in dilute, low pressure mixtures and introduced directly into the lamps on a rotary machine. However, the compound has a significant degree of dissociation at room temperature, which causes variable dosing. Neither does this method prevent color distortion since the hydrogen rapidly diffuses out of the vitreous fused silica envelope leaving the iodine as the element [35]. In a series of experiments performed by [32], no proper operating lamps could be obtained with HI pressures between 0 and 100 Torr. The lamps always blackened unacceptably within their lifetime, although the degree of blackening decreased somewhat at the higher pressures, see also "Tungsten" Suppl. Vol. A 7, 1987, pp. 79/80.

W–I–O Containing Systems

As has been mentioned above, the presence of oxygen is indispensable for starting and maintaining the cyclic process in iodine lamps which prevents lamp blackening. Tungsten removal from the lamp walls is assumed to be effected by the formation of gaseous WO_2I_2, which decomposes when diffusing towards the hotter regions near the incandescent filaments. A schematic representation of the gas phase composition in a W–I–O lamp as proposed by [11] is given in **Fig. 103**. According to this scheme, decomposition and formation of WO_2I_2 proceed via WO_2. High wall temperatures keep WO_2I_2 in the gaseous state; an excess of iodine prevents its decomposition which would result in irreversible WO_2 deposition at the lamp wall [11], see also [22]. Since WO_2I_2 decomposes at rather low temperatures, the iodide process is not able to heal hot spots of the W filament, but only to prevent blackening of the bulb. On the

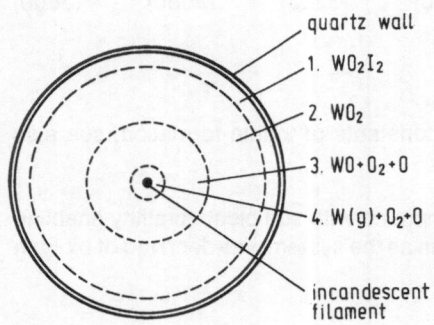

quartz wall
1. WO_2I_2
2. WO_2
3. $WO + O_2 + O$
4. $W(g) + O_2 + O$
incandescent filament

Fig. 103. Gas phase composition in the various reaction zones of a W–I–O lamp [11]. The ubiquitous iodine and inert gas are not indicated.

contrary, there is an axial transport of W from hot to less hot parts of the filament [6, 11], see also the experiments of [3, 4].

The temperature dependence of the total effective pressure $\Sigma p(W)$ is shown in **Fig. 104** for initial I_2 and O_2 pressures of 10^3 Pa (10^{-2} atm) and 10^{-1} Pa, respectively; for the definition of $\Sigma p(W)$ see p. 37. At an operating temperature of 3200 K the tungsten transport between neighboring turns of a W coil that have slightly differing temperatures takes place almost completely as tungsten vapor, and hence goes from the hotter to the cooler turn. At operating temperatures corresponding to the minimum in the $\Sigma p(W)$ vs. T curve, where tungsten is transported from cooler to the the hotter turns, the cool ends of the coils are attacked. This effect increases with the amount of oxygen (or water vapor, see p. 106) present [6, 24].

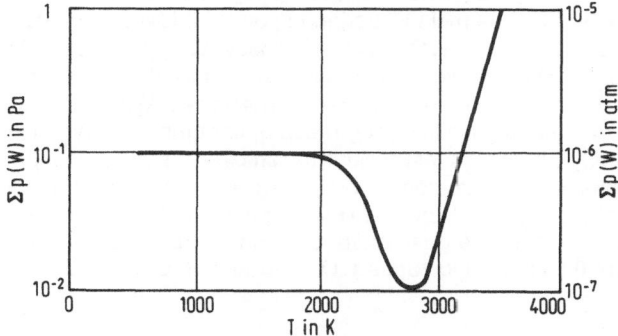

Fig. 104.   Temperature dependence of the total effective tungsten pressure $\Sigma p(W)$ in a W–I_2–O_2 lamp for initial pressures $p^0(I_2)=10^3$ Pa and $p^0(O_2)=10^{-1}$ Pa [6].

The temperature dependence of the gas phase composition at $p^0(I_2)=10^{-3}$ atm and $I_2:O_2=1:1$, derived from thermodynamic calculations, is given in Fig. 19, p. 40. The importance of WO_2I_2 is clearly perceivable. Several simple and polymeric W oxides appear as W-containing species in the gas phase at higher temperatures. At low temperatures, solid WO_2 forms (see Fig. 21, p. 42). The change of the mass balance ratio $W(I_2+O_2)$ with temperature is shown in **Fig. 105** for various values of $p^0(I_2)$ and $I_2:O_2=1:1$ [17, pp. 26/30]. Thermodynamic calculations of the gas phase composition for the ranges $500<T<3600$ K, $10^{-6}<p^0(I_2)<10^{-1}$ atm, and $I_2:O_2=10:1$ to $1:10$ were performed [25]. Again, at low and middle temperatures,

Fig. 105.   Temperature dependence of the mass balance ratio $W/(I_2+O_2)$ in the W–I_2–O_2(–inert gas) system for various initial pressures $p^0(I_2)$ (in atm) given as parameter at the curves and for $I_2:O_2=$ 1:1 [17].

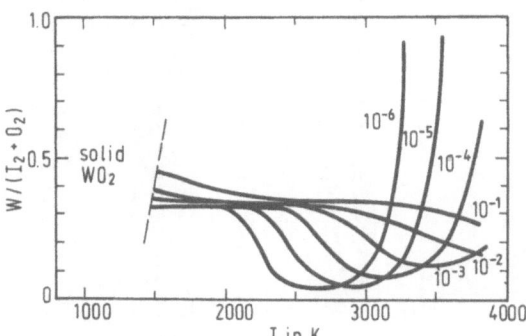

a pronounced formation of WO_2I_2 was noted, which is more stable than the binary iodides, but less than WO_2X_2 with X = F, Cl, Br. In the range from 1500 to 1700 K, solid WO_2 appears. In the region of thermal stability of WO_2I_2, the direction of W transport is from cold to hot [25].

W–I–O–H Containing Systems

The presence of H_2O promotes the tungsten transport in iodine lamps [3] (and also in a temperature gradient 800 (W) → 1000°C in a closed quartz glass ampule [23, pp. 163/4]). At the same time it shortens the lifetime of the lamps by initiation of the "water cycle" [1, p. 737]. Notably cold-end attack of the W coils is observed in the presence of H_2O [6].

In thermodynamic calculations of the gas phase composition, 23 species were considered: WO_2I_2, WI, WI_2, WI_4, W, H_2WO_4, WO, WO_2, WO_3, W_2O_6, W_3O_8, W_3O_9, W_4O_{12}, I, I_2, HI, H, H_2, O, O_2, H_2O, OH, and HO_2. The main W-bearing species at low temperatures is WO_2I_2; above 1500 K, the oxide species W_2O_6, WO_2, WO, and finally, W appear successively, see **Fig. 106**. In equilibrium with solid W, the gas phase "dissolves" more W at lower temperatures than at higher ones; thus, W is transported from cold to hot zones. When the partial pressure of O_2 becomes too low by either high initial I_2 : O_2 ratios or capturing of oxygen in stable oxides, the concentration of WO_2I_2 becomes insufficient to remove W from the walls of an iodine lamp. Excess oxygen causes condensation of WO_2, beginning at higher temperatures with an increasing O : H ratio. If oxygen is added as H_2O, deposition of WO_2 affects the H : O ratio, and, thus, the "solubility" of W in the gas phase, but only in the medium temperature range. For example, at p(tot) = 0.01 atm, and ratios I : O = 100 and H : O = 2 to 50, the W solubility has

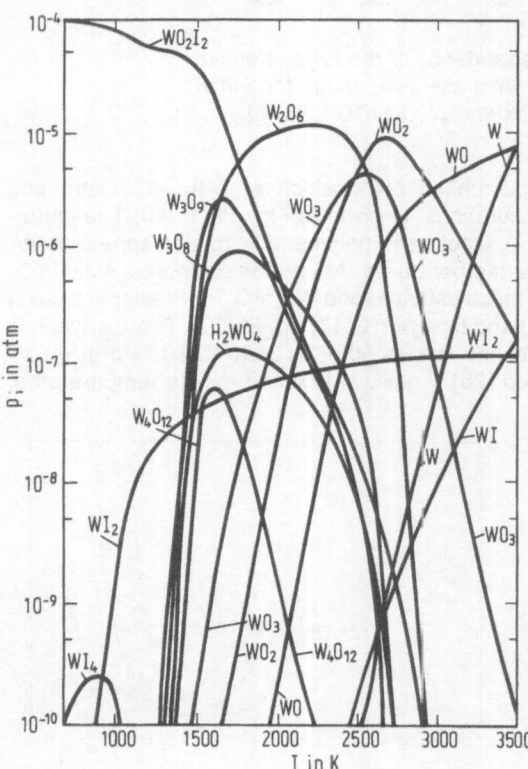

Fig. 106. Partial pressures p_i of W compounds in the W–I–O–H containing system as function of temperature for a total pressure of p(tot) = 0.01 atm, and ratios H : O = 2:1, I : O = 100:1 [24].

a minimum at about 1800 K. At low and very high temperatures, the W solubility is hardly changed by an increase in the H:O ratio. In a practical lamp the envelope will remain clear under the above conditions. At the two sides of the minimum, large solubility gradients will occur, leading to considerable tungsten transport along the filament. In accordance with this, the serious end attack of the W filaments observed in practical iodine lamps, coupled with whisker growth in the region between 1500 and 2000 K, is commonly ascribed to the presence of water vapor. If HI is introduced instead of I_2 at practicable I:O ratios, the W solubility at high temperatures is higher than at low temperatures. Thus lamp blackening can be expected, and has indeed been found in practice [24, pp. 161/4], see also [22].

Gas phase compositions for initial I_2 pressures $p^0(I_2)=10^{-1}$ to 10^{-6} atm at different oxygen:halogen ratios in the temperature range from 1000 to 3600 K were calculated [30]. Results obtained for $p^0(I_2)=10^{-3}$ atm and $I_2:H_2:O_2=1:1:0.5$ are shown in **Fig. 107**. The influence of water vapor and hydrogen on the temperature dependence of the mass balance ratio $W/(I_2+O_2)$ is shown in figures in the paper [30], for the influence of oxygen on $W/(I_2+O_2)$ in the gas phase at $I_2:H_2=1:1$ ($p^0(I_2)=10^{-3}$), see Fig. 22, p. 43. Equilibrium pressures of WO_2I_2, $WO_2(OH)_2$, H_2O, H_2, HI, I, and I_2 were calculated at 1100 and 1300 K, assuming pressures $p(H_2)$ and $p(I)$ between 10^{-3} and 10^{-5} and 10^{-1} and 2×10^{-3} atm, respectively [23].

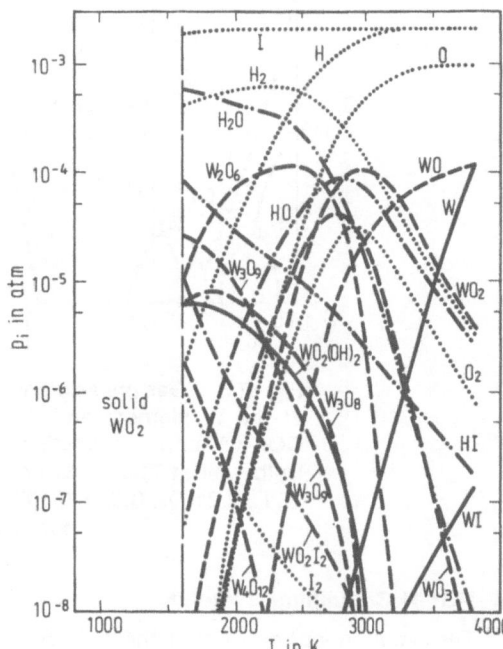

Fig. 107. Partial pressures of W compounds in the $W–I_2–O_2–H_2(–inert\ gas)$ system as a function of temperature for $p^0(I_2)=10^{-3}$ atm and $I_2:H_2:O_2=1:1:0.5$ [30].

W–I–O–C–H Containing Systems

A notable effect of C in the processes occurring in W–I lamps is only observed in the presence of oxygen. In the absence of oxygen, the gas phase composition is not changed appreciably. In oxygen-containing lamps, carbon present as impurity or addition counteracts the influence of oxygen by the formation of CO or CO_2 in competition to the formation of WO_2I_2. At higher temperatures, CO acts as an inert gas. The formation of W oxides is suppressed until the C:O ratio exceeds 1:1. Except for the capturing of oxygen, carbon has no significant influence on the transport reactions. As pointed out above, the presence of a certain amount of oxygen in necessary for the transport cycle to operate. Addition of oxygen to the lamp filling in the form of CO will be more advantageous than that in the form of elemental O_2. Due to its high stability CO does not permit undesired tungsten oxide formation and associated W removal at the incandescent filament. Near the relatively cool lamp wall CO is transformed into CO_2 via the Boudouard equilibrium, and CO_2 secures the required WO_2I_2 formation [17, pp. 30/7]; see also [26, 27].

Addition of graphite prevents the transport of W in an I-containing closed quartz glass ampule in a temperature gradient 800 (W)$\rightarrow$1000°C. The effect is ascribed to the removal of water traces required for the transport reaction: $W(s) + 2H_2O(g) + 6I(g)$ (or $3I_2(s, g)) \rightleftharpoons WO_2I_2(g) + 4HI(g)$; $C(s) + H_2O(g) \rightleftharpoons CO(g) + H_2(g)$ [23].

Measurements of the weight changes of a W coil in a closed system were the experimental basis for the dependence of the gas phase composition on the W filament temperature shown in **Fig. 108**. The glass bulb contained (at 700 K) 10 mbar I_2, 0.01 mbar O_2, 0.02 mbar CO, and 0.5 bar Kr [37].

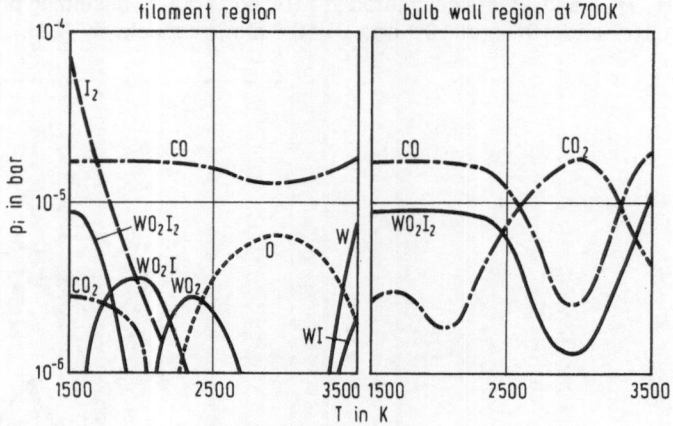

Fig. 108.   Gas phase composition (partial pressures p_i) at the W filament and the bulb wall of a W–I_2–O_2–CO(–Kr) lamp as a function of the filament temperature. Bulb wall at 700 K; fill pressures at 700 K (in mbar): 10 I_2, 0.01 O_2, 0.02 CO [37] ($p_i \approx$ constant $\approx$ 10 mbar is not shown).

W–I–Br–H Containing Systems

The simultaneous equilibria in the W–I_2–Br_2–H_2 system were studied by thermodynamic calculations at temperatures and pressures prevailing in halogen lamps, and, particularly at an ($I_2 + Br_2 + H_2$) pressure of 1.05×10^{-2} Torr and a total pressure of ~6 atm established by inert gases. The species H_2, I_2, Br_2, WI_2, WBr_2, WBr_5, WI, Br, H, HI, and HBr were considered as constituents of the gas phase, while the species WBr_4, WI_4, and WBr_6 were neglected. It was found that an increase in the H_2 concentration should suppress W halide formation without significantly changing the partial pressures of the atomic species I, Br, and H. When the H_2 concentration reaches 50% in the initial $I_2 + Br_2 + H_2$ mixture, WBr_5 practically disappears from the gas phase. For WI_2 the effect is much less pronounced and for WBr_2 it is appreciable only at low temperatures and high H_2 concentrations, but not at the temperatures at and near the incandescent filament. It is shown that under suitably chosen conditions the combined addition of iodine and bromine to the gas filling in halogen lamps can have considerable advantages over the use of one single halogen [28].

W–I Systems with Additional Sc, Na, Sn

The chemical reaction phenomena in a Sc–Na iodide gas discharge (arc) lamp with W electrodes, particularly those initiating tunsten transport, have been investigated [29].

The influence of SnI_4 on the processes occurring in tungsten-iodine incandescent filament lamps has been studied. The thermodynamic properties of SnI_4 induce a desired rather low iodine activity in the lamps. At the same time SnI_4 serves as a reservoir of iodine in case iodine should be removed from the cycle by impurity gettering. This brings about a significant increase in light output, minimizes color distortion, and improves the verical burning characteristics. The performance of a SnI_4-doped lamp, especially in a vertical burning position at higher than the rated voltage, is further improved by addition of I_2 or CH_2Br_2 to the lamp [5].

References:

[1] Zubler, E. G.; Mosby, F. A. (Illum. Eng. [N.Y.] **54** [1959] 734/40).
[2] Zubler, E. G.; Mosby, F. A. (U.S. 3160454 [1964] from [11]).
[3] Bayle, P.; Blanc, D.; Le Strat, J.; et al. (C.R. Hebd. Seances Acad. Sci. **258** [1964] 4710/2).
[4] Bayle, P.; Blanc, D.; Le Strat, J.; et al. (Lux **33** [1965] 211/4).
[5] Kulkami, A. D.; Martin, J.; Sell, H. G. (Illum. Eng. Soc. **6** [1977] 100/4).
[6] Dettingmeijer, J. H.; Dittmer, G.; Klopfer, A.; et al. (Philips Tech. Rev. **35** [1975] 302/6; Philips Tech. Rundsch. **35** [1975] 324/7).
[7] Schilling, W. (Lichttechnik **20** No. 12 [1968] 139A/142A).
[8] Price, D. H. (GEC J. Sci. Technol. **39** No. 3 [1972] 125/30).
[9] T'Jampens, G. (RGE Rev. Gen. Electr. **75** [1966] 990/6; Bull. Soc. R. Belge Electr. **83** No. 3 [1967] 247/56).
[10] Schilling, W. (ETZ Elektrotech. Z. B **13** [1961] 485/7).

[11] Rabenau, A. (Angew. Chem. **79** [1967] 43/9; Angew. Chem. Int. Ed. Engl. **6** [1967] 68/73).
[12] Kopelmann, B.; van Wormer, K. A. (Illum. Eng. [N.Y.] **63** [1986] 176/82).
[13] Brewer, L.; Bromley, L. A.; Gilles, P. W.; et al. (in: Quill, L. L.; The Chemistry and Metallurgy of Miscellaneous Materials, Thermodynamics, National Energy Series, Div. IV, Vol. 19B, McGraw-Hill, New York 1950, pp. 276/311).
[14] Neumann, G. M.; Knatz, W. (Z. Naturforsch. A **26** [1971] 863/9).
[15] Neumann, G. M.; Knatz, W. (Z. Naturforsch. A **26** [1971] 1046/53).
[16] Gottschalk, G.; Neumann, G. M. (Z. Metallk. **62** [1971] 910/5).
[17] Neumann, G. M. (Tech. Wiss. Abh. OSRAM-Ges. **11** [1973] 8/41).
[18] Tillack, J.; Eckerlin, P.; Dettingmeijer, J. H. (Angew. Chem. **78** [1966] 451; Angew. Chem. Int. Ed. Engl. **5** [1966] 421).
[19] Dettingmeijer, J. H.; Meinders, B. (Z. Anorg. Allg. Chem. **357** [1968] 1/10).
[20] Tillack, J. (Z. Anorg. Allg. Chem. **357** [1968] 11/24).

[21] Schäfer, H.; Giegling, D.; Rinke, K. (Z. Anorg. Allg. Chem. **357** [1968] 25/9).
[22] McHale, J. J. (Illum. Eng. [N.Y.] **66** [1971] 280/4, discussion pp. 284/6).
[23] Dettingmeijer, J. H.; Tillack, J.; Schäfer, H. (Z. Anorg. Allg. Chem. **369** [1969] 161/77).
[24] Dettingmeijer, J. H.; Meinders, B.; Nijland, L. M. (J. Less-Common Met. **35** [1974] 159/69).

[25] Neumann, G. M. (Z. Metallk. **64** [1973] 26/32).
[26] Collins, C. B.; Zubler, E. G. (U.S. 3132278 [1964]).
[27] Collins, C. B.; Holcomb, R. H. (U.S. 3364376 [1968]).
[28] Hangos, I.; Juhasz, I. (Acta Tech. Acad. Sci. Hung. **79** [1974] 101/3).
[29] Ishigami, T.; Saraki, H.; Honma, M. (J. Light Visual Environ. **7** No. 1 [1983] 1/6).
[30] Neumann, G. M. (Z. Metallk. **64** [1973] 379/85).

[31] Moore, J. A.; Jolly, C. M. (GEC J. Sci. Technol. **29** No. 2 [1962] 99/106).
[32] T'Jampens, G. R.; van de Weijer, M. H. A. (Philips Tech. Rev. **27** [1966] 173/9; Philips Tech. Rundsch. **27** [1966] 165/71).
[33] van Tijen, J. W. (Philips Tech. Rundsch. **23** [1961/62] 226/31).
[34] Shurgan, J. (U.S. 3091718 [1963], from [35]).
[35] Rees, J. M. (Light. Res. Tech. Rev. **2** [1970] 257/60).
[36] Strange, J. W.; Stewart, J. (Trans. Illum. Eng. Soc. [London] **28** [1963] 91/104; discussion pp. 104/9).
[37] Dittmer, G.; Niemann, U. (Philips J. Res. **36** [1981] 87/111).

3 Reactions with Sulfur

Phase Diagram. Products

A schematic phase diagram of the W–S system as proposed by [1] and redrawn by [2] is shown in **Fig. 109**. It is based on work done in conjunction with an investigation of the Fe–W–S system [3].

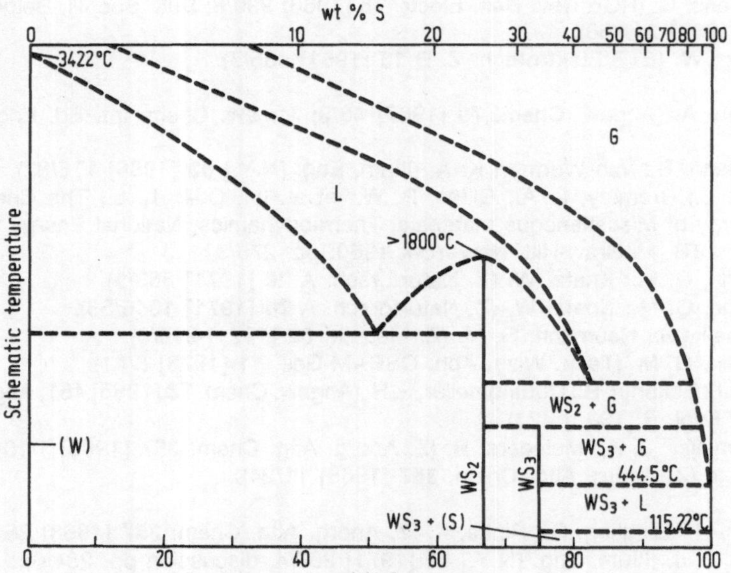

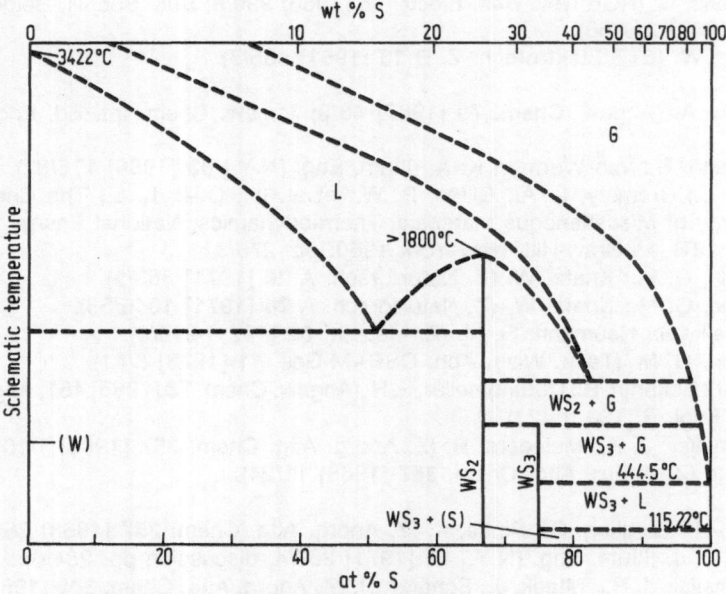

Fig. 109. Schematic phase diagram of the W–S system [2].

There is no information about the solubility of S in W, but by analogy to other metals (Cu, Fe, Co, Ni, Ti), it is estimated to be 10^{-3} to 10^{-2} wt% [4]. The lattice parameters of pure powder tungsten and tungsten saturated with sulfur (at 2100°C for 48 h) are 3.1644 ± 0.0003 and 3.1636 ± 0.0003 Å, respectively [5].

According to X-ray analyses [6 to 8], see also [9 to 11], the disulfide WS_2 is the only thermally stable intermediate phase in the system which can be prepared by direct synthesis from the elements [12]. It seems to exist in two modifications. Hexagonal (C 7 type) WS_2, as prepared by direct synthesis from the elements, has a homogeneity range from 66.1 to 66.67 at% ($WS_{1.95}$ to $WS_{2.0}$) [7], while samples obtained by decomposition of ammonium thiotungstate (($NH_4)_2WS_4$) were single-phase from $WS_{1.86}$ to $WS_{2.30}$ [13], see also [14 to 16]. Independent investigations of directly synthesized samples confirmed a lower phase limit of $WS_{1.96}$ and the small homogeneity range [6]. No perceptible region of homogeneity was found in X-ray studies and isopiestic experiments [11]. A rhombohedral modification, isotypic with rhombohedral MoS_2, was prepared by the reaction of WO_3 with S in molten Na_2CO_3 [17]. For crystallographic data for WS_2, see also [8, 16, 18 to 23].

The trisulfide WS_3 can be prepared in various indirect ways [6, 7, 9, 16, 17]. It is always amorphous [7, 16, 17]. The preparation of crystalline WS_3 by the decomposition of piperazinium thiotungstate with hydrochloric acid has been claimed [6], but is questionable [17, 23]. On heating, WS_3 decomposes into WS_2 and S. The decomposition temperatures reported range from 170 to 500°C and depend on pressure and atmosphere, see, e.g. [3, 6, 16, 17, 24].

No volatile sulfides of tungsten have been reported. For a calculated S_2 equilibrium pressure vs. $1/T$ diagram for the $W(s)/WS_2(s)$ interface at 1150 to 1450 K, see [25, p. 1680].

For discussions of W sulfides, see also [23, 26, 27].

For a high-temperature study of the W–S–O system, see [28].

Reaction Characteristics

Molten and even boiling sulfur attacks tungsten only very slowly. Passing sulfur vapor over tungsten at red heat produces WS_2 [29]. Early literature is quoted in "Wolfram", 1933, pp. 88, 184.

The synthesis of WS_2 and other W–S samples from the elements is usually carried out in sealed tubes, see, e.g. [6, 7, 11, 18]; preparation attempts in corundum and carbon crucibles in an Ar atmospheres yielded unsatisfactory results [3]. According to DTA studies, a mixture of the powdered elements begins to react at 400°C as indicated by the onset of a very strong exothermic peak. The exothermicity causes a sharp temperature increase and thereby a high pressure is created in the reaction vessel by the evaporation of still unreacted sulfur. The quartz ampules used for the investigation could not withstand this pressure and frequently broke. The synthesis of WS_2 from the mixed elements in sealed ampules therefore may be generally considered as dangerous and inefficient. This shortcoming can be eliminated if the reaction between the metal and sulfur, situated in different portions of the ampules at different temperatures, is carried out via the gas phase. The reaction of a 0.1 g sample of powdered W, which was heated from 300 to 700°C at a rate of 2°C/min and contacted with gaseous sulfur (corresponding to its vapor pressure at 210°C), reached measurable rates at 360°C. The maximum reaction rate was at 440°C, where the degree of conversion to WS_2 as the sole reaction product reached 62.5% [30].

A systematic thermogravimetric study of the reaction kinetics of both unannealed and annealed W sheet coupons with pure sulfur vapor was performed at temperatures between 290 and 560°C and sulfur pressures between 0.16 and 300 Torr. The results of weight-gain measurements are shown in **Fig. 110** and **Fig. 111**. For unannealed W sheet coupons the weight gain depends parabolically on the exposure time to sulfur when the temperature is low. At temperatures in the range 460 to 475°C the time dependence becomes linear and the rates of the

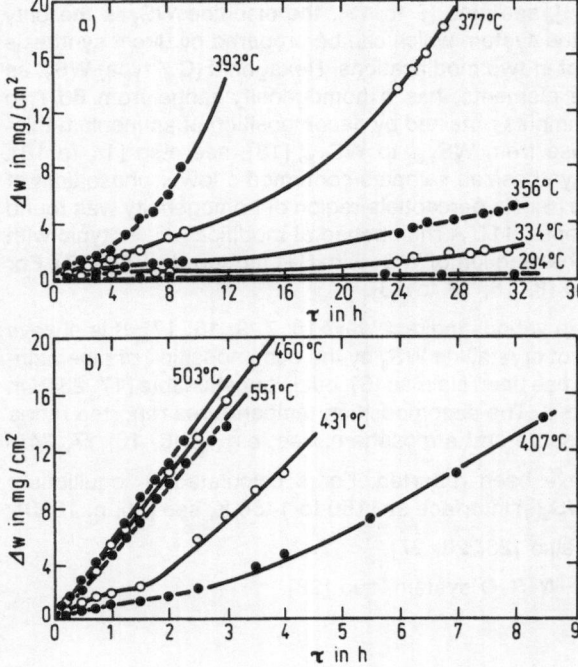

Fig. 110. Time-dependent weight gain, Δw, of unannealed tungsten sheet coupons exposed to sulfur vapor of 10.1 Torr at various temperatures a) below and b) above 400°C [31].

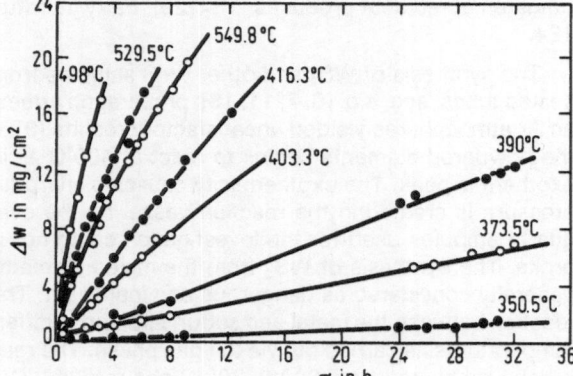

Fig. 111. Time-dependent weight gain, Δw, of annealed tungsten sheet coupons exposed to sulfur vapor of 9.5 Torr [31].

weight gain reach a maximum. At still higher temperatures the rates decrease again, see Fig. 110. Initial rates for T < 460 K and a sulfur pressure of 10.1 Torr were represented by an Arrhenius-type equation:

$$\log v \text{ (in mg} \cdot \text{cm}^{-2} \cdot \text{h}^{-1}) = 10.043 - 6934/T \text{ (10.1 Torr)}$$

with an apparent activation energy of 31.7 ± 2.8 kcal/mol. For the annealed metal, the weight gain is essentially linear with exposure time, though there is a slight tendency toward non-linearity at higher temperatures, see Fig. 111. Initial rates obtained below 475°C and for a sulfur pressure of 9.5 Torr were represented by

$$\log v \text{ (in mg} \cdot \text{cm}^{-2} \cdot \text{h}^{-1}) = 10.828 - 7448/T$$

with an apparent activation energy of 34.1 ± 3.2 kcal/mol. For both forms of metal, the sulfur-ization rates increase in a complex manner with increasing sulfur pressure for temperatures below the rate maximum, as is demonstrated by **Fig. 112** and **Fig. 113**, respectively. Above the transition temperature, the rates increase linearly with the partial pressure of S_2 with slopes of ~0.7 and ~0.3 mg·cm^{-2}·h^{-1}·Torr^{-1} for unannealed (521.5°C) and annealed (535°C) W, re-spectively. Tungsten disulfide was the only reaction product detected under all experimental

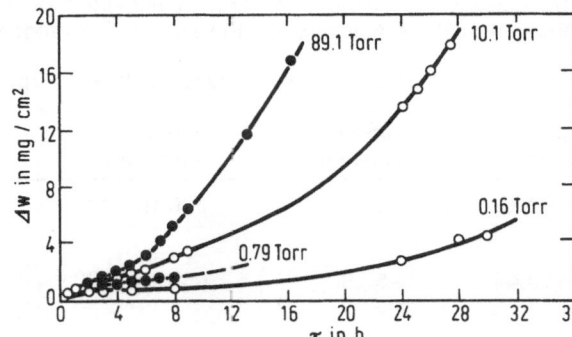

Fig. 112.   Effect of sulfur pressure on the time-dependent weight gain, Δw, of unannealed tungsten sheet coupons at 375°C [31].

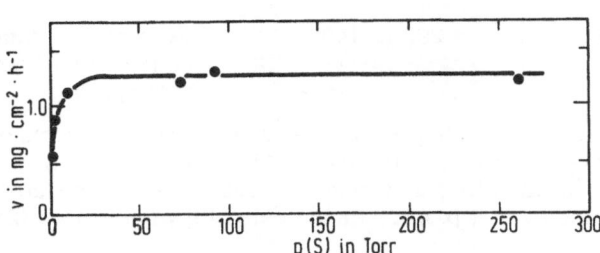

Fig. 113.   Effect of total sulfur pressure p(S) on the initial sulfurization rate v of annealed tungsten sheet coupons at 415°C [31].

conditions. The observed rate transition at 460 to 475°C is evidently associated with a change in the character of the WS$_2$ scale formed. At low temperatures this scale is brittle and nonprotective; in the case of the unannealed coupons, scale formation is accompanied by disintegration (splitting into lamellae) of the metal. At higher temperatures, the scale becomes more plastic and protective to the metal. Over the temperature range investigated, the rate of sulfurization is assumed to be controlled by some surface reaction and not by solid-state diffusion [31]. The area-corrected reaction rates of W powder with sulfur vapor of 3×10^{-2} Torr had ealier been found to be linear with time. The associated activation energy was 32 ± 3 kcal/mol; it dropped sharply to 11 kcal/mol above 460°C. The reaction started at 320°C and was accompanied by a large increase in surface [32]. In sulfur vapor of 2 atm pressure at 900°C,

W samples increased largely in volume and disintegrated completely. The sulfide layers did not tightly adhere to the base metal. They showed a double-layer structure with a dense inner layer and a porous outer one [33].

Solid tungsten specimens were sulfurized at 700 to 1200°C in a He-sulfur gas stream. The increase in the thickness of the surface coating with hexagonal WS_2 was measured. Plots of the squares of the weight increments per unit surface vs. time gave straight lines passing through the origin. The following values were determined for the diffusion coefficients, D, of S and W in the sulfide layer:

t in °C	700	900	1000	1100	1200
D_s in cm²/s	1.4	2.9	3.9	6.8	11
D_w in cm²/s	2.4	5.2	6.2	10	15

The activation energies of diffusion are 9.2 kcal/mol for W and 11.4 kcal/mol for S. Apparently diffusion of the metal through the sulfide film predominates [20].

Thermodynamic data for the reaction $W(s) + S_2(g) \rightarrow WS_2(s)$ are at variance:

$\Delta_r H$ in kcal/mol	$\Delta_r S$ in cal·mol⁻¹·K⁻¹	temperature range in K	comment	Ref.
−90.12	−35.9	600 to 770	formation of less-ordered hexagonal WS_2	[21]
−110.44	−57.9	650 to 1330	formation of ordered hexagonal WS_2	[21]
−80.4	−37.4	1370 to 1565	−	[34]
−62.36	−23.0	298 to 1400	experimental data [36] corrected for thermal segregation	[35]

The experimental basis for the data given above had been measurements of the equilibrium constants for $WS_2(s) + 2H_2(g) \leftrightarrow 2H_2S(g) + W(s)$ [21], for the formation of WS_2 from W in controlled H_2/H_2S gas mixtures [34], and for the reduction of WS_2/W mixtures by H_2 [36]. Data from [35] were adapted by [37, p. 429]. Earlier data [45] were revised [21].

The free-energy change for the reaction $W(s) + 2S$ (s, orthorhombic) $\rightarrow WS_2(s)$ was represented by $\Delta_r G$ (in cal/mol) $= -45840 + 3.68\,T \log T - 10.14 \cdot T$ for temperatures 300 K $<$ T $<$ 717.6 K [39].

For interactions of W and S at very high particle energies, see [40 to 42].

Diffusion

The penetration of S into polycrystalline tungsten at 1100 to 1200°C proceeds by grain boundary diffusion as well as by volume (bulk) diffusion. The penetration depth is only 15 to 20 µm after 5 h at 1100°C [4].

The diffusion of sulfur in monocrystalline tungsten at 1900 to 2200°C was studied by means of the β-emitting radioisotope ³⁵S. The diffusion direction was perpendicular to a plane with ⟨110⟩ orientation. The temperature dependence of diffusion coefficients was described by

$$D \text{ (in cm}^2\text{/s)} = (2.17^{+4.34}_{-1.45}) \times 10^{-5} \exp\left[-(69800 \pm 5000)/RT\right]$$

with the activation energy E_{diff} in cal/mol. The E_{diff} value is about half as large as the E_{diff} for the self-diffusion of tungsten. The entropy calculated from the preexponential factor has a negative value. There is a big difference from those D_0 and E_{diff} values derived from the theoretical formulae [43, 44] for an interstitial mechanism of diffusion. Therefore, a vacancy mechanism is considered as more probable, which is supported by the measurements of the lattice parameters of pure and S-saturated W [5].

For the diffusion of S and W in the sulfide layers during sulfurization experiments, see p. 114.

References:

[1] Elliott, R. P. (Constitution of Binary Alloys, 1st Suppl., McGraw-Hill, New York 1965, pp. 797/8).
[2] Massalski, T. D. (Binary Alloy Phase Diagrams, Vol. 1/2, American Society for Metals, Metals Park, Ohio, 1986, pp. 2011, 2013).
[3] Vogel, R.; Weizenkorn, H.-H. (Arch. Eisenhüttenwes. **32** [1961] 413/20).
[4] Pavlyuchenko, M. M.; Kononyuk, I. V. (Dokl. Akad. Nauk BSSR **8** No. 3 [1964] 157/60).
[5] Iovkov, V. P.; Panov, A. S.; Ryabenko, A. V. (Fiz. Met. Metalloved. **34** [1972] 1322/3; Phys. Met. Metallogr. [Engl. Transl.] **34** No. 6 [1972] 203/5).
[6] Glemser, O.; Sauer, H.; König, P. (Z. Anorg. Allg. Chem. **257** [1948] 241/6).
[7] Ehrlich, P. (Z. Anorg. Allg. Chem. **257** [1948] 247/53).
[8] Gardinier, C. F.; Chang, L. L. Y. (J. Less-Common Met. **61** [1978] 221/9).
[9] Beischer, D.; Öchsel, G. (Rev. Ger. Sci. Inorganic Chemistry, Pt. II, 1948, p. 33).
[10] Moh, G. H.; Udubasa, G. (Chem. Erde **35** [1976] 327/35).

[11] Shchukarev, S. A.; Morozova, M. P.; Damen, Kh. (Zh. Obshch. Khim **30** [1960] 2102/4; J. Gen. Chem. USSR [Engl. Transl.] **30** [1960] 2077/9).
[12] Hansen, M.; Anderko, K. (Constitution of Binary Alloys, McGraw-Hill, New York 1958, pp. 1170/1).
[13] Samoilov, S. M.; Rubinshtein, A. M. (Izv. Akad. Nauk SSSR Ser. Khim. **1959** 1905/12; Bull. Acad. Sci. USSR Div. Chem. Sci. [Engl. Transl.] **1959** 1819/24).
[14] Samoilov, S. M. (Izv. Akad. Nauk SSSR Ser. Khim. **1961** 1416/26; Bull. Acad. Sci. USSR Div. Chem. Sci. [Engl. Transl.] **1961** 1319/27).
[15] Samoilov, S. M.; Rubinshtein, A. M. (Izv. Akad. Nauk SSSR Ser. Khim. **1957** 1158/65; Bull. Acad. Sci. USSR Div. Chem. Sci. [Engl. Transl.] **1957** 1185/91).
[16] Rode, E. Ya.; Lebedev, B. A. (Zh. Neorg. Khim. **9** [1964] 2068/75; Russ. J. Inorg. Chem. **9** [1964] 1118/22).
[17] Wildervanck, J. C.; Jellinek, F. (Z. Anorg. Allg. Chem. **328** [1964] 309/18).
[18] van Arkel, A. E. (Recl. Trav. Chim. Pays-Bas **45** [1926] 437/44).
[19] Am. Soc. Test. Mater (Powder Diffr. File, ASTM 8-237).
[20] Koval'chenko, M. S.; Sychev, V. V.; Yurchenko, D. Z.; Tkachenko, Yu. G. (Izv. Akad. Nauk SSSR Met. **1974** No. 5, pp. 221/5; Russ. Metall. [Engl. Transl.] **1974** No. 5, pp. 180/4).

[21] Bartovská, L.; Černý, Č.; Kochanovská, A. (Collect. Czech. Chem. Commun. **31** [1966] 1439/52).
[22] Gait, R. I.; Mandarino, J. A. (Can. Mineral. **10** [1970] 729/31).
[23] Jellinek, F. (Arkiv Kemi **20** [1963] 447/80, 476/7).
[24] Sokol, L. (Collect. Czech. Chem. Commun. **21** [1956] 1140/5).
[25] Gulbransen, E. A.; Meier, G. S. (NBS Spec. Publ. [U.S.] No. 561-2 [1979] 1639/82; C. A. **92** [1980] No. 98072).

[26] Byalobzheskii, A. V.; Tsirlin, M. S.; Krasilov, B. I. (Vysokotemperaturnaya Korroziya i Zashchita Sverkhtugoplavkikh Metallov [High-Temperature Corrosion and Protection of Refractory Metals], Moscow 1977, pp. 1/224; C.A. **88** [1978] No. 179375).

[27] Shunk, F. A. (Constitution of Binary Alloys, 2nd Suppl. McGraw-Hill, New York 1969, p. 664).

[28] Anarbaev, A. A.; Bat'kaev, I. I.; Tleukulov, O. M. (Kompleksn. Ispol'z. Miner. Syr'ya **1988** No. 1, pp. 82/4 from C.A.**108** [1988] No. 190333).

[29] Riche, A. (Ann. Chim. Phys. [3] **50** [1857] 5/80, 13, 26).

[30] Opalovskii, A. A.; Lobkov, E. U.; Fedorov, V. E. (Zh. Prikl. Khim. [Leningrad] **47** [1974] 1200/3; J. Appl. Chem. USSR [Engl. Transl.] **47** [1974] 1241/3).

[31] Dutrizac, J. E. (J. Less-Common Met. **31** [1973] 281/97).

[32] Lambertin, M.; Colson, J. C.; Delafosse, D. (C. R. Seances Acad. Sci. C **270** [1970] 974/7).

[33] Gerlach, J.; Hamel, H. J. (Metall [Berlin] **23** [1969] 1006/11).

[34] Hager, J. P.; Elliott, J. F. (Trans. Metall. Soc. AIME **239** [1967] 513/20).

[35] Richardson, F. D.; Jeffes, J. H. E. (J. Iron Steel Inst. London **171** [1952] 165/75).

[36] Parravano, N.; Malquori, G. (Accd. Naz. Lincei Cl. Sci. Fis. Mat. Nat. Rend. [6] **7** [1928] 109/12, 189/92).

[37] Kubaschewski, O.; Evans, E. L.; Alcock, C. B. (Metallurgical Thermochemistry, 4th Ed., Pergamon, Oxford 1967).

[38] Ward, J. J.; Roy, J. P.; Herres, S. A. (R – 108 [1948] 1/97; N. S. A. **2** [1949] No. 928).

[39] Kelley, K. K. (Bull. U.S. Bur. Mines No. 406 [1937] from [38]).

[40] Keller, J. G.; Back, B. B.; Glagola, B. G.; et al. (Phys. Rev. [3] C **36** [1987] 1364/74).

[41] Abatzis, S.; Benayoun, M.; Beusch, W.; et al. (Proc. Rencontre Moriond **24** [1989] 491/5 from C.A. **112** [1990] No. 126755).

[42] Akesson, T.; Almehed, S.; Angelis, A. L. S.; et al. (Nucl. Phys. B **342** [1990] 279/301 from C.A. **113** [1990] No. 140071).

[43] Ferro, A. (J. Appl. Phys. **28** [1957] 895/900).

[44] Wert, C. A.; Zehner, C. (Phys. Rev. [2] **76** [1949] 1169/75).

[45] Černý, Č.; Habeš, M.; Zelená, M.; Erdös, E. (Collect. Czech. Chem. Commun. **24** [1959] 3836/42).

4 Reactions with Selenium

General. Products

No worked-out phase diagram is available.

The existence of the selenides WSe_2 [1, 2] and WSe_3 [1 to 3] has been reported [4], see also [5]. The diselenide is the only stable phase which can be obtained by direct synthesis from the elements at molar ratios W:Se between 1:0.5 and 1:3.0 [2] (see also [26] and the other publications on WSe_2 syntheses cited in the next section). WSe_2 is isostructural with MoS_2 (C7 type). Its range of homogeneity is similarly small as that of WS_2 [2], see also [25]. The absence of a noticeable homogeneity range of WSe_2 and of intermediate phases with Se contents lower than that of WSe$_2$ is generally confirmed by the thermal decomposition studies mentioned in the following, which in most cases included X-ray investigations of the residue at various degrees of decomposition. A homogeneity range with a limiting composition WSe$_{1.87}$ at 1534 K is postulated in [24]. At this composition the lattice parameter a of the (hexagonal) diselenide remains unchanged, but the parameter c increases.

Powdery WSe$_2$ is barely dissociated at 700 and 800°C in vacuum after residence times of 1 h. At higher temperatures, it decomposes rapidly into W and Se. The decomposition is virtually complete at 900 to 1000°C. Pressings of WSe$_2$ dissociate more slowly than the powdered form; a crust of tungsten powder forms on the pressing surface and becomes thicker with increasing time at temperature. A mixture of WSe$_2$ and W lies beneath this crust. The statement that WSe$_2$ is stable in vacuum up to 1350°C [6] was not confirmed [7, 8], see also [9]. The thermal decomposition of WSe$_2$ in a vacuum of $(1 \text{ to } 2) \times 10^{-2}$ Torr starts at 850 to 870°C and goes to completion at 1000°C [26]. Quantitative studies of the thermal decomposition of WSe$_2$ were carried out by a thermogravimetric technique [10], by Knudsen-effusion techniques [11, 12, 24], and by a simultaneous mass-loss, mass-spectrometric Knudsen-effusion technique [14]. The contradictory results from these measurements led to the conclusion that it might be appropriate to estimate the free energy changes for $1/2\,\text{WSe}_2(\text{s}) \to 1/2\,\text{W(s)} + \text{Se(g)}$ and $\text{WSe}_2(\text{s}) \to \text{W(s)} + \text{Se}_2(\text{g})$ and the respective Se and Se$_2$ pressures on the basis of a redetermined heat of formation of WSe$_2$(s) ($\Delta_f H^\circ_{298} = -185.3$ kJ/mol). For $1/2\,\text{WSe}_2(\text{s}) \to 1/2\,\text{W(s)} + \text{Se(g)}$ values of $\Delta_r G = 174$ and 120 kJ/mol and values of $p(\text{Se}) = 5 \times 10^{-4}$ and 6.5 Pa were estimated for 1100 and 1500 K, respectively. For $\text{WSe}_2(\text{s}) \to \text{W(s)} + \text{Se}_2(\text{g})$ values of $\Delta_r G = 135$ and 76 kJ/mol and values of $p(\text{Se}_2) = 0.04$ and 239 Pa were estimated for 1100 and 1500 K, respectively [15]. In all experimental decomposition studies, except [14], the gas phase was found to consist mainly of Se$_2$. This was also concluded from preliminary mass-spectrometric studies of the vapor phase over WSe$_2$ [16].

WSe$_3$ as obtained by decomposition of dissolved potassium selenotungstate with dilute sulfuric acid is in an amorphous state, which is maintained on heating with Se in a closed tube. Heating of WSe$_3$ in an evacuated closed tube results in decomposition at 220°C [2], see also [3].

Reaction Characteristics

W powder reacts with Se at 480 to 580°C with the evolution of a large amount of heat [25, 26].

First preparations of WSe$_2$ were carried out in closed tubes [1, 2]. Twenty-gram charges of the element mixture were enclosed in a quartz ampule at $\sim 10^{-5}$ Torr and fired for 10 to 15 h at 600 to 700°C. Then the loose, black product was remixed by shaking and fired for another 10 to 15 h at 1000 to 1200°C. The final product was metallic gray and could be transformed into single crystals by addition of some milligrams of iodine and by chemical vapor transport in a thermal gradient 900 → 700°C [17, p. 258], [18]. According to a second route, W powder was heated with a slight excess of Se from 20 to 900°C in 1.5 h and kept at this temperature for another 3 h. The product was homogenized at 800°C for 20 h in the presence of additional Se. Finally, free Se was distilled off into a cooled end of the quartz capsule which was used as the reaction vessel [7]. For further descriptions of syntheses in sealed ampules, see [14, 15] and, among others, also [19, 20, 25 to 28]. For the formation of WSe$_2$ from mixtures of the elements in the presence of niobium, see [13].

The electrically initiated synthesis of WSe$_2$ from W and Se in the crucible of a calorimetric bomb suffered from incomplete reaction. Between 10 and 15% of the Se did not combine with the W, and a post-reaction X-ray analysis revealed free W in addition to different crystallographic forms of WSe$_2$ [12].

Tungsten sheet coupons react quickly with selenium vapor, even at fairly low temperatures and pressures. The grey-black selenide WSe_2, formed as the only reaction product, is pulverulent and tends to spall off from the metal. The reaction kinetics of annealed tungsten coupons with pure selenium vapor of up to 55 Torr were investigated by a thermogravimetric technique at temperatures between 400 and 550°C. The weight gain of tungsten coupons exposed to two different selenium pressures at temperatures above 460°C is shown in **Fig. 114**. Prolonged measurements up to 30 h at 0.71 Torr selenium pressure and at 452 and 471°C showed some evidence of a plateau region after 6 to 8 h. However, at still longer times the weight gain

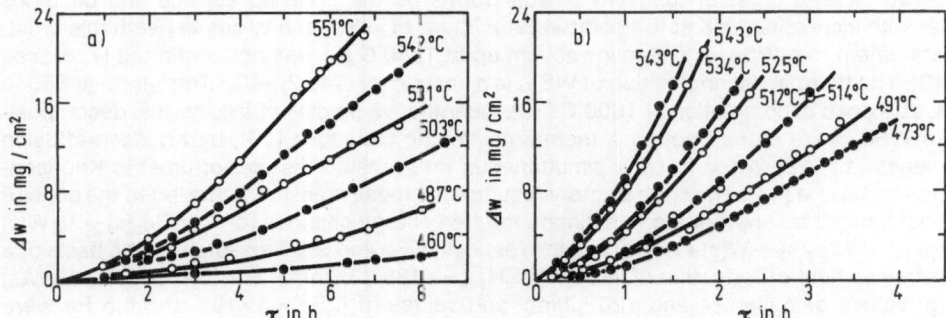

Fig. 114.   Time-dependent weight gain, Δw, of tungsten coupons in selenium vapor of a) 0.71 Torr and b) 11.1 Torr at various temperatures [21].

accelerated again. The time dependence of the weight gain at a given temperature and pressure was fitted to $\Delta w = a + b\tau + c\tau^2$. Initial rates ($\tau \to 0$) were erratic. Terminal rates arbitrarily derived at long observation times (the slope $\Delta w/\Delta\tau = b + 2c\tau$ is still a function of time), however, resulted in an apparent activation energy of about 26 kcal/mol; this value appeared to be essentially constant over the pressure range investigated. The selenization rate increased rapidly with increasing selenium pressure at low applied pressures, but was relatively insensitive to further pressure increases. **Fig. 115** shows the type of selenization curve obtained for different selenium pressures at 532°C. At this temperature, the vapor consists almost entirely of Se_2 molecules at all the pressures studied; also, the decomposition pressure of WSe_2 is negligible. The pressure dependence of the terminal rate constants as derived from the curves in Fig. 115 is displayed in **Fig. 116**. As can be seen, the initial rapid rate increase with pressure declines above about 10 Torr. The results suggest that the rate is possibly controlled by the adsorption of selenium on the surface of either the tungsten metal or the tungsten diselenide scale. The rate probably increases with elapsed reaction time, because the active surface also increases, either by the formation of loose WSe_2 particles, or by the roughening of the tungsten metal substrate. Autoheating could be excluded as the cause of the accelerating kinetics. The relatively high activation energy rules out rate control by Se mass transfer [21].

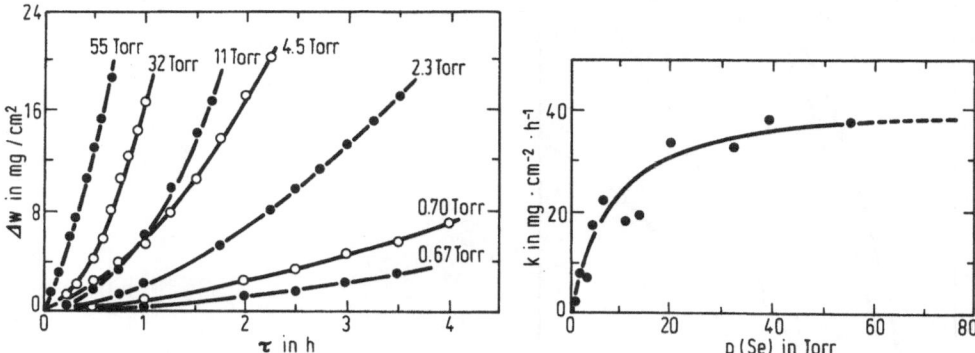

Fig. 115 (left).   Time-dependent weight gain, Δw, of tungsten coupons in selenium vapor at 532°C and various (total) selenium pressures [21].

Fig. 116 (right).   Effect of selenium pressure p(Se) on the "terminal" selenization rate of tungsten coupons at 532°C [21].

A different selenization behavior of tungsten was observed by [22] at 450 to ~650°C and 2×10^2 to 4×10^3 N/m² (~1.5 to 30 Torr) selenium pressure. The selenium vapor was transported by an Ar stream. **Fig. 117** presents weight gain vs. time curves for a constant selenium pressure of 4×10^3 N/m² and different temperatures, and **Fig. 118** does the same for a constant temperature of 500°C and different selenium pressures. The derived rates obey the relation

$$r \text{ (in kg} \cdot \text{m}^{-2} \cdot \text{s}^{-1}) = 1.3 \times 10^3 \ p(Se_2)^{0.5} \exp\left[-42\,000/RT\right]$$
$$(T = 723 \text{ to } 873 \text{ K}, \ p(Se_2) = 2 \times 10^2 \text{ to } 4 \times 10^3 \text{ N/m}^2)$$

where $p(Se_2)$ is the partial pressure of Se_2. The activation energy is given in J/mol. Diffusion of selenium through the WSe_2 scale is assumed, which forms a textured anti friction layer on the W. The reacting molecule species under the above conditions is mainly Se_2. The reaction usually starts after an incubation period, which decreases with increasing temperature and is required for the formation of nuclei of the selenide on active sites. Then growth proceeds by an island mechanism [22]. Thermogravimetric curves taken under static conditions at a selenium pressure corresponding to reservoir temperatures of 300 to 350°C indicate a conversion degree of ~45% after 12.5 h at 600°C and of nearly 98% after 1 h at 900°C (see the figure in the paper) [26].

When r.f.-sputtered W films on quartz substrates were exposed to Se vapor in a closed tube system, single-phase stoichiometric WSe_2 films with two types of orientations could be obtained depending on the selenium pressure. The basal plane of the WSe_2 crystallites was oriented either predominantly parallel or predominantly perpendicular to the substrate surface. The selenium pressure was adjusted by the temperature of an Se pellet stored separately from the W films in the colder end of 30 cm long evacuated and sealed silica glass tubes. The tungsten films (together with additional W foil as a getter for excess Se) were on the high-temperature side of the tubes, which were kept in a two-zone furnace with different temperature zones ranging from 543 to 1123 K. At selenium temperatures higher than 553 K a more rapid selenization started to take place, resulting in films which were either very porous or peeled completely off the substrate [23].

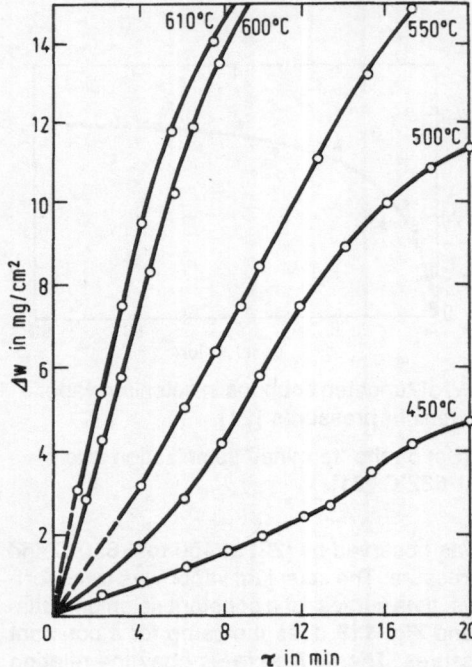

Fig. 117. Temperature-dependent weight gain, Δw, of tungsten plates in selenium vapor of 4×10^3 N/m² (~30 Torr) at various temperatures [22].

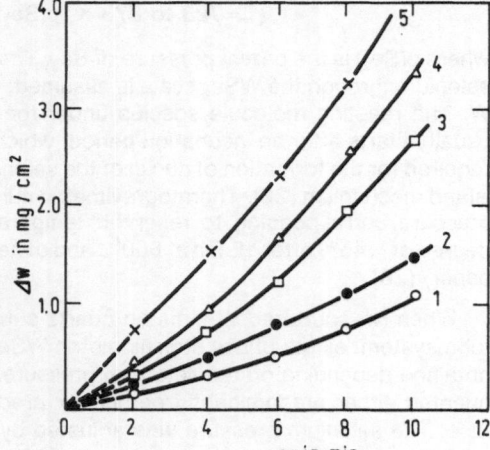

Fig. 118. Temperature-dependent weight gain, Δw, of tungsten plates in selenium vapor at 500°C and various pressures (in N/m²): 1) 2.3×10^2, 2) 5.3×10^2, 3) 1×10^3, 4) 2×10^3, 5) 4×10^3 [22].

References:

[1] Uelsmann, H. (Justus Liebigs Ann. Chem. **116** [1860] 122/7).
[2] Glemser, O.; Sauer, H.; König, P. (Z. Anorg. Allg. Chem. **257** [1948] 241/6).
[3] Moser, L.; Atynski, K. (Monatsh. Chem. **45** [1925] 235/50, 241).
[4] Hansen, M.; Anderko, K. (Constitution of Binary Alloys, McGraw-Hill, New York 1958, p. 1192).

[5] Kornilov, I. I.; Matveeva, N. M.; Pryakhina, L. I.; Polyakova, R. S. (Metallokhimicheskie Svoistva Elementov Periodicheskoi Systemy [Metal-Chemical Properties of the Elements of the Periodical System], Moscow 1966, pp. 179/82; C. A. **68** [1968] No. 16365).

[6] Magie, P. M. (Lubr. Eng. **22** No. 7 [1966] 262/9).

[7] Zelikman, A. N.; Kalikhman, V. L.; Krein, O. E.; et al. (Izv. Akad. Nauk SSSR Neorg. Mater. **6** [1970] 1930/4; Inorg. Mater. [Engl. Transl.] **6** [1970] 1696/9).

[8] Gladchenko, E. P.; Duksina, A. G.; Kalikhman, V. L.; et al. (Poroshk. Metall. [Kiev] **1970** No. 3, pp. 76/83; Sov. Powder Metall. Met. Ceram. [Engl. Transl.] **1970** 236/42).

[9] Sychev, V. V.; Yurchenko, D. Z.; Tkachenko, Yu. G.; et al. (Poroshk. Metall. [Kiev] **10** No. 9 [1971] 80/4; Sov. Powder Metall. Met. Ceram. [Engl. Transl.] **10** [1971] 744/7 from C. A. **76** [1972] No. 18564).

[10] Piskarev, N. V.; Mikhailov, E. S.; Chupakhin, M. S. (Zh. Neorg. Khim. **22** [1977] 2070/2; Russ. J. Inorg. Chem. [Engl. Transl.] **22** [1977] 1121/3).

[11] Mikhailov, E. S.; Glazunov, M. P.; Piskarev, N. V.; et al. (Zh. Fiz. Khim. **51** [1977] 722/3; Russ. J. Phys. Chem. [Engl. Transl.] **51** [1977] 422/3).

[12] Zelikman, A. N.; Kolchin, Yu. O.; Golutvin, Yu. M. (Zh. Fiz. Khim. **57** [1983] 856/8; Russ. J. Phys. Chem. [Engl. Transl.] **57** [1983] 519/21).

[13] Kalikhman, V. L. (Izv. Akad. Nauk SSSR Neorg. Mater. **16** [1980] 437/40; Inorg. Mater. [Engl. Transl.] **16** [1980] 290/2).

[14] Schiffman, R. A.; Franzen, H. F.; Ziegler, R. J. (High Temp. Sci. **15** [1982] 69/78).

[15] O'Hare, P. A. G.; Lewis, B. M.; Parkinson, B. A. (J. Chem. Thermodyn. **20** [1988] 681/91).

[16] Mikhailov, E. S.; Piskarev, N. V.; Chupakhin, M. S. (in: Tez. Dokl. 4th Vses. Semin. Khim. Primen. Khal'kogenidov (Abstr. Pap. 4th All-Union Semin. Chem. Use Chalcogenides), Ushgorod 1975 from [10]).

[17] Brixner, L. H. (J. Inorg. Nucl. Chem. **24** [1962] 257/63).

[18] Brixner, L. H. (J. Electrochem. Soc. **110** [1963] 289/93).

[19] Kershaw, R.; Vlasse, M.; Wold, A. (Inorg. Chem. **6** [1967] 1599/602).

[20] Upadhyayula, L. C.; Loferski, J. J.; Wold, A.; et al. (J. Appl. Phys. **39** [1968] 4736/40).

[21] Dutrizac, J. E. (J. Less-Common Met. **33** [1973] 341/53).

[22] Kolchin, Yu. O.; Lobova, T. A. (Nauchn. Tr. Mosk. Inst. Stali Splavov No. 131 [1981] 55/8; C. A. **96** [1982] No. 92450).

[23] Jaeger-Waldau, A.; Bucher, E. (Thin Solid Films **200** [1991] 157/64).

[24] Viksman, G. Sh.; Gordienko, S. P.; Yanaki, A. A.; et al. (Poroshk. Metall. [Kiev] **25** No. 1 [1986] 75/8; Sov. Powder Metall. Met. Ceram. [Engl. Transl.] **25** [1986] 64/6).

[25] Opalovskii, A. A.; Fedorov, V. E.; Erenburg, B. G.; et al. (Izv. Sib. Otd. Akad. Nauk SSSR Ser. Khim. Nauk **1969** No. 5, pp. 62/5).

[26] Opalovski, A. A.; Fedorov, V. E.; Lobkov, E. Yu.; et al. (in: Khal'gogenidy, Vol. 2, Naukova Dumka, Kiew 1970, pp. 86/92).

[27] Champion, I. A. (Br. J. Appl. Phys. **16** [1965] 1035/7).

[28] Hicks, W. T. (J. Electrochem. Soc. **111** [1964] 1085/65).

5   Reactions with Tellurium

The phase diagram of the W–Te system as proposed by [1] and redrawn by [2] is shown in **Fig. 119**. The orthorhombic (space group Pmmn) ditelluride has been identified as the only intermediate phase [3 to 5], see the reviews [6 to 8]. The formation of WTe_4 as a thin coating on tungsten in Te melts at high temperatures (>700°C) is assumed in [9]. The solubility of W in Te is low (<0.3 at%), as microscopic studies show [3].

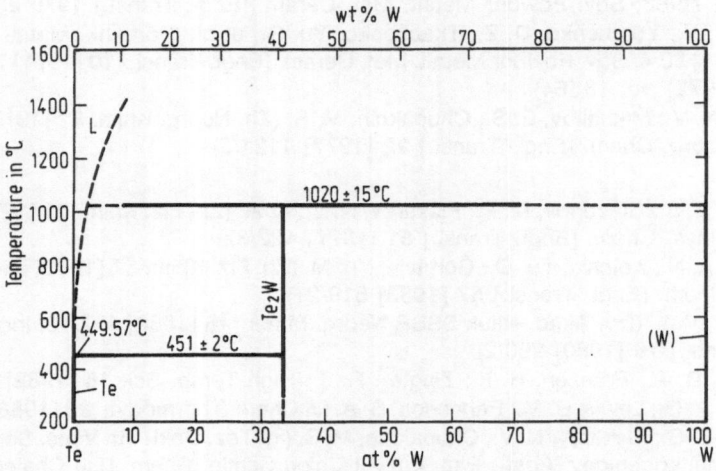

Fig. 119.   Phase diagram of the W–Te system
according to [1, 2].

Metallic gray WTe_2 was obtained by heating stoichiometric mixtures of the elements in evacuated and sealed quartz ampules. The mixture was first fired for 10 to 15 h at 600 to 700°C, then remixed by shaking, and again fired for another 10 to 15 h at 1000 to 1200°C. Single crystals can be prepared by adding a few mg of iodine (or bromine) and transporting the product in a thermal gradient from 900 to 700°C [5]. The direct synthesis from elemental mixtures in heated, closed quartz tubes has also been used repeatedly as a preparation method [3, 4, 10, 11].

The thermal decomposition of WTe_2 in a vacuum of $(1 \text{ to } 2) \times 10^{-2}$ Torr starts at 600 to 620°C and goes to completion at 800 to 850°C [10].

Alloying of tungsten with tellurium in vacuum starts at 640°C [10]. According to [9], the attack by Te vapor at elevated temperatures is minimal. A weight increase was noted between 700 and 800°C, but on further heating to 945°C the weight decreased, and, upon cooling the sample, a thin coating peeled off which was assumed to be WTe_4 [9].

References:

[1] Moffatt, W. G. (Handbook of Binary Phase Diagrams, Business Growth Services, Schenectady, New York, 1976).
[2] Massalski, T. D. (Binary Alloy Phase Diagrams, Vol. 1/2, American Society for Metals, Metals Park, Ohio, 1986, pp. 2118/9).
[3] Morette, A. (C. R. Hebd. Seances Acad. Sci. **216** [1943] 566/8; Ann. Chim. [Paris] [11] **19** [1944] 130/43).

[4] Knop, O.; Haraldsen, H. (Can. J. Chem. **34** [1956] 1142/5).
[5] Brixner, L. H. (J. Inorg. Nucl. Chem. **24** [1962] 257/63).
[6] Hansen, M.; Anderko, K. (Constitution of Binary Alloys, McGraw-Hill, New York 1958, p. 1231).
[7] Shunk, F. A. (Constitution of Binary Alloys, 2nd Suppl., McGraw-Hill, New York 1969, p. 700).
[8] Kornilov, I. I.; Matveeva, N. M.; Pryakhina, L. I.; Polyakova, R. S. (Metallokhimicheskie Svoistva Elementov Periodicheskoi Systemy [Metal-Chemical Properties of the Elements of the Periodical System], Moscow 1966, pp. 179/82; C.A. **68** [1968] No. 16365).
[9] Matson, L. K. (MLM-1720 (TID-4500) [1970] 1/14; N.S.A. **24** [1970] No. 29912).
[10] Opalovskii, A. A.; Fedorov, V. E.; Lobkov, E. Yu.; et al. (in: Khal'gogenidy, Vol. 2, Naukova Dumka, Kiew, 1970, pp. 86/92).

[11] Champion, I. A. (Br. J. Appl. Phys. **16** [1965] 1035/7).

6 Reactions with Polonium

Tungsten did not react with polonium vapor at up to 700°C [1]. This is to be expected considering the already rather low reactivity of tellurium towards W [2].

References:

[1] Witteman, W. G.; Giorgi, A. L.; Vier, D. T. (J. Phys. Chem. **64** [1960] 434/40).
[2] Matson, L. K. (MLM-1720 (TID-4500) [1970] 1/14, 3; N.S.A. **24** [1976] No. 29912).

7 Reactions with Boron

Phase Diagram

The phase diagram of the W–B system as proposed by [1] and redrawn by [2] is shown in **Fig. 120**. It is mainly based on the work of [3]. For earlier phase diagrams, see [4, 5, 69]. For the W–B system as a special case of the ternary systems W–B–M (M = Y, La, Ce, Pr, Gd, and Er), see [6]. A calculation of the W–B phase diagram was attempted [79], see also [80] for the thermochemical data.

The identity of the most B-rich intermediate phase has not been definitely established, see, e.g. [10]. Formulas close to WB_4 [7 to 9, 11 to 13, 39, 65, 67, 69] and to WB_{12} [5, 14 to 16] were proposed; the formula $W_{2-x}B_9$ (x ~1/6) was postulated [76]. W_2B_5 has been reported to be the phase richest in B [17 to 22, 24, 25, 32, 33, 77]. Carbon impurities cause decomposition of WB_{12} into W_2B_5 and boron carbide [5, 16]. WB_4 forms W_2B_5 and disappears when heated with 8 wt% C for 1 h at 1400°C [7]. For the influence of C on the phase relations at high B contents, see also [31]. The existence of the phases W_2B_3 [20, 25], W_3B_4 [20, 25], and WB_2 [20, 25 to 28, 73, 74] (see also [11, 30]) has not been confirmed. The monoboride WB occurs in two modifications (α and β) [29, 31, 70]. A low-temperature modification of W_2B_5 prepared from WO_2 and B and reportedly stable at temperatures below 1300 to 1500°C was described [71]. This finding conflicts with other reports [17, 72].

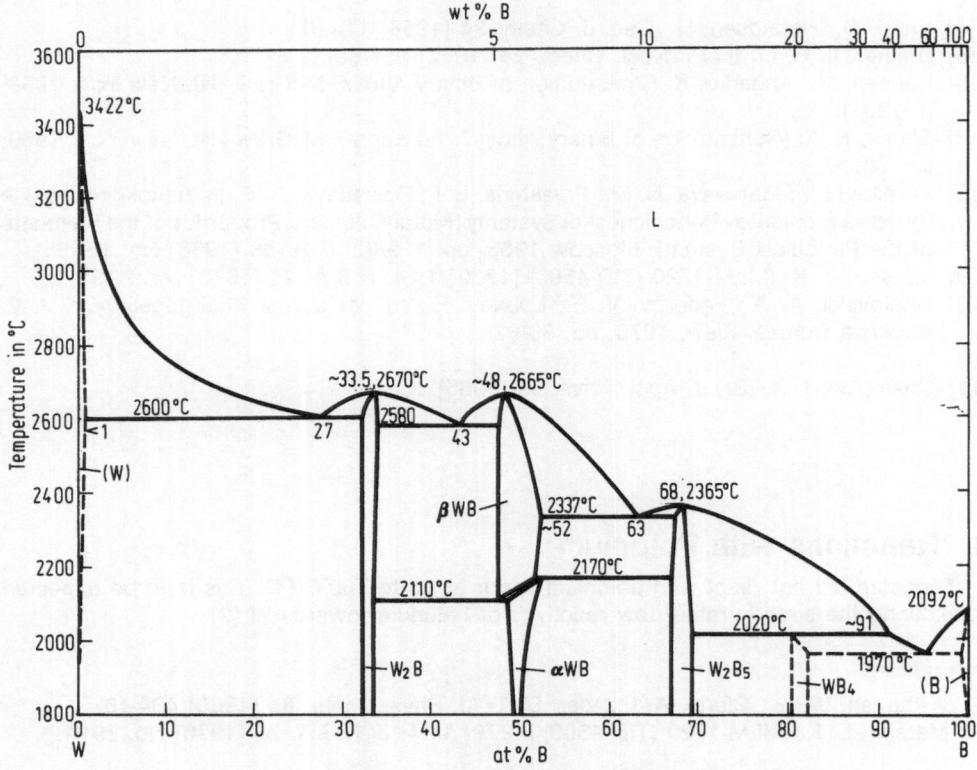

Fig. 120. Phase diagram of the W–B system according to [1, 2].

Melting characteristics of the intermediate phases:

phase	nature of melting	temperature in °C	Ref.
W₂B (~33.5 at%)	congruent	2670±16	[3]
	congruent	2780	[9]
	congruent	~2740	[5]
	congruent	2770±80	[4, 34]
WB(β) (~48 at% B)	congruent	2665±16	[3]
	congruent	~2800	[5]
	—	2860±80	[4, 34]
?ᵃ⁾	—	2920±50	[35]
W₂B₅ (68±1 at%)	congruent	2365±15	[3]
	congruent	2980	[31]
	incongruent	~2370	[5]
	?ᵇ⁾	2200	[30]

phase	nature of melting	temperature in °C	Ref.
WB_4	incongruent	2020 ± 30	[3]
	congruent	≥ 2200	[69]
WB_{12}	congruent	~ 2440	[5]

a) Attribution of the datum to WB by [36]. – b) The authors suggest that W_2B_5, in analogy to Mo_2B_5, may transform to WB_2 near the melting point; see also [36].

WB_4 decomposes reversibly above 1600°C into W_2B_5 and B [7], see also [39]. WB_{12} decomposes above 1800°C [14].

Homogeneity ranges of the intermediate phases:

phase	range in at% B	Ref.
W_2B	narrow	[5, 14, 17, 18, 20, 25]
WB	48.0 to 50.5	[17, 18]
	44.4 to 50 to 55	[20, 25]
W_2B_5	68 to 75	[20, 25]
	66.7 to 68	[17]
	66.7 to 69.4	[69]
	$\sim$64 to 70 at $\leq$2200°C a)	[5]
WB_4	narrow	[69]
WB_{12}	negligible	[5]

a) With rising temperature, the region of homogeneity extends in the direction of compositions enriched in boron. At 2250°C, the homogeneity extends to ~76 at% B.

According to electron microprobe analyses, the W content of the B-saturated W_2B_5 phase was 88.2 wt% (theoretical 87.2 wt%) [11]. Microprobe and chemical analyses as well as measured densities indicated that WB_4 may contain more than stoichiometric amounts of B (up to the composition WB_5) [11]. At any reactant ratio in the range $4.5 \leq B/W \leq 12$ a boride WB_{-4} is obtained whose composition lies withing the range $WB_{3.6 \text{ to } 4.3}$ as indicated by X-powder diffraction data and chemical analysis [65].

Reported temperatures for the transition α-WB into β-WB scatter: 1950 to 2400°C [25], >1900°C [69, 70], 2300 to 2400°C [5], and 2110 to 2170 ($\pm$70)°C [3]. A much lower transition temperature of 1850 ± 50°C was reported earlier [29]. No α-WB → β-WB transition was found at the temperature of the assumed α-WB + C eutectic, i.e. at 2270°C [37], see also [5].

The mutual solubilities of W and B in the solid state are low. The solubility of B in W is <1 at% [3], <0.9 at% [20, 25], ~0.2 at% at 2500°C and ~0.1 at% at 1000°C [9]. For the formation of W_2B in strain-aged tungsten alloyed with 0.002 and 0.01 wt% B, see [63]. Boron dissolves up to 0.3 to 0.5 at% W at 1800°C [38]. Boron alloys with 1 and 6 at% W, prepared by induction melting, were both found to be two-phase systems (with no evidence for a eutectic) [23] as were zone-melted alloys with 0.31 to 1.45 at% W [16].

Characteristics of the eutectica:

phase composition	at% B	temperature in °C	Ref.
(W) + W$_2$B	27±0.5	2600±12	[3]
	23	2600	[9]
	~25	~2650	[5]
W$_2$B + β WB	43±1	2580±12	[3]
	~42	~2550	[5]
β WB + W$_2$B$_5$	63±1	2337±7	[3]
W$_2$B$_5$ + WB$_{12}$	~83	~2270	[5]
WB$_4$ + (B)	~95	1970	[3]
WB$_{12}$ + (B)	~95	~2100	[5]

Reaction Characteristics

Various W borides were synthesizec by a solid-state reaction between tungsten and amorphous boron powders. Mixed powders with compositions varying between $0.4 \leq B/W \leq 13.0$ were pretreated at 500°C for 60 min in an H$_2$ stream and subsequently heated at 800 to 1500°C up to 120 min in an Ar stream. The formation of the individual borides depended on the B/W ratio in the powder mixture, the reaction temperature, and the holding time at the reaction temperature. W$_2$B, WB, W$_2$B$_5$, and WB$_4$ were successively obtained within 60 min at 1400°C, when the B/W ratio increased within the range given above, see **Fig. 121**. The boride WB$_{12}$ was not observed. At B/W = 0.4 and a holding time of 15 min, no boride formation occurred below 1000°C. Formation of W$_2$B started at ~1000°C and continued up to 1500°C. At B/W = 7.0 and

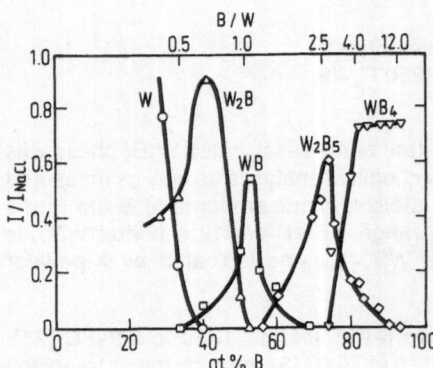

Fig. 121.   Relative amounts of W borides formed from W + B powder mixtures within 60 min at 1400°C as a function of B content or B/W ratio. The relative amounts of the borides are represented by the normalized relative intensities of selected X-ray diffraction lines: 110, 211, 021, 004, and 101, respectively, for W, W$_2$B, WB, W$_2$B$_5$, and WB$_4$. The normalized relative intensities were determined by the peak height ratio I/I$_{NaCl}$, where I$_{NaCl}$ is the intensity of the 200 line of sodium chloride [39].

the same holding time, the formation of WB, W$_2$B$_5$, and WB$_4$ was initiated at 800, 950, and 1200°C, respectively. WB was the only crystalline phase at 950°C, W$_2$B$_5$ at ~1200°C; a maximum amount of WB$_4$ was formed besides a small amount of W$_2$B$_5$ at 1400 to 1500°C. Above 1550°C the amount of WB$_4$ decreased, wnile the amount of W$_2$B$_5$ increased again. At B/W = 1.1 and 1400 to 1500°C, WB formed besides small amounts of W$_2$B within holding times of ≤60 min and as the only phase after longer holding times. At 1300°C, W$_2$B did not appear. At B/W = 2.75 and 1300 to 1500°C, W$_2$B$_5$ was the only phase formed after holding times of ≤120 min. The amounts of WB and W_2B_5 increased with holding time and temperature. Single-phase WB_4 could only be obtained with large excess of B (B/W≥9.0) It decomposes into W_2B_5 and B above 1400 to 1500°C, but is stabilized by excess B [39]; see also [40, 41]. This is in line with earlier investigations. When pressed powder mixtures of W and B with ratios 1:3≤W:B≤1:10 were heated at >1100°C, the tetraboride WB_4 formed besides W_2B_5 at lower W:B ratios and free from W_2B_5 at higher W:B ratios. The proportion of the WB_4 phase increased with increasing B content of the element mixture and increasing temperature up to 1400°C. At lower (1000°C) and higher (1600°C) temperatures, only W_2B_5 was obtained [7], [8, pp. 1342/5]. According to [11], WB_4 free from W_2B_5 forms only at W:B ratios lower than 1:8. The requirement for excess Ar pressure in the synthesis of WB_4 is stressed to avoid B losses by evaporation. Otherwise, the equilibrium will be shifted towards W_2B_5. Under 2 atm Ar WB_4 formed at 1700 and 1900°C within the range 4.5≤B/W≤12; specimens of composition $WB_{4.5}$ had a single-phase structure with only very minor amounts of free boron. Formation of WB_4 started at 1650°C, but at this temperature W_2B_5 and WB also started to form [65].

When W and B powders, mixed in the proportion of W_2B_5, were pressed and heated in sealed quartz ampules for 350 h at 800 and 900°C, the products consisted mainly of α-WB and B. The X-ray powder diagram had only a few of the lines belonging to the low-temperature modification of W_2B_5 [71].

Boronizing compact W with B powder for 1 at 1400°C in vacuum produces mostly W_2B and W_2B_5 besides very small amounts of α-WB. The weight increase after 1 h at 1300°C was 0.16 mg/cm² [21]; see also [42, 66, 75]. On heating W pieces with amorphous B at temperatures ≥900°C and a pressure of 10^{-3} mbar, WB layers were obtained on the W up to a thickness of 20 μm. These were free from visible cracks and showed a smooth transition to the base metal. Boronizing to layer thicknesses of 80 μm (10 h at 100°C) or 110 μm (>8 h at 1100°C) resulted in WB and W_2B_5 layers which showed brittle cracking in bending tests. In the initial stage, the growth of the boride case into the metal is influenced by epitaxial effects. The thickness of the scale generally grows according to a $\sqrt{\tau}$-time dependence when oxygen is totally excluded. Diffusion of B is rate-controlling [22, pp. 216, 225/7], [32, 33]. On cylindrical W specimens, pack-boronized in 84 wt% B+16 wt% borax in graphite cartridges under flowing hydrogen at 110 to 1400°C, an inner layer of W_2B forms below an external layer of WB + W_2B_5. The boride layers are separated from each other by a pronounced boundary and differ from each other in microhardness, structure, and etchability. The inner layer consists of columnar crystallites elongated in a direction perpendicular to the tungsten surface. The boundary between the inner layer and parent metal has a twisted shape with sharp projections and recesses, while that between the inner and outer boride layers is relatively straight. The thickness d of the (entire) boride layer grew as follows:

T in K	1373		1473		
time in h	6	8	1	2	4
d in μm	45	64	60	90	120

In accordance with the $\sqrt{\tau}$-time dependence it is assumed that B atoms diffuse via vacancies on B sublattice sites of the boride layer (for diffusion constants, see below) [24, 66]. When tungsten was pack-boronized in amorphous boron and 3% NH_4Cl, a single-phase W_2B layer formed [55]. For surface-boronizing of W in amorphous B powder contained in a C crucible, see also [43].

Boride formation by heating sputter-deposited B films on a W tip at various temperatures was studied by field ion microscopy. Stable W_2B_5 was formed at ≥1200 K. Frequently, the formation of W_2B_5 {0001} occurred on {011} (not the central W(011)). Characteristics in the He ion image of W_2B_5 obtained in this study showed a resemblance to that previously observed at the core of boron fibers. These fibers had been prepared by chemical vapor deposition on thin (diameter 10 to 20 µm) W wires at 1400 to 1600 K and could also have contained WB_4 in their core [44]. The occurrence of WB, W_2B_5, and WB_4 in the core of such fibers is reported [45].

Chemical vapor deposition of B on hot W at 900 to 1200°C and 10^{-7} to 10^{-6} Torr leads to the formation of two boride layers. The inner one consists of W_2B_5, the outer one of WB_4 and B. Boron deposition is effected by reduction of BCl_3 with H_2 [15, p. 3]. At 1050 and 1200°C vapor-deposited boron diffuses rapidly into the tungsten. X-ray diffraction studies reveal the formation of α-WB, W_2B_5, and WB_4 besides metallic W (base metal) and amorphous B [13, 67]. The borides W_2B, WB, and W_2B_5 were obtained at 850 to 1200°C within test periods of ≤120 min. The higher borides prevailed (or were formed exclusively) at higher temperatures and longer exposure times [46]. Layers of B a unidentified W borides formed within 15 min at 1500°C [47]; see also [48, 49]. No boride formation was noted when B was deposited on W wires by thermal decomposition of BBr_3 at 1000 to 1600°C after 30 to 60 min [50].

For epitaxy in boride layers grown on tungsten, see also [68].

Bombardment Effects

Irradiation of amorphous W films with 20- and 40-keV B ions at intensities of 8.2×10^{16} to 1.62×10^{17} ions/cm^2 resulted in β-WB formation, crystallization, and formation of circular defects in the microcrystalline W matrix [64].

Boron (^{11}B) atoms, which were implanted into tungsten single crystals at an energy of 80 keV and at doses of 10^{15} cm^{-2} settled on octahedral interstitial sites [51], see also [52]. An earlier study by a similar technique led to the same conclusion [53].

Diffusion

Diffusion of B into W was investigated repeatedly by Russian workers [24, 54 to 59, 66]. Early experimental results obtained at 1170 to 1570 K [58] were reviewed [60, 78]. Later measurements in the 1270 to 2170 K range [55] were reevaluated and revised to yield an increased acitivation energy of diffusion [54] (for comparison: $E_A = 17.2$ [58], 20.7 [55], and 30.6 kcal/mol [54]. A value as high as 64 kcal/mol has also been reported [24, 66]. It was assumed [54, 55] and there was also experimental evidence [26, 66] that B diffusion into W occurs via reactive diffusion and that the diffusion in the W_2B layer, i.e., the inner layer adjoining the tungsten, is the rate-controlling step. An empirical relation for diffusion of interstitial elements in transition metals with a bcc lattice predicts on activation energy of 54 kcal/mol for diffusion of B into W [61].

Diffusion of B into W from boride coatings on W (produced by CVD processes) started at ca. 900°C [49, p. 502].

The diffusion of W impurities in zone-refined, β-rhombohedral B at 1173 to 2373 K can be described by

$$D(W) = 2.5 \times 10^{-8} \exp[-7800/RT] + 2 \times 10^2 \exp[-44200/RT]$$

with the activation energy in cal/mol. The presence of two linear sections on the D vs. 1/T curve is probably due to superimposition of grain-boundary diffusion on bulk diffusion. Diffusion at grain boundaries is very pronounced at low temperatures, where it predominates over bulk diffusion [62].

References:

[1] Moffatt, W. G. (Handbook of Binary Phase Diagrams, Business Growth Services, Schenectady, New York, 1976).

[2] Massalski, T. D. (Binary Alloy Phase Diagrams, Vol. 1/2, American Society for Metals, Metals Park, Ohio, 1986, pp. 395/397).

[3] Rudy, E. (AFML-TR-65-2 Pt. V; AD-689843 [1969] 1/689, 214/5).

[4] Kieffer, R.; Benesovsky, F. (Hartstoffe, Springer, Wien 1963, pp. 429/34).

[5] Portnoi, K. I.; Romashov, V. M.; Levinskii, Yu, V.; Romanovich, I. V. (Poroshk. Metall. [Kiev] **1967** No. 5, pp. 75/80; Sov. Powder Metall. Met. Ceram. [Engl. Transl.] **1967** 398/402).

[6] Kuz'ma, Yu. B.; Mikhalenko, S. I.; Chaban, N. F. (Issled. Primen. Splavov Tugoplavkikh Met. **1983** 5/11; C.A. **100** [1984] No. 72496).

[7] Chrétien, A.; Helgorsky, J. (C. R. Hebd. Seances Acad. Sci. **252** [1961] 742/4).

[8] Helgorsky, J. (Ann. Chim. [Paris] **6** [1961] 1339/81).

[9] Goldschmidt, H. J.; Catherall, E. A.; Ham, W. M.; Oliver, D. A. (ASD-TDR-62-25 Pt. II [AD-418033] [1963], pp. 1/36) from [10, 11].

[10] Shunk, F. A. (Constitution of Binary Alloys, 2nd Suppl., McGraw-Hill, New York 1969, pp. 101/2).

[11] Romans, P. A.; Krug, M. P. (Acta Crystallogr. **20** [1966] 313/6).

[12] Shtein, L. M. (Legir. Obrab. Legk. Splavov **1981** 117/9; C.A. **96** [1982] No. 56374).

[13] Galasso, F.; Pinto, J. (Trans. Metall. Soc. AIME **242** [1968] 754/5).

[14] Rudy, E.; Benesovsky, F.; Toth, L. (Z. Metallk. **54** [1963] 345/53).

[15] DiCarlo, J. A. (NASA-TM-79077, E-9894 [1979] 1/23; C.A. **91** [1979] No. 161538).

[16] Avlokhashvili, D. Zh.; Tavadze, G. F.; Gabuniya, D. L. (Metalloved. Korrz. Met. **5** [1977] 21/5; C.A. **88** [1978] No. 195847).

[17] Kiessling, R. (Acta Chem. Scand. **1** [1947] 893/916).

[18] Brewer, L.; Sawyer, D. L.; Templeton, D. H.; Dauben, C. H. (J. Am. Ceram. Soc. **34** [1951] 173/9).

[19] Kornilov, I. I.; Matveeva, N. M.; Pryakhina, L. I.; Polyakova, R. S. (Metallokhimicheskie Svoistva Elementov Periodicheskoi Systemy [Metal-Chemical Properties of the Elements of the Periodical System], Moscow 1966, pp. 179/82; C.A. **68** [1968] No. 16365).

[20] Samsonov, G. V. (Izv. Sekt. Fiz.-Khim. Anal. Inst. Obshch. Neorg. Khim Akad. Nauk SSSR **27** [1956] 97/125; Ref. Zh. Khim. **1957** No. 11177).

[21] Minkevich, A. N.; Rastorguev, L. N.; Andryushechkin, V. I. (Izv. Vyssh. Uchebn. Zaved. Chern. Metall. **1960** No. 7, pp. 171/9; C.A. **1961** 11231).

[22] Vetters, H. (Sitz. Arbeitskreises Rastermikrosk. Materialprüf. **10** 1981 [1982] 215/28; C.A. **99** [1983] No. 57383).

[23] Seybolt, A. U. (Trans. Am. Soc. Met. **52** [1960] 971/89).

[24] Epik, A. P. (Poroshk. Metall. [Kiev] **3** No. 5 [1963] 21/7; Sov. Powder Metall. Met. Ceram. [Engl. Transl.] **1963** 361/5).

[25] Samsonov, G. V. (Dokl. Akad. Nauk SSSR **113** [1957] 1299/301; Dokl. Chem. [Engl. Transl.] **113** [1957] 417/9).

[26] Tucker, S. A.; Moody, H. R. (Proc. Chem. Soc. London **17** [1901] 129/30).

[27] Wedekind, E. (Ber. Dtsch. Chem. Ges. **46** [1913] 1198/207, 1206).

[28] Halla, F.; Thury, W. (Z. Anorg. Allg. Chem. **249** [1942] 229/37).

[29] Post, B.; Glaser, F. W. (J. Chem. Phys. **20** [1952] 1050/1).

[30] Post, B.; Glaser, F. W.; Moskowitz, D. (Acta Metall. **2** [1954] 20/5).

[31] Glaser, F. W. (Trans. Am. Inst. Min. Metall. Pet. Eng. **194** [1952] 391/6).

[32] Vetters, H.; Mayr, P. (Scanning Electron Microsc. **1983** No. 1, pp. 105/11; C. A. **99** [1983] No. 199128).

[33] Vetters, H.; Mayr, P. (HTM Härterei-Tech. Mitt. **40** No. 5 [1985] 198/210).

[34] Kieffer, R.; Benesovsky, F.; Honak, E. R. (Z. Anorg. Allg. Chem. **268** [1952] 191/200).

[35] Moers, K.; Agte, C. (Z. Anorg. Allg. Chem. **198** [1931] 233/75, 243/75).

[36] Hansen, M.; Anderko, K. (Constitution of Binary Alloys, McGraw-Hill, New York 1958, pp. 264/5).

[37] Levinskii, Yu. V.; Salibekov, S. E.; Levinskaya, M. Kh. (Poroshk. Metall. [Kiev] **5** No. 12 [1965] 56/62; Sov. Powder Metall. Met. Ceram. [Engl. Transl.] **1965** 1004/9).

[38] Jimenez-Crespo, A.; Tergenius, L. E.; Lundström, T. (J. Less-Common Met. **77** [1981] 147/50).

[39] Itoh, H.; Matsudaira, T.; Naka, S.; et al. (J. Mater. Sci **22** [1987] 2811/5).

[40] Matsudaira, T.; Itoh, H.; Naka, S.; et al. (Yogyo Kyokaishi **95** No. 2 [1987] 248/52 from C. A. **106** [1987] No. 124531).

[41] Matsudaira, T.; Itoh, H.; Naka, S.; et al. (Zairyo **36** [1987] 1167/71 from C. A. **108** [1988] No. 80553).

[42] Minkevic, A. N. (Härterei-Tech. Mitt. **17** No. 3 [1962] 141/7).

[43] Matsuda, F.; Nakata, K. (1st Int. Conf. Surf. Eng., Brighton, UK, 1985 [1986], Vol. 3, pp. 109/20 from C. A. **107** [1987] No. 138902).

[44] Omae, N.; Nakamura, A.; Koike, S.; Umeno, M. (J. Vac. Sci. Technol. A **5** [1987] 1367/70).

[45] Shtein, L. M. (Legir. Obrab. Legk. Splavov **1981** 117/9; C. A. **96** [1982] No. 56374).

[46] Deiss, W. J.; Andrieux, J. L. (Bull. Soc. Chim. Fr. **1959** 178/82).

[47] Champbell, I. E.; Powell, C. F.; Nowicki, D. H.; Gonser, B. W. (Trans. Electrochem. Soc. **96** [1949] 318/33, 331).

[48] Boman, M.; Carlsson, J. O. (Surf. Technol. **24** [1985] 173/90).

[49] Bonetti, R.; Comte, D.; Hintermann, H. E. (Proc. Conf. Chem. Vap. Deposition 5th Int. Conf., Slough, Engl., 1975, pp. 495/508; C. A. **84** [1976] No. 154195).

[50] v. Naray-Szabo, S.; Tobias, C. W. (J. Am. Chem. Soc. **71** [1949] 1882/3).

[51] Beck, K.; Kopitzki, K.; Krauss, G.; Mertler, G. (Radiat. Eff. **77** [1983] 79/87).

[52] Andersen, J. U.; Laegsgaard, E.; Feldman, L. C. (Radiat. Eff. **12** [1972] 219/23).

[53] Skakun, N. A.; Matyash, P. P.; Dikii, N. P. (Ukr. Fiz. Zh. **19** [1974] 1610/3).

[54] Samsonov, G. V.; Epik, A. P. (Fiz. Met. Metalloved. **14** [1962] 479/80; Phys. Met. Metallogr. [Engl. Transl.] **14** No. 3 [1962] 144/5).

[55] Samsonov, G. V.; Latysheva, V. P. (Fiz. Met. Metalloved. **2** [1956] 309/19; C. A. **1957** 6258).

[56] Samsonov, G. V.; Latysheva, V. P. (Dokl. Akad. Nauk SSSR **109** [1956] 582/5; C. A. **1957** 9249).

[57] Samsonov, G. V. (Bor Tr. Konf. Khim. Bora Ego Soedin., Moscow 1955 [1958], pp. 74/89; C. A. **1960** 23564).

[58] Samsonov, G. V. (Dokl. Akad. Nauk SSSR **93** [1953] 859/61).

[59] Samsonov, G. V. (Izv. Sekt. Fiz.-Khim. Anal. Inst. Obshch. Neorg. Khim. Akad. Nauk SSSR **27** [1956] 97/125; Ref. Zh. Khim. **1957** No. 11177).

[60] Klopp, W. D.; Barth, V. D. (DMIC-Memo-50 [1960] 1/10; C. A. **57** [1962] 6984).

[61] Spivak, I. I. (Fiz. Met. Metalloved. **22** [1966] 859/64; Phys. Met. Matallogr. [Engl. Transl.] **22** No. 6 [1966] 52/7).

[62] Panteleeva, G. V.; Kharimov, R. Kh. (Izv. Akad. Nauk SSSR Neorg. Mater. **11** [1975] 943/4; Inorg. Mater. [Engl. Transl.] **11** [1975] 806/7).

[63] Tolstobrov, Yu, O.; Povarova, K. B. (Fiz. Khim. Obrab. Mater. **187** No. 5, pp. 121/4 from C. A. **108** [1988] No. 42281).

[64] Khitrova, V. I.; Tikhonova, A. A.; Titov, V. V.; et al. (Izv. Akad. Nauk Kaz. SSR Ser. Fiz. Mat. **1981** No. 6, pp. 50/6; C. A. **96** [1982] No. 113729).

[65] Bodrova, L. G.; Koval'chenko, M. S.; Serebryakova, T. I. (Poroshk. Metall. [Kiev] **1974** No. 1, pp. 1/4; Sov. Powder Metall. Met. Ceram. [Engl. Transl.] **13** [1974] 1/3).

[66] Samsonov, G. V.; Epik, A. P. (Coatings of High-Temperature Materials, Pt. 1, Plenum, New York 1966, pp. 7/111, 25/9).

[67] Galasso, F.; Paton, A. (Trans. Metall. Soc. AIME **236** [1966] 1751/2).

[68] Gebhardt, M. (in: Bardsley, W.; Hurle, D. T. J.; Mullin, J. B.; Hartman, P.; Crystal Growth: An Introduction, North-Holland, Amsterdam 1973, pp. 105/42, 127).

[69] Lundström, T. (Arkiv Kemi **30** [1968] 115/27, 118/20).

[70] Boller, H.; Rieger, W.; Nowotny, H. (Monatsh. Chem. **95** [1964] 1497/501).

[71] Kuz'ma, Yu. B.; Serebryakova, T. I.; Plakhina, A. M. (Zh. Neorg. Khim **12** [1967] 559/60; Russ. Inorg. Chem. [Engl. Transl.] **12** [1967] 288/9).

[72] Peshev, P.; Bliznakov, G.; Leyarovska, L. (J. Less-Common Met. **13** [1967] 241/7).

[73] Woods, H. P.; Wawner, F. E., Jr.; Fox, B. G. (Science **151** [1966] 75).

[74] Moissan, H. (C. R. Hebd. Seances Acad. Sci. **123** [1896] 13/6).

[75] Minkevich, A. N. (Metalloved. Term. Obrab. Met. **1961** No. 8, pp. 9/15; Met. Sci. Heat Treat. Met. [Engl. Transl.] **1961** No. 8, pp. 347/51).

[76] Nowotny, H.; Haschke, H.; Benesovsky, F. (Monatsh. Chem. **98** [1967] 547/54).

[77] Kiessling, R. (Acta Chem. Scand. **4** [1950] 209/27).

[78] Seith, W. (Diffusion in Metallen, Springer, Berlin 1955, p. 66).

[79] Kaufman, L.; Uhrenius, B.; Birnie, D.; Taylor, K. (CALPHAD: Comput. Coupling Phase Diagrams Thermochem. **8** [1984] 25/66 from [80]).

[80] Rogl, P.; Schuster, J. C. (Phase Diagrams of Ternary Boron Nitride and Silicon Nitride Systems, ASM International, Materials Park, Ohio, 1992, pp. 117/20).

8   Reactions with Carbon

8.1   Phase Diagram

A schematic phase diagram of the W–C system as proposed by [1] and redrawn by [2] is shown in **Fig. 122**. It is based mainly on the work of [3]. A thermodynamic evaluation of the W–C system has been made by use of a two-sublattice regular solution model for the interstitial solution phases and an ordinary subregular solution model for the liquid phase [52]. Literature data on the system up to and including 1969 were reviewed [64 to 67]. For earlier reviews see [4 to 9].

A saturation concentration of 40 and 50 at% C in molten W at 3450 and 3800°C, respectively, was found [10], which, however, seems to be too low with regard to the phase diagram [70].

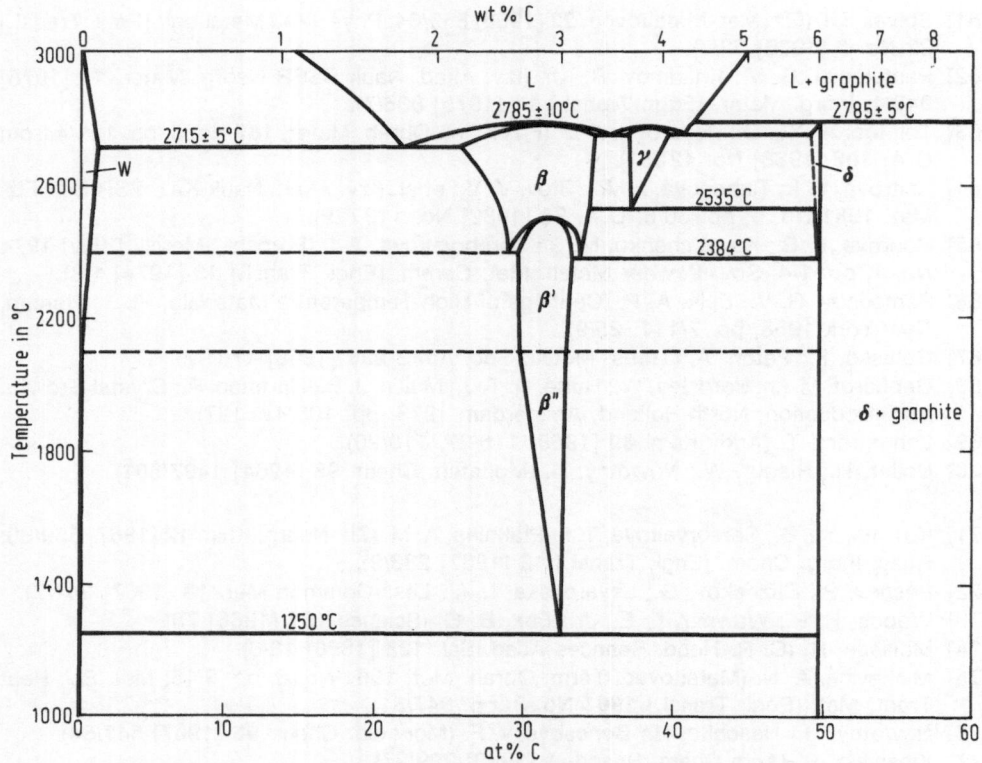

Fig. 122   Schematic phase diagram of the W–C system according to [1, 2].

Intermediate Phases

The following intermediate phases have been identified: W_2C in the modifications β, β′, and β″, WC_{1-x} (termed γ in Fig. 122), and WC (termed δ in Fig. 122) [2, 3]; for earlier identifications of these phases, see the reviews cited above. The existence of a phase W_3C besides WC and another, unidentified carbide between W_3C and WC has been reported [11]. A carbide with a C content between W_2C and WC was considered as being possible [12, 13]. The formation of one or perhaps two unstable carbides (W_5C_2 or W_3C_2?) was suggested [14], based on carburizing and decarburizing tests, see [7]. The existence of W_3C_2 at high temperatures was also supposed [13], but could not be substantiated up to 2600°C. A cubic phase W_5C_3, stable only above 2540°C and being formed by a peritectic reaction, was found [15]. It is probably identical with the high-temperature modification of WC [16, 17] (according to Fig. 122, it should be identical with the γ phase). No evidence was found for the existence of a high-temperature modification of W_2C by DTA, metallography, or X-ray diffractometry [16, 17].

W_2C Phase. A polymorphic transformation of W_2C occurs at ~2400°C [12, 13, 18, 19]. A cubic phase obtained by sparking W foil or powder with C electrodes as well as by the self-sparking of W_2C is assumed to be the high-temperature polymorph of W_2C [20], see [4]. The same phase was prepared by sparking WC electrodes in petroleum [68]. This phase [20, 68] was also regarded as WC_{1-x} [66, 67]. A face-centered cubic W_2C modification, perhaps also corresponding to the high-temperature form of W_2C according to [12, 13, 18, 19], was found in W layers prepared by decomposition of W carbonyl [21]. Mechanical treatment changes the unstable high-temperature modification in the low-temperature polymorph. The β, β', and β'' modifications (see Fig. 122) crystallize in the PbO_2, hexagonal L'3, and hexagonal C6 types, respectively [2]. The transition of the β' into the β modification at 32.6 at% C occurs at 2490 ± 30°C (maximum). The $\beta'' \rightarrow \beta'$ transition occurs between about 2100 and 2140°C [3]. An orthorhombic low-temperature (≤ 800°C) modification of W_2C was found in X-ray diffraction studies of the W–Re–C system, while an intermediate phase with an NaCl structure could not be detected in the binary W–C system [22].

There is general agreement that W_2C melts congruently. Data for the melting point (in °C) are: 2880 [23, 55], 2860 ± 50 [24], ~2820 (maximum) [15], 2795 (maximum at 30 at% C) [16, 17], 2785 ± 10 (maximum) [2], 2776 ± 12 (maximum at 39 at% C) [3], 2750 (maximum) [25], 2730 ± 15 [26], 2730 [27]. W_2C decomposes in a eutectoid reaction at 1215°C into W and WC [28]. Thermodynamic calculations suggest a eutectoid decomposition at 1327°C [29]. The eutectoid temperature is given as 1300°C [30], 1302 ± 5°C [31], see also [32]. The $\beta' + \delta$ eutectoid (Fig. 122) lies at 2384°C [2] or 2380 ± 30°C and ~33.4 at% C [3].

In contrast to WC, the W_2C phase has an appreciable range of homogeneity [25, 33, 34]. According to X-ray work [35 to 38], the homogeneity range above 1400°C extends by ~2 at% to lower than stoichiometric C contents, i.e., lies within the limits $WC_{0.5}$, and $WC_{0.45}$. According to [15 to 17], the W-rich limit shifts from 28 at% C at 2460 to 2475°C to 26 at% C at 2710 to 2735°C. The C-rich limit at 2460 to 2675°C corresponds to the stoichiometric composition. At 2525°C, the eutectoid decomposition temperature of WC_{1-x} (see below), some excess C may be dissolved in W_2C [16, 17]. Units of 25.6 at% C at 2710°C (W–W_2C eutectic) and ~35 at% C (read from figure) at 2735°C (W_2C–WC_{1-x} eutectic) are reported [3]. The homogeneity range of the β' modification at 2200°C extends from ~29.6 to ~32.7 at% C [3]. A C deficiency of the W_2C phase was also noted [20].

WC_{1-x} (γ) Phase. A face-centered cubic phase, which was considered to be a high-temperature modification of WC, was first detected in rapidly quenched alloys with compositions in the range W_2C–WC [16, 17, 39]; see, however, [37]. Preliminary data [40] suggested that a cubic modification of WC can be retained by quenching under high pressure. In fact, a cubic phase could be prepared at 35 to 60 kbar pressure and ~1800°C in the presence of TiC as stabilizer [41]. Formation of WC_{1-x} in tungsten/carbon multilayers was reported [69]. Detailed studies showed that the C content of the phase in question never comes near that of stoichiometric WC [3]. Therefore, in accordance with [1, 2], the phase is treated here separately from WC (or W_2C).

WC_{1-x} crystallizes in the B 1 type [2]. It forms from the melt at 2747 ± 12°C with 39 at% C [3] and decomposes at 2530 ± 20°C [3] or 2535°C [1, 2] via a eutectoid reaction into W_2C (β) and WC (δ). The eutectoid composition lies at ~38 at% C. According to [16, 17], the "high-temperature modification" of WC forms peritectically at 2785°C with 50 at% C and decomposes eutectoidally with ~37.5 at% C at 2525°C. The "W_5C_3" phase, which is possibly also identical with the γ phase, is reported to form from the melt and W_2C at ~2790°C and to decompose into W_2C and C at 2530°C (40 at% C) [15].

WC Phase. The phase crystallizes in the hexagonal B_h type [2]. A cubic "high-temperature modification" is probably identical with the WC_{1-x} phase, see above. WC has only a negligible homogeneity range.

A congruent melting point of WC is only reported in early work: 2870 ± 50°C [24], 2780°C [23], 2720 ± 20°C [42], 2600 to 2700°C [11]. According to most other studies, WC decomposes when or before melting. The nature of the decomposition is contested. More recent studies established a peritectic decomposition into melt and C at 2776 ± 10°C (peritectic melt 42 at% C) [3] or 2785 ± 5°C [1, 2]. The same decomposition reaction had earlier been found at

~2600°C [25] and at 2880°C [43]. According to [16, 17], WC decomposes into WC_{1-x} (γ) and C at 2755°C. A decomposition into W_5C_3 and C at 2730°C was suggested; the heat effect noted at 2730°C, however, could be caused either by peritectoid decomposition or by a phase change [15]. Thermodynamic calculations and reaction studies indicate a peritectoid decomposition into W_2C and C at 1325°C [28] or 1760°C [30]. WC is considered to be metastable above this temperature [28].

Eutectics

The following eutectics have been observed in the W–C system:

phase composition	temperature in °C	at% C	Ref.	phase composition	temperature in °C	at% C	Ref.
$W + W_2C$	2425	—	[20]	$W + W_2C$	2710 ± 12	22	[3, 46]
	2475	~18.9	[25]		2715 ± 5	~22	[1, 2]
	2475	20	[44]	$W_2C + WC_{1-x}$[a]	2760	36	[16]
	~2480	—	[45]	$W_2C + WC_{1-x}$[a]	2735 ± 12	36.5	[3]
	2690	~17.9	[11]	$W_5C_3 + C$	2765	~39	[15]
	2735	—	[15]	$W_2C + WC$	2525	~41.9	[8, 25]
	2732 ± 22	—	[42]			39.6	[47]
	2710	25	[16, 17]	WC_{1-x}[a] $+ WC$	2720 ± 12	41	[3]

[a] "High-temperature modification" of WC.

Solid Solubilities of Carbon in Tungsten

The saturation concentration, c(sat) in at%, of carbon in tungsten was determined in the 1400 to 2000°C range by measurement of the residual electric resistivity of carburized (CH_4 atmosphere) W wires which had been quenched to 4 K, $\ln c(sat) = 5.2 - 1.71 \times 10^4/T$ [60]. These data and the results obtained from an isothermal carburizing method ($\ln c(sat) = 4.67 - 1.5 \times 10^4/T$ for 1795 to 2640°C [46], see also [59]) were combined to a common equation: $\ln c(sat) = 5.0 - 1.57 \times 10^4/T$ [60]. The enthalpy change for the transition of 1 mol C from W_2C into W thus amounts to 31 ± 2 kcal/mol [60]. The maximum solubility at the $W–W_2C$ eutectic temperature of ca. 2710°C is around 0.7 to 0.8 at% C [46, 60]. Earlier measurements of the carbon solid solubility by X-ray diffraction techniques [20] (see also [31]) indicated lower solubilities (and a lower eutectic temperature). Above the eutectic temperature, measured solubilities decreased rapidly [20]. Several solubilities reported for individual temperatures [11, 25, 31, 38, 48 to 51, 53, 54, 56 to 58] deviate in part significantly.

For carbide precipitation in W-rich alloys, see also [61 to 63].

References:

[1] Metals Handbook, Metallography, Structures, and Phase Diagrams, 8th Ed., Vol. 8, American Society for Metals, Metals Park, Ohio, 1973.

[2] Massalski, T. D. (Binary Alloy Phase Diagrams, Vol. 1/2, American Society for Metals, Metals Park, Ohio, 1986, pp. 599/600).

[3] Rudy, E. (AFML-TR-65-2 Pt. V; AD-689843 [1969] 1/689).

[4] Shunk, F. A. (Constitution of Binary Alloys, 2nd Suppl., McGraw-Hill, New York 1969, pp. 167/9).

[5] Elliott, R. P. (Constitution of Binary Alloys, 1st Suppl., McGraw-Hill, New York 1965, pp. 236/7).

[6] Kieffer, R.; Benesovsky, F. (Hartstoffe, Springer, Wien 1963, pp. 187/90).

[7] Hansen, M. (Der Aufbau der Zweistofflegierungen, Springer, Berlin 1936, 386/92).

[8] Schwarzkopf, P.; Kieffer, R.; Leszynski, W.; Benesovsky, F. (Refractory Hard Metals, McMillan, New York 1953, pp. 138/61, 153/6).

[9] Hansen, M.; Anderko, K. (Constitution of Binary Alloys, McGraw-Hill, New York 1958, pp. 391/3).

[10] Kashin, V. I.; Klibanov, E. L.; Tsilosani, A. G.; Tylkina, M. A.; Konieva, L. Z. (Izv. Akad. Nauk SSSR Met. **1971** No. 2, pp. 39/45; Russ. Metall. [Engl. Transl.] **1971** No. 2, pp. 25/30).

[11] Ruff, O.; Wunsch, R. (Z. Anorg. Allg. Chem. **85** [1914] 292/328).

[12] Becker, K. (Z. Physik **51** [1928] 481/9).

[13] Becker, K. (Z. Metallkd. **20** [1928] 437/41).

[14] Schenck, R.; Kurzen, F.; Wesselkock, H. (Z. Anorg. Allg. Chem. **203** [1932] 177/83).

[15] Dolloff, R. T.; Sara, R. V. (WADD-TR-60-143-Pt. II [1961] 1/19; N.S.A. **16** [1962] No. 7898).

[16] Sara, R. V. (J. Am. Ceram. Soc. **48** [1965] 251/7).

[17] Sara, R. V.; Dolloff, R. T. (WADD-TR-60-143-Pt. III; AD-277794 [1962] 1/38; N.S.A. **16** [1962] No. 24798).

[18] Skaupy, F.; Becker, K. (Z. Elektrochem. Angew. Phys. Chem. **33** [1927] 512/3).

[19] Becker, K. (Z. Elektrochem. Angew. Phys. Chem. **34** [1928] 640/2).

[20] Goldschmidt, H. J.; Brand, J. A. (J. Less-Common Met. **5** [1963] 181/94).

[21] Lander, J. J.; Germer, L. H. (Trans. Am. Inst. Min. Metall. Pet. Eng. **175** [1948] 648/92).

[22] Kuz'ma, Yu. B.; Lakh, V. I.; Markiv, V. Ya.; Stadnyk, B. I.; Gladyshevskii, E. I. (Poroshk. Metall. [Kiev] **3** No. 4 [1963] 40/8; Sov. Powder Metall. Met. Ceram. [Engl. Transl.] **1963** 286/92).

[23] Andrews, M. R. (J. Phys. Chem. **27** [1923] 270/83).

[24] Agte, C.; Alterthum, H. (Z. Tech. Phys. **11** [1930] 182/91).

[25] Sykes, W. P. (Trans. Am. Soc. Steel Treat. **18** [1930] 968/92).

[26] Barnes, B. T. (J. Phys. Chem. **33** [1929] 688/91).

[27] Brewer, L.; Bromley, L. A.; Gilles, P. W.; Lofgren, N. L. (in: Quill, L. L., The Chemistry and Metallurgy of Miscellaneous Materials: Thermodynamics, National Nuclear Energy Series, Div. IV, Vol. 19B, McGraw-Hill, New York 1950, pp. 40/59).

[28] Orton, G. W. (Trans. Metall. Soc. AIME **230** [1964] 600/2; Diss. Ohio State Univ., Columbus 1961, pp. 1/55 from Diss. Abstr. **22** [1961] 527).

[29] Worrell, W. L. (Trans. Metall. Soc. AIME **233** [1965] 1173/7).

[30] Cunningham, G. W. (private communication to [28]).

[31] Gupta, D. K.; Seigle, L. L. (Metall. Trans. A **3** [1975] 1939/44).

[32] Gupta, D. K. (Diss. State Univ. New York, Stony Brook 1973, pp. 1/97 from Diss. Abstr. Int. B **35** [1974] 1274).

[33] Gupta, D. K.; Seigle, L. L. (N-74-29983 [1975] 1/25; C.A. **83** [1975] No. 153582).

[34] Samsonov, G. V.; Latysheva, V. P. (Dokl. Akad. Nauk SSSR **109** [1956] 582/5; C.A. **1957** 9249).

[35] Rautala, P.; Norton, J. T. (Plansee Proc. 1st Semin., Reutte/Tyrol, Austria, 1952 [1953], pp. 303/16 from [6]).

[36] Nowotny, H.; Parthé, E.; Kieffer, R.; Benesovsky, F. (Z. Metallkd. **45** [1954] 97/101).

[37] Rudy, E.; Benesovsky, F.; Rudy, E. (Monatsh. Chem. **93** [1962] 693/707).

[38] Rudy, E.; Benesovsky, F.; Toth, L. E. (Z. Metallkd. **54** [1963] 345/53).

[39] Kirner, K. (private communication 1957 to [6]).

[40] Radcliffe, S. V.; Clougherty, E.; Schatz, M.; Harvey, J. S.; Kaufman, L.; Tenner, L. E.; Kulin, S. A. (ADS-TDR-63-90 [AD-417034] [1963] 1/90, 21/5).

[41] Sevast'yanova, L. G.; Velikodnyi, Yu. A.; Zubova, E. V.; Kovba, L. M.; Krutskikh, V. M.; Kudrya, N. A. (Dokl. Akad. Nauk SSSR **229** [1976] 357/9; Dokl. Chem. [Engl. Transl.] **226/231** [1976] 476/7).

[42] Nadler, M. R.; Kempter, C. P. (J. Phys. Chem. **64** [1960] 1468/71).

[43] Friederich, E.; Sittig, L. (Z. Anorg. Allg. Chem. **144** [1925] 184/5).

[44] Elyutin, V. P.; Sosedov, V. P.; Maurakh, M. A.; et al. (Konstr. Mater. Osn. Ugleroda No. 9 [1974] 183/90).

[45] Lally, F. J.; Hiltz, R. H. (J. Met. **14** [1962] 424/8).

[46] Gebhardt, E.; Fromm, E.; Roy, U. (Z. Metallkd. **57** [1966] 732/6).

[47] Dawihl, W. (FIAT Rev. Ger. Sci. 1939–1946, General Metallurgy **1948** p. 94 from [4]).

[48] Savitskii, E. M.; Burkhanov, G. S.; Kirillova, V. M. (Fiz. Khim. Obrab. Mater. **1978** No. 6, pp. 21/7; C.A. **90** [1979] No. 59543).

[49] Kornilov, I. I.; Matveeva, N. M.; Pryakhina, L. I.; Polyakova, R. S. (Metallokhimicheskie Svoistva Elementov Periodicheskoi Systemy [Metal-Chemical Properties of the Elements of the Periodical System], Moscow 1966, pp. 179/82; C.A. **68** [1968] No. 16365).

[50] Savitskii, E.; Burkhanov, G. S.; Kopetskii, Ch. V. (Izv. Akad. Nauk SSSR Metall. Gorn. Delo **1963** No. 6, pp. 12/33; abbr. transl.: Russ. Metall. Min. **1963** No. 6, pp. 12/7; C.A. **60** [1964] 6572).

[51] Andrews, M. R.; Dushman, S. (J. Phys. Chem. **29** [1925] 462/72).

[52] Gustafson, P. (Mat. Sci. Tech. **2** [1986] 653/8).

[53] Allen, B. C.; Maykuth, D. J.; Jaffee, R. I. (AFSWC-TR-60-6 [1959] 101 pp.; N.S.A. **14** [1960] No. 19409; J. Inst. Met. **90** [1961] 120/8).

[54] Klopp, W. D.; Barth, V. D. (DMIC-Memo-50 [1960] 1/10; C.A. **57** [1962] 6984).

[55] Andrews, M. R.; Dushman, S. (J. Franklin Inst. **192** [1921] 545/6).

[56] Shchelkonogov, V. Ya. (Elektron Svoistva Tverd. Tel Fazovye Prevrashch. **1978** 115/21; C.A. **94** [1981] No. 19276).

[57] Kozlova, M. V.; Lashko, N. F.; Bogina, N. Kh.; et al. (Zavod. Lab. **43** [1977] 413/6; Ind. Lab. [Engl. Transl.] **43** [1977] 491/4).

[58] Kovenskii, I. I. (Diffusion Body-Cent. Cubic Met. Pap. Int. Conf., Gatlinburgh, Tenn., 1964 [1965], pp. 283/7).

[59] Fromm, E.; Jehn, H. (Metall. Trans. **3** [1972] 1685/92).

[60] Kuhlmann, H.-H. (Tech.-Wiss. Abh. OSRAM-Ges. **11** [1973] 328/32).

[61] Savitskii, E. M.; Povarova, K. B.; Makrov, P. V.; Ugaste, Yu. E. (Izv. Akad. Nauk SSSR Met. **1972** No. 6, pp. 177/85; Russ. Metall. [Engl. Transl.] **1972** No. 6, pp. 132/45).

[62] Touboul, J.-P.; Langeron, J.-P. (Compt. Rend. Seances Acad. Sci. C **267** [1968] 1285/7).

[63] Dekhtyar, I. Ya.; Patoka, V. I.; Silant'ev, V. I.; et al. (Ukr. Fiz. Zh. [Russ. Ed.] **30** [1985] 940/6; C.A. **103** [1985] No. 79722).

[64] Sale, F. R. (J. Less-Common Met. **100** [1984] 277/97).

[65] Hultgren, R.; Desai, P. D.; Hawkins, D. T.; et al. (Selected Values of the Thermodynamic Properties of Binary Alloys, Metals Park, Ohio, 1973, p. 538).

[66] Storms, E. K. (The Refractory Carbides, Academic, New York 1967, p. 145).

[67] Yih, S. W. H.; Wang, C. T. (Tungsten, Plenum New York – London 1979, pp. 385/7).

[68] Lautz, G.; Schneider, D. (Z. Naturforsch. **16a** [1961] 1368/72).

[69] Carim, A. H.; De Jong, A. F.; Houdy, P. (Thin Solid Films **176** [1989] L177/L182).

[70] Jehn, H. (in: Fromm, E.; Gebhardt, E., Gase und Kohlenstoff in Metallen, Springer, Berlin – Heidelberg – New York 1976, pp. 552/63).

8.2   Reaction Characteristics

General

Tungsten and carbon react at higher temperatures to form tungsten carbides. The temperatures reported for the start of the reaction usually range between 1050 and 1500°C, see, e.g. [1 to 7]. They depend on the form of the reactants and the reaction atmosphere. The carburization of polycrystalline W wires with carbon powder starts at 1550°C. With monocrystalline wires no carbide formation is seen up to 1900°C [8]. Tungsten wires embedded directly into graphite (in holes drilled into a graphite block) and heated up to 982°C in an He+H$_2$ atmosphere became severely embrittled although the thermoelectric shift remained small [9]. Thin layers of W and C (the latter deposited by decomposition of gaseous hydrocarbons) react at temperatures as low as 800°C to form W$_2$C and WC [10]. In sputter-deposited multilayer W/C films, carbide formation has been observed even at room temperature, see p. 143. Carbon evaporated from a graphite rod reacts rapidly with W foils at 1000°C to form a mixed carbide-carbon film. Evidently, small vapor phase species of C are much more reactive with W than the relatively massive particles of carbon black [11].

Some evidence for WC formation was found already at room temperature, when a W–C powder mixture was ground for a prolonged time (e.g., 24 h) in a ball mill [86, 87]. Formation of W$_2$C under these conditions was established and the factors promoting the reaction were discussed in detail [88], see also [108].

Reaction Products and Reaction Extent

The formation and preparation of tungsten carbides by direct combination of the elements, described in the literature up to about 1958 [12 to 31], has been reviewed in some detail [32, 33]. For surveys on tungsten carbides and their formation see also [34, 35].

Carbides Formed in Melts. The classical method of preparing tungsten carbides is the fusion process, either in the electric arc or the carbon-tube furnace. The process permits the ready preparation of W$_2$C as well as eutectic W$_2$C–WC mixtures (~ 4 to 4.5 wt% C). In samples containing WC, also free graphite occurs due to the peritectic decomposition of the mono-carbide. The production of fused W–C alloys containing about 3.5 to 4 wt% C has become important for the large-scale manufacture of castings and of W carbide-based hard-facing alloys [32, 33]; for early work see [12 to 19, 31]. In the fusion process, always more C evaporates than W [16, 17], see, however [21]. Only arc melting and quenching produces a pure W$_2$C phase, which cannot be obtained without small amounts of other phases by powder metallurgy techniques. Arc melting of WC produces, in part, W$_2$C and C [36]. For the preparation of W–C compounds by arc melting, see also [37], [38 p. 182].

Carbides Formed in Heated Powder Mixtures. The most common laboratory method for the preparation of tungsten carbides on a laboratory or technical scale is the reaction of the metal or of the hydride with solid carbon. This technique is also the best for producing relatively pure compounds. The preparation of high-purity homogeneous samples for research purposes is, however, a difficult task. Very high temperatures and good vacuum conditions or highly purified gases are generally required. Typical temperatures for reacting the powder mixtures are 1400 to 1600°C. These temperatures are not the lowest possible ones but reactions proceed at a reasonable rate [32, 33, 39]. The choice of the reaction temperature depends on the composition of the powder mixtures or the desired end product, the particle size, the furnace type, and the protective atmosphere. Probably the most usual method is to carry out the operation in a hydrogen atmosphere in some kind of tube furnace. Another common technique is to heat large batches contained in cylindrical graphite crucibles in a simple form of a high-frequency furnace. By using a closed crucible, it is possible to work without any specially provided protective atmosphere. The action is essentially one of gas carburizing and diffusion. Carbon monoxide or methane is formed depending upon the atmosphere in the furnace and this acts as the carbon carrier [40, 90, 91]. For more recent preparation descriptions than those considered in [32, 33], see, e.g. [36, 37, 41 to 48, 95].

During the first run of a carburization attempt, usually much carbon remains uncombined and the operation has to be repeated by grinding and reheating. C amounts exceeding the composition of the monocarbide do not react [32, 33]. In some cases the powder mixtures were either cold-pressed before heating [21, 37, 42, 44, 49] or hot-pressing procedures were employed [36, 43, 44, 47]. A pre-shock treatment of the powder mixture promoted the W carburization significantly. W powder was mixed with an equimolar amount of acetylene black in a ball mill and the mixture packed into a stainless steel tube with a 0.5 mm thick wall. The tube was then subjected to a shock wave generated by an underwater explosion. No carbide could be detected in the charge after this treatment. However, on subsequent firing for 1 h at 1300°C in a vacuum, the amount of combined C reached 90.2%, while in a hydraulically compressed (3000 kg/cm²) sample only 72.3% of the C had reacted. Also, the shock-compressed sample consisted of almost pure WC, while the conventionally compressed sample also contained W and W_2C, as shown by X-ray analyses [45]. Shock waves were also generated by means of a high-current relativistic electron beam, which not only produced a pressure pulse but also heated the 1:1 (molar ratio) powder mixture contained in a special reaction chamber. Formation of W_2C as well as WC was observed [89].

As mentioned above, the presence of a gas having a definite partial pressure of carbon is required for the fomation of carbides: $2W + C \leftrightarrow W_2C$ and $W + C \leftrightarrow WC$. In the carbon-tube furnace, the carburizing atmosphere (hydrocarbon) can be produced by the reaction of hydrogen with the hot carbon tube and carbides are readily formed, while in the carbon-free tungsten-tube furnace, carbide formation proceeds slowly [32, 33]. This was exemplified by systematic studies during which carbide formation was examined in differently composed charges under H_2 and vacuum in various furnace types [16 to 18, 22, 23]. The end products, identified by X-ray diffraction, are given in Table 3.

Table 3

Products from Carburization of Tungsten with Carbon Black under Different Conditions [16 to 18, 22, 23] from [32, 33].

furnace and conditions of carburization	charge		
	W	$2W + C$	$W + C$
carbon-tube furnace, H_2, 1400°C, 90 min	W_2C	WC	WC
tungsten-wire furnace, H_2, 1400°C, 90 min	W	W, little W_2C	W, little W_2C
tungsten-tube furnace, vacuum (10^{-4} Torr), 1400°C, 90 min	—	—	W, W_2C, WC
tungsten-tube furnace, vacuum (5×10^{-3} Torr), 2000°C, 8 min	—	—	W_2C, WC, and free C

When a black mix (intimate mixture of W and C powders) with nominal 20 μm W powder and presumably equimolar composition was heated in purified Ar for 10 min at 1608°C, the product consisted of 73 wt% W_2C and only 2.5 wt% WC (referred to the total amount of W-containing components). The sample was contained in an open graphite boat, and heating was performed in a graphite muffle-tube furnace. In 90 min at 1608°C, 6 wt% WC formed in the product. The same percentage of WC was obtained in H_2 in less than 1 min. When pure tungsten powder, not a black mix, was heated to 1608°C in a tungsten boat in Ar, there was no evidence of carbide formation. Surface diffusion seems to be responsible for the C transport in the black mix samples heated in Ar. Transport of C by residual traces of CO could be excluded, since the addition of up to 0.5% CO to the Ar did not affect the carburization rate. Surface diffusion, however, is slow compared to the C transport enhanced by H_2 gas. Hydrogen reacts with the carbon parts of the carburization apparatus to form CH_4 which transports C to samples in open boats below ~1600°C (and causes some soot deposition on the boats and their contents). Above 1600°C, C_2H_2 formation becomes important and carbon loss occurs from samples in open receptacles. The loss is especially severe at temperatures $\geq$1830°C. The change of the product composition with the carburization time in H_2 at 1119°C is shown in **Fig. 123** for samples made with nominal 1.3 µm W powder. It is typical of nearly all the results obtained with W particle sizes of 1.3 to 20 µm in the temperature range of 1056 to 1833°C. The W metal always disappears more of less rapidly by the formation of a large amount of W_2C which is slowly converted to WC. At the highest temperatures studied, complete conversion to W_2C occurred before carburization to WC took place. Formation of W_2C also occurs below 1250°C, i.e., its eutectoid decomposition temperature. The rate of nucleation for W_2C apparently overwhelms that for WC in the initial stages of carburization, and W_2C forms as a metastable compound in large amounts at a temperature as low as 949°C [90].

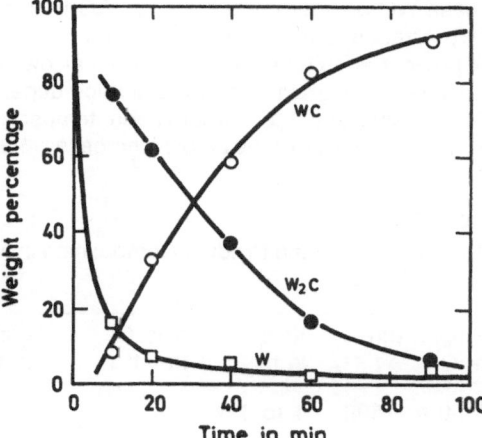

Fig. 123.   Reaction products from nominal 1.30 µm tungsten in W–C powder mixtures heated in H_2 at 1119°C for different periods of time [90].

The influence of temperature on the end products formed within 90 min in equimolar W–C mixtures in H_2 and vacuum is demonstrated by Table 4. The change of the free and bound C in the products is shown in **Fig. 124**. The mean diameter of the W grains was 3.12 µm; the starting mixtures were homogenized for 4 h in a ball mill [49 to 51].

Table 4

Influence of Temperature on the Products Formed from Equimolar W–C Mixtures in an H_2 Atmosphere or in Vacuum (heating time 90 min) [49 to 51].

temperature in °C ...	900	1000	1100	1200	$\geq$1300
in H_2 atmosphere	W	W	W, WC, W_2C	W, WC, W_2C	WC
in vacuum	W, WC, W_2C	W, WC, W_2C	W	WC	WC

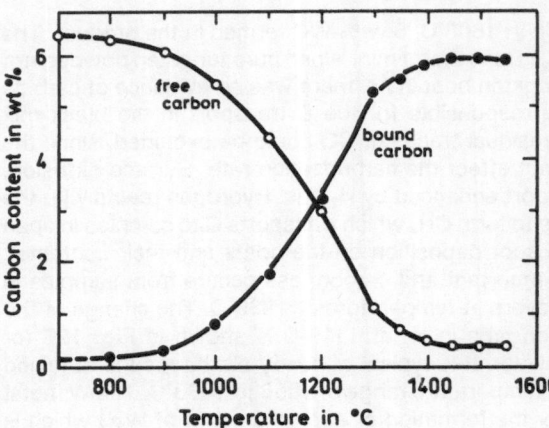

Fig. 124. Variation of the free and bound C of equimolar W–C mixtures with the carburization temperature in H_2 and vacuum. The holding time at each temperature was 90 min, the mean diameter of the W grains was 3.12 µm [49].

The reaction extent and product composition found in nearly equimolar powder mixtures (14.6 µm W) after heating in the absence of H_2, the C transport being procured by CO of unspecified origin, are given in Table 5. The coarser the tungsten particles, the higher is the temperature required for complete carburization, and the longer the time required. If single phase WC powder containing no W_2C is to be produced from coarse tungsten powder, the carburization temperature should not remain under 1800°C and the size distribution of the W powder should be restricted to a narrow range, from which particle sizes of ≥50 µm are excluded. The grain size of the carbide depends mainly on the carburization process parameters, especially the carburization temperature. When the temperature is ≥1700°C, the average grain size of the carbide increases sharply, but the bulk density remains constant [91].

Table 5

Reaction Extent and Product Composition in Nearly Equimolar W–C Mixtures Heated at 1300 to 1900°C [91].

temperature in °C	time in h	free C in wt%	combined C in wt%	composition of product	appearance of product
1300 to 1500	1 to 1.5	>2	—	WC, W_2C, W	loose black powder
1600	1	≤0.98	5.10	WC, W_2C[1]	loose dark gray powder
1700	1, 1.5	0.43, 0.28	5.66, 5.99	WC, W_2C	slightly sintered dark gray powder
1800	1	0.12	6.08	WC[2]	slightly sintered gray powder
1900	0.5	0.035	6.12	WC[2]	slightly sintered gray powder

[1] Center of extra coarse particles still tungsten. – [2] Carburization of extra coarse particles still uncomplete.

W powders with mean diameters of 1.1 µm are reported to be completely carburized to WC already in the temperature ranges 1000 to 1300°C in H_2 and 900 to 1000°C in vacuum [50]. According to [52], the carburization of tungsten with excess solid C proceeds via the steps

$$2W(s)+C(s) \rightarrow W_2C(s) \text{ and } W_2C(s)+C(s) \rightarrow 2WC(s).$$

Under pure H_2 at 800 to 1000°C, free residual C and 27 to 34.5% of the carbon bound in WC are removed. In H_2 containing as little as 1% CH_4, only the excess (free) carbon is removed without any change of the carbide [52]. The effect of hydrogen halide additions on the carburization of W powder in an H_2–0.5% propane gas mixture at atmospheric pressure had earlier been studied [6]. Intimately mixed equimolar W–C (carbon black) charges were heated for 4 h at 1100°C in an alumina-tube furnace in the different gas atmospheres or, for comparison, in a vacuum of 8 Torr. The extent of the reaction was determined by carbon analyses; the theoretical C content of WC is 6.13%. Results are presented in Table 6. As can be seen, the vacuum had a pronounced inhibiting effect on the reaction, which was evidently proceeding via the gaseous phase. The gas mixture containing HI was the most effective in promoting the reaction. However, the differences in reaction extent found with the different gas mixtures were comparatively small. This indicates that the promotion of the reaction by the gas phase is not the rate-determining process. The carburization of each individual W powder particle from the outside to the interior is apparently slower and thus rate-determining [6], see also [32, 33]. For reactions of W with hydrocarbons and CO, see the latter monographs and "Tungsten" Suppl. Vol. A7, 1987, pp. 334/78, 95/7.

Table 6

Effect of Hydrogen Halides on the Carbide Formation in an H_2–Propane Atmosphere. Equimolar W–carbon black mixture (reaction time 4 h) [6].

reaction atmosphere	total carbon in %	free carbon in %	combined carbon in %	weight loss in %	extent of reaction in %
vacuum	6.03	3.2	2.83	0.8	46.7
H_2 + 0.5% propane	6.03	0.85	5.18	0.15	81
H_2 + HCl (6:1.3) + 0.5% propane	6.1	0.75	5.35	1.2	87.6
H_2 + HBr (6:1.3) + 0.5% propane	6.1	0.71	5.39	1.3	88.5
H_2 + HI (6:1.3) + 0.5% propane	6.05	0.4	5.65	0.9	93.5

The stepwise formation of W_2C and WC was confirmed by several authors. Spherical tungsten particles obtained by plasma atomization with a particle size of 50 to 150 µm reacted with lamp black at 1600 to 1700°C to form thick internal layers of W_2C and, later on, also external layers of WC. The thickness of the W_2C layer increased markedly with both time and temperature, while the WC layers grew only very slowly. When the carbide layer thickness exceeded 70 to 100 µm, cracks occurred. After complete impregnation of the W particles with C, the particle shape was preserved, but the final carbide particles had deep cracks [53]. When monocrystalline W powder with (mostly rhombodecahedral) particles of 30 to 150 µm diameter was carburized with a large excess of carbon black (exceeding by several times the stoichio-metric amount required of WC at 1450°C, incomplete reaction occurred. Again, a layer of WC formed on the outside of the W paricles (isolated from each other by the excess C). This was followed by a thicker W_2C layer penetrated by WC crystals. Unreacted W remained in the core. At 1700°C, the reaction was complete and the original tungsten monocrystals formed polycrystalline grains of WC which approximately retained the shape of the W grains. At 2000°C, the formed WC recrystallized [54]. A mixed layer of WC and W_2C, about 1 μm thick, formed on 15 μm W particles carburized at 1200°C for a few hours (evidently under H_2). At 1000°C no reaction occurred. At 1550°C the reaction proceeded nearly to completion; an outer zone of the original grains consisted of WC crystals, and some W_2C remained in the core. Finer W powders with grain sizes of 3.6 or 0.6 μm were completely transformed into WC at 1600 to 1900°C or 1400 to 1600°C, respectively [55].

Phase studies on tungsten-covered graphite (W vapor phase deposited from a WF_6–H_2 mix-ture) after annealing for 0.5 to 2 h at 2300°C and 5×10^{-5} Torr revealed three phases, well separated by boundaries, as shown by microhardness tests [56].

Carbides Formed from Dense W Samples. Short heating of mono- and polycrystalline W samples in graphite crucibles containing graphite powder at ≥1600°C results in the formation of an outer carbide case which is dissolved into the tungsten matrix during subsequent annealing at 1900 to 2200°C in a vacuum. In the polycrystalline samples, the carbide phase, which is assumed to be W_2C, is then not only dispersed within the grains, but has also precipitated at the grain boundaries [57]. An outer carbide layer and carbide (W_2C) segrega-tions at the grain boundaries and in the interior of the larger grains of the W core were also found on the cross sections of tungsten cylinders carburized at 1795 to 2640°C for 930 to 5 min. The core metal was saturated with C [58]. The carburized zone on polycrystalline tungsten rods, heated with carbon black in a stream of H_2 at 1500 to 1800°C, consisted of an inner layer of W_2C and an outer layer of WC. The layers were separated by a sharp boundary. The layer thickness of W_2C was always larger than that of WC, which is formed first. Grain boundary diffusion of C did not occur [26]. According to earlier diffusion studies on electrically fused tungsten beads at 1535 to 1805°C, the carburized zone consists mainly of WC, with a very small amount of residual W. No preferential penetration along the grain boundaries was observed. The W beads had been tightly packed in a carbon tube with pure degassed carbon black [24]. Primary formation of W_2C, followed by formation of WC in the later stages of carburization was postulated [59].

Tungsten tubing of 0.25 mm wall thickness heated with graphite for 60 min under H_2 is completely penetrated by W_2C [60, 61].

Carbide Formation in W–C Films and Multilayers. Carbon evaporated from a graphite rod forms films which contain WC particles on tungsten foils at 1000±25°C. The lath-like WC crystals appear to be oriented perpendicular to one another. The atmosphere was pure argon at 5×10^{-5} Torr. At residual H_2O or O_2 (partial) pressures as low as about 10^{-7} Torr, the residual gas reacts with carbon and thus disturbs the results [11].

A thin layer of graphite was sprayed at 500°C onto evaporated tungsten films of ~500 Å thickness on a micra carrier. After separation of the C-covered W film from the carrier, carbide formation was investigated at temperatures up to 1500°C under a vacuum of 10^{-5} Torr. Both W_2C and WC were found after 1 h at 800°C [10].

WC, W_2C, or mixtures of both were obtained by dual ion beam deposition. Tungsten and graphite targets were alternately sputtered by a broad beam of 1.5 keV Ar^+ ions. The resulting thin films on glass, single crystal silicon, or stainless steel substrates were simultaneously bombarded by 15 keV Ar^+ from another ion source to induce carbide formation. All depositions were at room temperature. The composition of the films and type of carbides formed could be controlled by adjusting the W and C arrival ratio via the relative time of sputtering. XPS examinations showed the existence of free C even for W/C ratios >1 [92].

Carbide formation has been observed when separate layers of W and C are sputter- or ion-beam-deposited successively on each other at room temperature and then are heated. Such multilayer systems composed of thin (several nm) alternating layers of W and C on various substrates have invoked interest for their utility as (soft) X-ray optical elements, e.g., as reflectors and monochromators. A major limitation on the practical application of multilayer structures for this purpose is the requirement that they withstand significant heat loads in the working environment. A great number of studies, e.g. [98 to 105], have shown that dramatic changes occur after heat treatment. At higher temperatures the layer structure is destroyed. Carbide formation is in some cases found already at, or slightly above room temperature [97, 103, 106, 107] and generally increases on annealing, although, under certain conditions, it may fail to appear up to 770 [105, 109, 110] or 1000°C [104], see also [99, 101]. The occurrence of carbide formation seems to depend decisively on the thickness of the W layer and the W/C thickness ratio [98 to 102]. The formation of W_2C [98 to 102], $\beta\ WC_{1-x}$ [106], and WC [97, 107] has been reported.

Carbides Formed in an Argon Plasma and by Spark-Treatment. Tungsten powder (grain size 1 to 2 μm) and graphite powder (grain size 44 μm) were separately, but simultaneously injected with Ar as carrier gas into a transferred arc Ar plasma (150 A, 55 V). Tungsten evaporated completely and reacted with part of the carbon forming WC and (hexagonal) W_2C condensates on thin (002)-oriented graphite flakes as the plasma jet decayed [62].

Spark-treating tungsten foils using a graphite electrode yielded high-temperature W_2C (cubic) (in addition to unconsumed W). The atmosphere was pure Ar. The possibility that the high-temperature phase might have been stabilized by interstitials other than carbon, namely nitrogen or oxygen, was eliminated by side experiments [38].

Kinetics

For starting temperatures of tungsten carburization, see p. 137.

The dissolution rate of C (graphite) in arc-molten W is high at the beginning, but declines rapidly within 30 s from $\sim 0.5 \times 10^{-2}\ g \cdot cm^{-2} \cdot s^{-1}$ to zero when an arc current of 150 A and an Ar atmosphere are used. For the initial stage, the kinetic law is $c = K \cdot \tau^{1/2}$ with c = carbon concentration in wt%, τ = time in s, $K = 112.5 \cdot c_0 \cdot s \cdot D^{1/2}$. In the latter term, c_0 is given as $\rho \cdot c_{sat}/100$ with ρ = density of the melt, s = area of contact between metal and graphite surfaces, and D = diffusion coefficient [63].

Zone-refined tungsten single crystals readily pick up carbon during cementation in graphite in a vacuum furnace (residual pressure 3×10^{-4} Pa). At 1300°C, the C content increases in 6 h from 0.0012 to 0.054 wt% [64]. The depth of the carburized layer on arc-melted W beads heated in carbon black for 1 h is 0.025 to 0.039 mm at 1535 to 1545°C, 0.054 to 0.057 mm at 1640°C, 0.069 to 0.076 mm at 1700 to 1710°C, and 0.108 to 0.117 mm at 1800 to 1805°C [24]. In diffusion studies with ^{14}C as a radiotracer, carbon was found to penetrate into tungsten to a depth of 5 μm in 3.5 h at 1100 to 1450°C [85]. For detailed data on the thickness of the carburized zone in tungsten powder particles under various conditions, see, e.g. [90, 91]. An activation energy of 115 ±10 kcal/mol is derived for the dissolution of C in bulk W from the activation energy of volume diffusion and the segregation energy of C onto the W(100) surface [65].

Tungsten chemical vapor-deposited by WF_6 decomposition on a graphite substrate is stable up to 1500°C. At 2300 and 2400°C, it is carburized at rates of 0.05 and 0.08 mm/min, respectively [56].

Three regimes of different carburization rates are apparent from the temperature dependence of the amounts of free and bound C in equimolar W–C powder mixtures heated for 90 min in vacuum or H$_2$ (see Fig. 124, p. 140) [49 to 51]. The kinetics of the carburization reaction was studied by heating a mixture of 5 to 10 μm W powder with lamp black in a crucible furnace with a N$_2$+CO atmosphere at 1000 to 1500°C. The contents of both free and bound C in the products were determined by chemical analyses. The rate of conversion found was compared to a model assuming spherical particles. The degree of conversion α was given by the equation

$$\alpha = 1 - \{1 - [16\,800 \; \tau \; \exp(-25\,043/T)]^{1/2}\}^3$$

with τ = carburization time [46].

Mechanism. The formation process of tungsten carbide from tungsten and carbon depends on: (a) the supply of carbon atoms to the surface of the tungsten; (b) the diffusion of C through the layer of tungsten carbide; and (c) the chemical reaction rate at the interface. Since the tungsten carbide forms a nonporous zone, it is probable that (b) is the rate-determining step at all events in experiments where contact is permitted between the solid carbon and the tungsten [24, p. 393]. This assumption is confirmed for compacted W by studies of [24, 26, 59, 66]. The promoting effect of a carburizing gas yielding a definite partial pressure of C on the carburization reaction has, however, been established by many experiments with powder mixtures. CH$_4$ formed with H$_2$ in the protective atmosphere, or CO formed with residual O may act as carbon carriers and decisively contribute to the C supply to the W surface even though C diffusion into the interior of the W remains rate controlling in many cases. For powder mixtures, W (and C) transport via the gas phase has to be assumed to explain the extraordinary grain growth in the last stages or at the end of carburization at higher temperatures, which exceeds that due to the difference in densities of tungsten and tungsten carbide. This W transport may be enhanced by volatile W suboxides. A detailed discussion of the various processes influencing the final grain size of tungsten carbide is given in [40, 48, 55]. The chemical reaction at the interface (see the reaction scheme above) is always assumed to proceed fast.

An analysis of the carburization kinetics of tungsten powders mixed with carbon and heated in H$_2$ has been attempted with a mathematical model which presumes C diffusion through a shell of WC growing into the W particles as the rate-determining step. The W particles are taken as spheres. The activation energy for the carburization (i.e., the above kind of diffusion) is derived as 58 kcal/mol in the temperature range 1056 to 1833°C. The rate of carburization is found to be inversely proportional to the square of the particle size (1.3 to 20 μm). Although a wide range of particle sizes is present in each sample, the initial „spherical" rate plot is linear up to 80 to 90% of the carburization. Eventually the rate begins to slow because of an increasing proportion of incompletely carburized large particles, and, ultimately, the largest sized particle determines the time for complete carburization. Hydrogen gas is important to transport carbon as methane or acetylene, but increased hydrogen pressure increases the rate of carburization so little that an adsorbed species such as CH probably controls the carbon concentration at the particle surface [90]. In another more recent paper the carburization of tungsten powder is also shown to be a reaction diffusion process. The experiments were performed in absence of hydrogen. CO of unspecified origin is assumed to be responsible for the initial surface reaction. The carburization then proceeds from the outside to the inside of the polycrystalline tungsten particles or the sintered agglomerates of tungsten, which function as reaction units. It is believed that at first WC is formed as a thin layer at the outside of the W particles, and that subsequently inner layers of W$_2$C develop by the reaction WC + W → W$_2$C (see also pp. 141/2). The coarser the tungsten particles, the higher is the temperature required for complete carburization, and the longer the time required [91].

Thermodynamics

A thermodynamic calculation of the equilibria in the W–C–H and W–C–O systems, based on literature data, was carried out [52].

The enthalpy change for the transition of 1 mol C from the W_2C phase into solid solution in W is $\Delta H = 29.8$ [58] or 34 ± 2 kcal/mol [67], combined to 31 ± 2 kcal/mol [67]. The corresponding change in free energy (in cal/mol) for

$$W_2C \rightarrow C(in\ W(s)) + 2\,W(s)$$

is $\Delta G_T = 29\,800 - 0.14\,T$ (T = 2073 to 2873 K) [68, 69]. Using the solubility data of [58], the relative partial molar free energy of C in the dilute solid solution was calculated to be $\Delta \overline{G}_C = (23\,000 \pm 3000) - [0.68 - R\ ln\ x_C]T$ cal/mol with $x_C =$ mole fraction of C in the saturated solid solution. The heat of solution obtained of C in W was $\Delta \overline{H}_C = 23\,000 \pm 5000$ cal/mol and the excess entropy for the interstitial solid solution, assuming that the C atoms are in the octahedral sites, was $\Delta \overline{S}_{C,exc} = -1.5 \pm 2$ cal·mol^{-1}·K^{-1} [70]. Enthalpy and excess entropy from a second evaluation of the same solubility data obviously contain an error, compare fig. 2 and table II of [71].

The standard free energy of formation of W_2C according to

$$2\,W(s) + C(graphite) \rightarrow W_2C(s)$$

was calculated from the activity of carbon in the two-phase region $W + W_2C$. The activity was obtained from the C content of iron rods equilibrated with mixtures of the metal and carbide powders:

$$\Delta_f G° = -(7300 \pm 300) - (0.56 \pm 0.2)T \text{ cal/mol}$$

for 1575 K $\leq$ T $\leq$ 1660 K [70, 72], see also [73, p. 35]. Equations $\Delta_f G° = -6970 - 0.735\,T$ [74] and $\Delta_f G° = -6400 - 1.0\,T$ [76], see also [68, p. 552], lead to similar results in the appropriate temperature range. For earlier $\Delta_f G_T°$ values of W_2C, see [77 to 79]. A value $\Delta_f H_{298}° = -6.3 \pm 0.6$ kcal/mol was obtained for the heat of formation of W_2C by combustion calorimetry [81]. A much more negative $\Delta_f H_{298}° = -11 \pm 4$ kcal/mol had previously been estimated [82]. Values $\Delta S_{298}° = 1.3$ and 2.5 ± 1.5 cal·mol^{-1}·K^{-1} are estimated by [76] and [82], respectively, and combined to a recommended $\Delta S_{298}° = 2.0 \pm 1.5$ cal·mol^{-1}·K^{-1} [76].

The standard free energy of formation of WC according to

$$W(s) + C(graphite) \rightarrow WC(s)$$
$$\Delta_f G° = -(10\,100 \pm 200) + (1.19 \pm 0.1)T \text{ cal/mol}$$

for 1150 K $\leq$ T $\leq$ 1575 K was obtained by the same method quote above for W_2C [70, 72], see also [68, p. 35]. Slightly deviating equations are: $\Delta_f G° = -10\,020 + 1.17\,T$ [74], $-9600 + 1.0\,T$ [76], and $-9000 + 0.4\,T$ [75, p. 429], [96, p. 384]. For earlier $\Delta_f G_T°$ values, see [77, 80]. Small negative $\Delta_f S_{298}°$ values of -2.1 and -1.4 cal·mol^{-1}·K^{-1} depend on whether $\Delta_f H_{298}°$ is -9670 or $-10\,600$ cal/mol [70]. Small but positive $\Delta_f S_{298}°$ values were estimated before [76, 82].

Influence of Impurities and Additions

For the importance of H_2 and CO in the environmental atmosphere for the reaction of W with C, see the preceding chapters.

The influence of common impurities in tungsten on the rate and extent of carburization as well as on the final grain size of the carbides formed in powder mixtures has been thoroughly studied [48]. Impurities like alkali and alkaline earth metals, B, Al, Si, and P generally impede the grain growth and recrystallization of the carbide. The alkali metals furthermore hinder the diffusion of C into the W particles, an effect which is partly remedied by B, Si, or P. Of the latter elements B and Si, and also Al, form thermodynamically rather stable oxides, which can be reduced by C during carburization, but at the expence of C losses changing the C balance in the products. The study also showed that the initial content of some of these impurities in the W is substantially reduced during the carburization operation [48]. A free-energy-minimization computer program was used to study the evaporation of impurities from W during the carburization. The evaporation temperatures for most impurities were found to be strongly dependent of the amount of sulfur present [84].

Interlayers of TiC, ZrC, TiN, ZrN, or SiC hinder the carbon diffusion into W [56].

TiC acts as a stabilizer for cubic WC_{1-x} (see p. 133) in promoting its synthesis in high pressure-high temperature experiments. ZrC and VC have no positive effect [47].

Bombardment Effects

The size distribution of vacancy clusters in a tungsten FIM tip was studied after irradiation with 200 keV C^+ ions at 65 and 800°C. The C^+ flux was 3.2×10^{13} $cm^{-2} \cdot s^{-1}$; the irradiation doses ranged between 6.8×10^{14} and 3.4×10^{17} C^+/cm^2. Helium was used as the imaging gas; photographs were taken every time when three (110) plane layers had been field-evaporated. At 65°C, the density of vacancy clusters at each particular size increased with the irradiation dose, but vacancy clusters with diameters >15 Å were not observed. At 800°C, the density of clusters with ≤15 Å diameter was saturated and remained constant, but that of clusters >15 Å increased with the dose. The total vacancy concentration was proportional to the square root of the dose, $d^{1/2}$, at both temperatures. The concentration of single vacancies at 65°C was also proportional to $d^{1/2}$; at 800°C it was approximately constant. This difference was shown to be due to vacancy migration and cluster growth at the higher temperature. The vacancy clusters in the size range <15 Å, which come from depleted zones, apparently act as nuclei of voids [93].

C atoms on W surfaces (supplied, for example, by low pressures of CH_4 in an AES system) can recoil into surface regions of W during sputter cleaning due to energy transfer from the sputtering ions. The implanted C forms carbides with W and this decreases the sputter removal rate of C [94].

References:

[1] Kingery, W. D. (Proc. Int. Symp. High Temp. Technol., Asilomar, Calif., 1959 [1960], pp. 76/89).

[2] Johnson, P. D. (J. Am. Ceram. Soc. **33** [1950] 168/71).

[3] Kieffer, R.; Benesovsky, F. (Metallurgia [Manchester] **58** [1958] 119/24; Planseeber. Pulvermetall. **5** [1957] 56/71).

[4] Schley, R.; Metauer, G.; Gentil, J. (CEA-Conf-2935 [1974] 1/16; C.A. **85** [1976] No. 128967).

[5] Lugscheider, E.; Eck, R.; Ettmayer, P. (Radex-Rundsch. **1983** No. 1/2, pp. 52/84).

[6] Hüttig, G. F.; Fattinger, V.; Kohla, K. (Powder Metal. Bull. **5** [1950] 30/7).

[7] Foulias, S. D.; Rawlings, K. J.; Hopkins, B. J. (J. Phys. C: Solid State Phys. **14** [1981] 5403/9).

[8] Geiss, W.; van Liempt, J. A. M. (Z. Metallkd. **16** [1924] 317/8).

[9] Benett, R. L.; Hemphill, H. L.; Rainey, W. T., Jr.; Keilholtz, G. W. (ORNL-TM-746 [1964] 1/7; N.S.A. **18** [1964] No. 10473).

[10] Zaitsev, A. A.; Ignatov, D. V.; Fedorchuk, N. M.; Lazarev, E. M. (Fiz. Khim. Obrabot. Mater. **1976** No. 6, pp. 90/3; C.A. **86** [1977] No. 125531).

[11] Irving, S. M.; Walker, P. L., Jr. (Carbon **5** [1967] 399/402).

[12] Moissan, H. (C. R. Hebd. Seances Acad. Sci. **116** [1893] 1225/7, **123** [1896] 13/6, **125** [1897] 839/44).

[13] Ruff, O.; Wunsch, R. (Z. Anorg. Allg. Chem. **85** [1914] 192/328).
[14] Westgren, A.; Phragmén, G. (Z. Anorg. Allg. Chem. **156** [1936] 27/36).
[15] Hultgren, A. (Metallographic Study of Tungsten Steels, Wiley, New York 1920).
[16] Becker, K. (Z. Elektrochem. **34** [1928] 640/2).
[17] Becker, K. (Z. Metallkd. **20** [1928] 437/41).
[18] Becker, K. (Z. Physik **51** [1928] 481/9).
[19] Sykes, W. P. (Trans. Am. Soc. Steel Treat. **18** [1930] 968/91).
[20] Swalin, R. A. (Acta Crystallogr. **10** [1957] 473/4).

[21] Friederich, E.; Sittig, L. (Z. Anorg. Allg. Chem. **144** [1925] 169/89).
[22] Skaupy, F. (Z. Elektrochem. **33** [1927] 487/91).
[23] Becker, K.; Hölbling, R. (Z. Angew. Chem. **40** [1927] 512/3).
[24] Pirani, M.; Sandor, J. (J. Inst. Met. **73** [1947] 385/95; discussion pp. 807/8).
[25] Krainer, H.; Konopicky, K. (Berg- u. Hüttenmänn. Monatsh. Montan. Hochsch. Leoben **92** [1947] 166/78).
[26] Kreimer, G. S.; Efros, L. D.; Voronkova, E. A. (Zh. Tekh. Fiz. **22** [1952] 858/73; C. A. **1956** 7536).
[27] Samsonov, G. V.; Latysheva, V. P. (Dokl. Akad. Nauk SSSR **109** [1956] 582/5; C. A. **1957** 23564).
[28] Samsonov, G. V.; Latysheva, V. P. (Fiz. Met. Metalloved. **2** [1956] 309/19; C. A. **1957** 6258).
[29] Samsonov, G. V. (Izv. Sekt. Fiz. Khim. Anal. Inst. Obshch. Neorg. Khim. Akad. Nauk SSSR **27** [1956] 97/125).
[30] Samsonov, G. V. (Bor Tr. Konf. Khim. Bora Ego Soedin., Moscow 1955 [1958], pp. 74/89; C. A. **1960** 23564).

[31] B.I.O.S. (Final Rep. No. 1076 from [32, p. 187], [33, p. 153]).
[32] Kieffer, R.; Benesovsky, F. (Hartstoffe, Springer, Wien 1963, pp. 170/87).
[33] Schwarzkopf, P.; Kieffer, R. (Refractory Hard Metals; Borides, Carbides, Nitrides, and Silicides, MacMillan Comp., New York 1953, pp. 138/53).
[34] Storms, E. K. (The Refractory Carbides, Academic, New York 1967, pp. 143/54).
[35] Yih, S. W. H.; Wang, C. T. (Tungsten, Plenum, New York – London 1979, pp. 385/403).
[36] Rudy, E.; Benesovsky, F.; Rudy, E. (Monatsh. Chem. **93** [1962] 693/707).
[37] Rudy, E.; Benesovsky, F.; Todt, L. E. (Z. Metallkd. **54** [1963] 345/53).
[38] Goldschmidt, H. J.; Brand, J. A. (J. Less-Common Met. **5** [1963] 181/94).
[39] Toth, L. E. (Transition Metal Carbides and Nitrides, Academic, New York – London 1971, pp. 11/4).
[40] Lardner, E. (Powder Metall. **13** [1970] 394/428).

[41] Meinhardt, D.; Krisement, O. (Arch. Eisenhüttenw. **33** [1962] 493/9).
[42] Alekseev, V. I.; Shvartsman, L. A. (Izv. Akad. Nauk SSSR Metall. Gorn. Delo **1963** No. 1, pp. 91/6; Russ. Metall. Min. **1963** No. 1, p. 41 [Abstract only]).
[43] Stecher, P.; Benesovsky, F.; Nowotny, H. (Planseeber. Pulvermetall. **12** [1964] 89/95).
[44] Sara, R. V. (J. Am. Ceram. Soc. **48** [1965] 251/7).
[45] Horiguchi, Y.; Nomura, Y. (Chem. Ind. [London] **1965** 1791/2).
[46] Mirzaev, A. M.; Davidson, A. M.; Alkatsev, M. I. (Izv. Vyssh. Uchebn. Zaved. Tsvetn. Metall. **1975** No. 6, pp. 89/91; C. A. **84** [1976] No. 183500).
[47] Sevast'yanova, L. G.; Velikodnyi, Yu. A.; Zubova, E. V.; et al. (Dokl. Akad. Nauk SSSR **229** [1976] 357/9; Dokl. Chem. [Engl. Transl.] **226/231** [1976] 476/7).
[48] Lassner, E.; Schreiner, M.; Lux, B. (Proc. 10th Plansee Semin., Reutte, Austria, 1981 [1982], Vol. 2, pp. 761/93; C. A. **98** [1983] No. 76599).

[49] Storin, A.; Zamfirescu, C. (Cercet. Metal. Inst. Cercet. Metal. [Bucharest] **16** [1975] 531/40; C.A. **85** [1976] No. 147459).

[50] Storin, A.; Zamfirescu, C. (Tirages Prelim. 4th Symp. Eur. Metall. Poudres, Grenoble 1975, Vol. 1, Paper No. 9, pp. 1/7; C.A. **88** [1978] No. 10607).

[51] Storin, A. (Cercet. Metal. **14** [1974] 321/31).

[52] Vaskevich, N. K.; Senchikin, V. K.; Tret'yakov, V. I. (Poroshk. Metall. [Kiev] **1985** No. 6, pp. 69/73; Sov. Powder Metall. Met. Ceram. [Engl. Transl.] **24** [1985] 489/92).

[53] Krasnov, A. N.; Burykina, A. L. (Zashch. Pokrytiya Met. **2** [1968] 261/8; Prot. Coat. Met. [Engl. Transl.] **2** [1970] 208/12).

[54] Pikunov, D. V.; Tret'yakov, V. I. (Tsvetn. Met. [Moscow] **1978** No. 1, pp. 68/9; Tsvetn. Met. [N.Y.] **19** No. 1 [1978] 75/6).

[55] Hara, A.; Miyake, M. (Monatsh. Chem. **103** [1972] 1465/82).

[56] Emyashev, A. V.; Slavgorodskaya, Z. V.; Martynov, S. Z. (Konstr. Mater. Osn. Ugleroda No. 13 [1978] 70/8; C.A. **91** [1979] No. 111107).

[57] Sell, H. G.; Keith, G. H.; Schnitzel, R. H.; Cerulli, N. F. (WADD-TR-60-37 (Pt. IV) [AD-423841] [1963] 1/103; N.S.A. **18** [1964] No. 32154).

[58] Gebhardt, E.; Fromm, E.; Roy, U. (Z. Metallkd. **57** [1966] 732/6).

[59] Andrews, M. R.; Dushman, S. (J. Phys. Chem. **29** [1925] 462/72).

[60] Fries, R. J.; Cummings, J. E.; Hoffman, C. G.; Daily, S. A. (J. Nucl. Mater. **32** [1969] 171/3).

[61] Fries, R. J.; Cummings, J. E.; Hoffman, C. G.; Daily, S. A. (Plansee Proc. 6th Semin., Reutte, Austria, 1968 [1969], LA-3795 [1967] 32 pp.; N.S.A. **22** [1968] No. 38622).

[62] Ronsheim, P.; Toth, E.; Pfender, E. (J. Mater. Sci. Lett. **1** [1982] 343/6).

[63] Kashin, V. I.; Klibanov, E. L.; Tsilosani, A. G.; et al. (Izv. Akad. Nauk SSSR Met. **1971** No. 2, pp. 39/45; Russ. Metall. [Engl. Transl.] **1971** No. 2, pp. 25/30).

[64] Savitskii, E. M.; Burkhanov, G. S.; Kirillova, V. M. [Savitsky, E. M.; Burhanov, G. S.; Kirillowa, V. M.] (Prakt. Metallogr. **15** [1978] 453/9).

[65] Rawlings, K. J.; Foulias, S. D.; Hopkins, B. J. (Surf. Sci. **109** [1981] 513/21).

[66] Gerasimov, A. F.; Konev, V. N.; Timofeeva, N. F. (Fiz. Met. Metalloved. **11** [1961] 596/600; Phys. Met. Metallogr. [Engl. Transl.] **11** No. 4 [1961] 106/10).

[67] Kuhlmann, H.-H. (Tech.-Wiss. Abh. OSRAM-Ges. **11** [1973] 328/32).

[68] Jehn, H. (in: Fromm, E.; Gebhardt, E., Gase und Kohlenstoff in Metallen, Springer, Berlin – Heidelberg – New York 1976, pp. 552/63).

[69] Fromm, E.; Jehn, H. (Metall. Trans. **3** [1972] 1685/92).

[70] Gupta, D. K.; Seigle, L. L. (Metall. Trans. A **6** [1975] 1939/44).

[71] Mozzhukhin, E. I. (Izv. Vyssh. Uchebn. Zaved. Chern. Metall. **1976** No. 1, pp. 151/6; C.A. **85** [1976] No. 66852).

[72] Gupta, D. K.; Seigle, L. L. (N-74-29983 [1975] 1/25; C.A. **83** [1975] No. 153582).

[73] Jehn, H.; Speck, H.; Fromm, E.; Hörz, G. (Phys. Daten/Phys. Data No. 5-10 [1980] 1/64, 34/9).

[74] Gupta, D. K. (Diss. State Univ. New York, Stony Brook 1973, pp. 1/97 from Diss. Abstr. Int. B **35** [1974] 1274).

[75] Kubachewski, O.; Evans, E. L.; Alcock, C. B. (Metallurgical Thermochemistry, 4th Ed., Pergamon, Oxford 1967).

[76] Worrell, W. L. (Trans. Metall. Soc. AIME **233** [1965] 241/2, 1173/7).

[77] Orton, G. W. (Trans. Metall. Soc. AIME **230** [1964] 600/2).

[78] Cunningham, G. W. (private communication 1963 to [77]).

[79] Cunningham, G. W.; Ward, J. J.; Alexander, C. A. (BMI-1601 [1962] 1/34; N.S.A. **17** [1963] No. 3470).

[80] Gleiser, M.; Chipman, J. (Trans. Metall. Soc. AIME **224** [1962] 1278/9).

[81] Mah, A. D. (USBM-RI-6337 [1963] 1/9).

[82] Krikorian, O. H. (UCRL-2888 [1955] 136 pp. from [76, 81]; N.S.A. **9** [1955] No. 4704).

[83] McGraw, L. D.; Seltz, H.; Snyder, P. E. (J. Am. Chem. Soc. **69** [1947] 329/31).

[84] Qvick, J.; Noläng, B. I.; Richardson, M. W.; Snell, P.-O. (High Temp. – High Pressures **14** [1982] 171/81).

[85] Aleksandrov, L. N.; Shchelkonogov, V. Ya., (Poroshk. Metall. [Kiev] **1964** No. 4, pp. 28/32; Sov. Powder Metall. Met. Ceram. [Engl. Transl.] **1964** 288/91).

[86] Matteazzi, P.; Le Caer, G. (J. Am. Ceram. Soc. **74** [1991] 1382/90).

[87] Matteazzi, P. (Ceramurgia **20** [1990] 227/33 from C.A. **115** [1991] No. 55173).

[88] Butyagin, P. Yu.; Davydkin, V. Yu.; Trusov, L. I.; et al. (Dokl. Akad. Nauk SSSR **308** [1989] 405/9; Dokl. Phys. Chem. [Engl. Transl.] **304/309** [1989] 746/9).

[89] Batsanov, S. S.; Demidov, B. A.; Ivkin, M. V.; et al. (Izv. Akad. Nauk SSSR Neorg. Mater. **26** [1990] 2100/2; Inorg. Mater. [Engl. Transl.] **26** [1990] 1799/801).

[90] McCarty, L. V.; Donelson, R.; Hehemann, R. F. (Metall. Trans. A **18** [1987] 969/74).

[91] Tao, Zhengji (Int. J. Refract. Hard Met. **6** [1987] 221/5).

[92] Min, Zhang; Li, Wenzhi; Li, Hengde (Nucl. Instrum. Methods Phys. Res. B **59/60** Pt. 2 [1991] 1358/61).

[93] Igata, N.; Shibata, K.; Sato, S. (Radiat. Eff. **41** [1979] 251/60).

[94] Ingrey, S.; Johnson, M. B.; Streater, R. W.; Sproule, G. I. (J. Vac. Sci. Technol. **20** [1982] 968/70).

[95] Yamamoto, Ryoji; Nukui, Takatoshi; Tokyo Tungsten Co., Ltd. (Jpn. 03-75214 [1989/91] from C.A. **115** [1991] No. 32092).

[96] Kubaschewski, O.; Alcock, C. B. (Metallurgical Thermochemistry, 5th Ed., Pergamon, New York 1979, pp. 320, 384 from [97]).

[97] Jankowski, A. F.; Schrawyer, L. R.; Wall, M. A.; et al. (J. Vac. Sci. Technol. A **7** [1989] 2914/8).

[98] Nguyen, Tai D. (LBL-29206 [1990] 97 pp. from C.A. **114** [1991] No. 217634).

[99] Takagi, Y.; Flessa, S. A.; Hart, K. L.; et al. (Proc. SPIE Int. Soc. Opt. Eng. **563** [1985] 66/9; C.A. **104** [1986] No. 61071).

[100] Lamble, G. M.; Heald, S. M.; Sayers, D. E.; et al. (Physica B **158** [1989] 672/3).

[101] Lamble, G. M.; Heald, S. M.; Sayers, D. E.; et al. (J. Appl. Phys. **65** [1989] 4250/5).

[102] Lamble, G. M.; Heald, S. M.; Sayers, D. E.; et al. (Mater. Res. Soc. Symp. Proc. **103** [1988] 101/7 from C.A. **109** [1988] No. 159597).

[103] Kortright, J. B.; Denlinger, J. D. (Mater. Res. Soc. Symp. Proc. **103** [1988] 95/100 from C.A. **109** [1988] No. 159596).

[104] Jiang, Zuimin; Jiang, Xiaoming; Liu, Wenhan; Wu, Ziqin (J. Appl. Phys. **65** [1989] 196/200).

[105] Ziegler, E.; Lepêtre, Y.; Schuller, I. K. (Proc. SPIE Int. Soc. Opt. Eng. **640** [1986] 145/8; C.A. **106** [1987] No. 54285).

[106] Carim, A. H.; De Jong, A. F.; Houdy, P. (Thin Solid Films **176** [1989] L177/L182).

[107] Petford-Long, A. K.; Stearns, M. B.; Chang, C.-H.; et al. (J. Appl. Phys. **61** [1987] 1422/8).

[108] Kolbanov, I. V.; Davydkin, V. Yu.; Butyagin, P. Yu. (Mekhanokhim. Sint. Dokl. Vses. Nauchno-Tekh. Konf., Vladivostok 1990, pp. 50/2 from C.A. **114** [1991] No. 190939).

[109] Ziegler, E.; Lepêtre, Y.; Schuller, I. K.; Spiller, E. (Appl. Phys. Lett. **48** [1986] 1354/6).

[110] Lepêtre, Y.; Ziegler, E.; Schuller, I. K.; Rivoira, R. (J. Appl. Phys. **60** [1986] 2301/3).

8.3 Diffusion

Carbon Diffusion Through Tungsten. The temperature dependence of the diffusion coefficient D_C of C in W found by radiotracer studies with ^{14}C is shown in **Fig. 125**. The following values were derived for the constants D_0 and E_{diff} in the equation $D_C = D_0 \exp(-E_{diff}/RT)$:

temperature range in °C	D_0 in cm²/s	E_{diff} in kcal/mol	specimens	Ref.
300 to 400	—	45.0	tablets	[1]
900 to 1450	2.24×10^{-2}	48.0 ± 4.5	tablets	[1]
900	3.0×10^{-1} (assumed)	61.5 ± 1.5	polycrystalline wires	[2]
1200 to 1600	8.91×10^{-2}	53.5	rectangular specimens of polycrystalline arc-melted W; purity 99.51%	[3]
1250 to 1450	3.0×10^{-1}	49.5	0.8 mm wires	[4]
1500 to 1800	$(3.45 \pm 0.12) \times 10^{-3}$	37.8 ± 0.14	disks of randomly oriented single crystals; purity 99.99%	[5, 6]
1800 to 2800	9.22×10^{-3}	40.4	polycrystalline wires; purity 99.86%	[7]

The above results are commented in the following text on pp. 150/2.

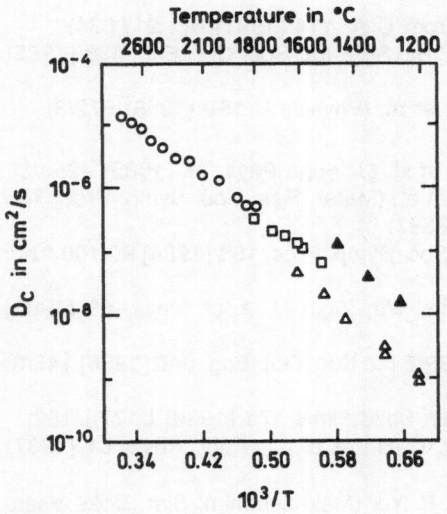

Fig. 125.   Temperature dependence of the diffusion coefficient D_C of carbon in tungsten: △ [3], ▲ [4], □ [5], ○ [7] (from [5]).

For further data on C diffusion in W see [13, 14, 32].

Some carburization of the tungsten by a ^{14}C-containing graphite lubricant was noted at temperatures as low as 480°C (4 h). The data given in the table are based on an experimental value of $D_C = 2.15 \times 10^{-12}$ cm²/s at 900°C [2].

Diffusion parameters determined by other than radiotracer techniques:

temperature range in °C	D_0 in cm²/s	E_{diff} in kcal/mol	method	specimens	Ref.
300 to 400	3.15×10^{-3}	41.1 ± 3.6	internal friction	poly-crystalline wires	[12]; see also [1, 35]
927 to 1127	—	$\leq 59.8 \pm 3.6$[1]	analysis of C segregation kinetics to (110) face	W(110) ribbon	[8]
1077 to 1147	3	$\leq 59 \pm 8$[1]	analysis of C segregation kinetics to (100) face	W(100) crystal	[9]
1056 to 1833	—	58	carburization kinetics	powder particles	[33]
~1800 to 2100	—	72	decarborization rate of carburized specimens in low-pressure O_2 atmosphere	poly-crystalline filaments	[10]
~1800 to 2500	1.6×10^{-6}	50.7	CO generation rate on C-containing samples in O_2	poly-crystalline ribbons	[34]
1823 to 2223	—	≤ 55[1]	decrease of surface coverage	FEM tip	[11]

[1] Values probably too high due to potential barrier just below the surface.

The concentration of C as a function of the penetration depth shows the usual sigmoid curve [1]. In the disks of randomly oriented crystals a zone of carbon depletion adjacent to the carbide/tungsten interface formed is observed as a result of C migration to the interface. The extent of the depleted zone is a function of the cooling rate and reaches up to 60 μm. The driving force for the interface enrichment effect is assumed to be the supersaturation of the solid solution with C as the specimen temperature is decreased, resulting in C migration to the interface. Computer-calculated concentration profiles with supersaturations of 1.42 and 1.44 agreed well with experimental data for two tungsten samples cooled at different rates. For the data presented in the table above, only the experimental concentration profiles in the distance range above 60 μm from the interface were evaluated [5, 6].

A significant dependence of the C diffusion parameters on the penetration depth was also noted: $D_C (\leq 2\ \mu m) = 4 \times 10^{-2} \exp[-27\,000/T]$ cm²/s, $D_C (>2\ \mu m) = 0.3 \exp[-25\,000/T]$ cm²/s [4]. It is assumed that microinclusions of W carbide are formed at the lower penetration depths which impede the C penetration [4], see also [15]. The diffusion mobility of C is strongly increased by lattice defects, which, in turn, facilitate C segregation around the defects. Different diffusion mechanisms were established at room temperature to 800°C, 800 to 1850°C, 1850 to 2400°C, and ≥ 2400°C. Intense carbide formation, especially on structure defects, occurred at all stages. Local carbide formation at $\leq 800°C$ occurred by C segregation on dislocations [1]. A distinct temperature dependence of the activation energy of diffusion was observed above 2600°C [7]. The deviations of the experimental D values from the selected $D_C = D_0$ exp $(-E_{diff}/RT)$ line amounted to 15, 16.3, and 17.5% at 2600, 2700, and 2800°C, respectively, and to only 4 to 11% at lower temperatures [7]. The deviations from the Arrhenius law at higher temperatures were explained by a model presuming diffusion of interstitial C atoms via a vacancy mechanism [16].

Local segregation of C in W by diffusion processes was studied. A mechanism for C segregation on grain boundaries was proposed [17]; see [17] for additional literature. Segregation of C in the mounds (associated with eruptions of liquid metal) that occur on closely spaced tungsten electrodes subjected to repeated 35 A arcs of 1.44 ms duration was investigated [18]. For LEED and electron microscopic investigations of carbon segregation on a W(100) surface, see [19]; for AES studies of C segregation of the surface of a polycrystalline W ribbon, see [20].

The activation energy for C diffusion is roughly proportional to the melting point T_m of the base material when the data for a series of bcc metals (W, Mo, Cr, Ta, Nb, V, Zr, and Ti) are compared; the coefficient of proportionality is ~ 10 to 13 cal·mol^{-1}·K^{-1}. The preexponential factor D_0 is largely determined by the activation energy according to $D_0 = A \exp(B \cdot E_{diff})$, where $A = 3.2 \times 10^{-4}$ cm²/s and $B = 10^{-4}$ cal^{-1}·mol^{-1}. For W, the calculated value for D_0 is 1.69×10^{-2} cm²/s, as compared to an experimental $D_0 = 8.91 \times 10^{-2}$ cm²/s [3]. Assuming an interstitial vacancy mechanism of diffusion, the activation energy is given by $E_{diff} = K \, T_m/r_0$ with $K =$ constant factor, $r_0 =$ radius of the metal atom. The value of 51.0 kcal/mol obtained for W [21] is in good agreement with an experimental value of 53.5 kcal/mol [3]. The relationship $E_{diff} = 11300 + 8.4 \, T_m$ cal/mol [12] predicts an activation energy at the lower end of measured values.

Permeation of carbon through tungsten was measured between 1535 and 1805°C. The activation energy of 59 ± 5 kcal/mol was taken representative for diffusion [30]. An activation energy of 34 kcal/mol [23] was recalculated from permeation measurements reported before [26 to 28], see also [13].

Carbon Diffusion Through Tungsten Carbide Layers. At the higher C concentrations attained in the carburization of W, carbide layers are formed on the outside of the metal. The thickness of these outer carbide shells increases at a rate governed by the diffusion of C through the carbide (reactive diffusion). Diffusion coefficients for reactive diffusion in the WC phase, D_r (in cm²/s) $= 1.56 \times 10^{-3} \exp[-52000/T]$ [25], were reported without giving further details. Activation energies of 112 ± 3 kcal/mol [24] and 108 kcal/mol [10] were derived from permeation through the W_2C phase measured around 2000 K.

Coefficients of carbon diffusion (or permeation) through the tungsten metal and the appropriate activation energies were correlated to simple electronic parameters of W which were thought to be characteristic for W_2C formation too [23, 26 to 29].

Carbonitriding experiments at 1000 to 1200°C showed that nitrogen has a retarding effect on the diffusion of carbon in tungsten. It is suggested that the diffusion of C and N takes place mainly through the reaction products of the metal [36].

For the influence of alkali and alkaline earth metals, Al, P, Si, and B on the diffusion processes operative in the carburization of W and the grain growth of the carbides formed, see [22].

Grain Boundary Diffusion. When the porosity of tungsten samples exceeds $\sim 3\%$, grain boundary diffusion becomes significant. The activation energy for boundary diffusion of C on W grains has been evaluated as E_{diff} (bound) $= 34.2$ kcal/mol [15]. No grain boundary diffusion was observed in arc-melted W beads at 1535 to 1805°C [31].

The coefficients of grain (bulk) and boundary diffusion at 900°C were calculated as $D_{grain} = 5.2 \times 10^{-12}$ and $D_{bound} = 4.0 \times 10^{-10}$ cm²/s, when a packing of spherical grains was assumed. Values $D'_{grain} = 10.4 \times 10^{-12}$ and $D'_{bound} = 8.5 \times 10^{-10}$ are obtained, when a cyclindrical-grain model was used taking account of the actual columnar structure of wire test specimens [2]. Coefficients were also calculated for temperatures of 1250 to 1450°C [15].

Carbon Diffusion in Molten Tungsten. From the dissolution rate of C in liquid tungsten, the diffusion coefficient D at 3450 and 3800°C was calculated as 0.39×10^{-3} and 0.54×10^{-3} cm²/s, respectively. The activation energy of C diffusion in molten W is ~30 kcal/mol (error ≤30%), close to that in the solid metal [31].

References:

[1] Shchelkonogov, V. Ya. (Elektron. Svoistva Tverd. Tel Fazovye Prevrashch. **1978** 115/211; C.A. **94** [1981] No. 19276).

[2] Aleksandrov, L. N. (Inzh.-Fiz. Zh. **5** No. 9 [1962] 53/8; Int. Chem. Eng. **3** No. 1 [1963] 108/11; C.A. **58** [1963] 3172).

[3] Nakonechnikov, A. I.; Pavlinov, L. V.; Bykov, V. N. (Fiz. Met. Metalloved. **22** [1966] 234/8; Phys. Met. Metallogr. [Engl. Transl.] **22** No. 2 [1966] 73/7).

[4] Aleksandrov, L. N.; Shchelkonogov, V. Ya. (Poroshk. Metall. [Kiev] **1964** No. 4, pp. 28/32; Soviet Powder Metall. Met. Ceram. [Engl. Transl.] **1964** 288/91).

[5] Shepela, A. (J. Less-Common Met. **26** [1972] 33/43).

[6] Shepela, A. (Diss. Cornell Univ., Ithaca 1970, pp. 1/274 from Diss. Abstr. Int. B **31** [1970/71] 224/5).

[7] Kovenskii, I. I. (Diffus. Body Cent. Cubic Met. Pap. Int. Conf., Gatlinburgh, Tenn., 1964 [1965], pp. 283/7).

[8] Foulias, S. D.; Rawlings, K. J.; Hopkins, B. J. (J. Phys. C **14** [1981] 5403/9).

[9] Rawlings, K. J.; Foulias, S. D.; Hopkins, B. J. (Surf. Sci. **109** [1981] 513/21).

[10] Andrews, M. R.; Dushman, S. (J. Phys. Chem. **29** [1925] 462/72).

[11] Piquet, A.; Pralong, G.; Roux, H.; et al. (Vide [Suppl.] No. 185 [1977] 376/86).

[12] Shchelkonogov, V. Ya.; Aleksandrov, L. N.; Piterimov, V. A.; Mordyuk, V. S. (Fiz. Met. Metalloved. **25** [1968] 80/4; Phys. Met. Metallogr. [Engl. Transl.] **25** [1968] No. 1 [1968] 68/72).

[13] Klopp, W. D.; Barth, V. D. (DMIC-Memo-50 [1960] 1/10; C.A. **57** [1962] 6984).

[14] Joffreau, P. O.; Haubner, R.; Lux, B. (Int. J. Refract. Hard Met. **7** [1988] 186/94).

[15] Shchelkonogov, V. Ya. (Uch. Zap. Mord. Gos. Univ. **50** No. 1 [1966] 3/9; C.A. **68** [1968] No. 33455).

[16] Smirnov, L. I.; Alekseev, E. R. (Deposited Doc. VINITI 323-83 [1982] 1/12; C.A. **100** [1984] No. 145367).

[17] Anastasiadi, G. P.; Girshov, V. L.; Gulyaev, B. B.; Sheyanova, E. V. (Fiz. Met. Metalloved. **44** [1977] 345/9; Phys. Met. Metallogr. [Engl. Transl.] **44** No. 2 [1977] 99/103).

[18] Ostermayer, F. W.; Koch, F. B. (Appl. Phys. Lett. **36** [1980] 266/8).

[19] Bas, E. B. (Proc. 7th Int. Vac. Congr., Vienna 1977, Vol. 2, pp. 1233/6; C.A. **88** [1978] No. 95334).

[20] Joyner, R. W.; Rickman, J.; Roberts, M. W. (Surf. Sci. **39** [1973] 445/9).

[21] Gruzin, P. L.; Makarov, N. M. (Fiz. Met. Metalloved. **37** [1974] 895/6; Phys. Met. Metallogr. [Engl. Transl.] **37** No. 4 [1974] 204/5).

[22] Lassner, E.; Schreiner, M.; Lux, B. (Proc. 10th Plansee-Semin., Reutte, Austria, 1981 [1982], Vol. 3, pp. 761/93; C.A. **98** [1983] No. 76599).

[23] Samsonov, G. V.; Epik, A. P. (Fiz. Met. Metalloved. **14** [1962] 479/80; Phys. Met. Metallogr. [Engl. Transl.] **14** No. 3 [1962] 144/5).

[24] Kreimer, G. S.; Efros, L. D.; Voronkova, E. A. (Zh. Tekh. Fiz. **22** [1952] 858/73; C. A. **1956** 7536).

[25] Samsonov, G. V.; Burykina, A. L.; E'pik, A. P. (Zashch. Pokrytiya Met. **6** [1974] 5/24; Transl.: FDT-MT-24-1341-117 [1974] 3/33, 7).

[26] Samsonov, G. V.; Latysheva, V. P. (Fiz. Met. Metalloved. **2** [1956] 309/19; C. A. **1957** 6258).

[27] Samsonov, G. V.; Latysheva, V. P. (Dokl. Akad. Nauk SSSR **109** [1956] 582/5; C. A. **1957** 9249).

[28] Samsonov, G. V.; Solonnikova, L. A. (Fiz. Met. Metalloved. **5** [1957] 565/6; Phys. Met. Metallogr. [Engl. Transl.] **5** No. 3 [1957] 177).

[29] Samsonov, G. V. (Bor Tr. Konf. Khim. Bora Ego Soedin., Moscow 1955 [1958], pp. 74/89; C. A. **1960** 23564).

[30] Pirani, M.; Sandor, J. (J. Inst. Met. **73** [1947] 385/95; Discussion (correspondence) 807/8).

[31] Kashin, V. I.; Klibanov, E. L.; Tsilosani, A. G.; et al. (Izv. Akad. Nauk SSSR Met. **1971** No. 2, pp. 39/45; Russ. Metall. [Engl. Transl.] **1971** No. 2, pp. 25/30).

[32] Jehn, H. (in: Fromm, E.; Gebhardt, E., Gase und Kohlenstoff in Metallen, Springer, Berlin – Heidelberg – New York 1976, pp. 552/63).

[33] McCarty, L. V.; Donelson, R.; Hehemann, R. F. (Metall. Trans. A **18** [1987] 969/74).

[34] Becker, J. A.; Becker, E. J.; Brandes, R. G. (J. Appl. Phys. **32** [1961] 411/23).

[35] Schnitzel, R. H. (Trans. Metall. Soc. AIME **233** [1965] 186/922).

[36] Gerasimov, A. F.; Konev, V. N.; Timofeeva, N. F. (Fiz. Met. Metalloved. **11** [1961] 596/600; Phys. Met. Metallogr. [Engl. Transl.] **11** No. 4 [1961] 106/10).

9   Reactions with Silicon

9.1   Phase Diagram

The most recent, experimentally determined version of the phase diagram for the W–Si system is shown in **Fig. 126**. It is based on differential thermal analysis (DTA), X-ray, and microscopic studies as well as hardness measurements on W–Si alloy samples prepared by arc melting single-crystal Si with metal-ceramic (99.6% pure) or electron-beam-melted W [1]. However, the data points in the W rich region are consistently lower than the results of earlier investigations [2 to 9]. These results are the main basis of the assessed phase diagram in **Fig. 127** from the recent review [10]. For previously reported phase diagrams and reviews see [11 to 22]. Thermochemical data and lattice stability values of the pure elements were coupled to compute a phase diagram for, and an analytical description of, the W–Si system [23]. A thermodynamic evaluation of the W–Si system was carried out using a one-lattice substitutional model to describe the Gibbs energy of the solution phases in the system. It was found that the phase diagram according to [1] could be represented by simple thermodynamic models and seemed to be consistent with the existing thermodynamic properties of the compounds, while the other set of phase diagram data presented great difficulties in this respect [24]; see also the table on p. 156. Second-order polynomial equations for the solidus and liquidus at the W-rich end of the phase diagram were derived from existing phase diagram data and measured equilibrium distribution coefficients [25].

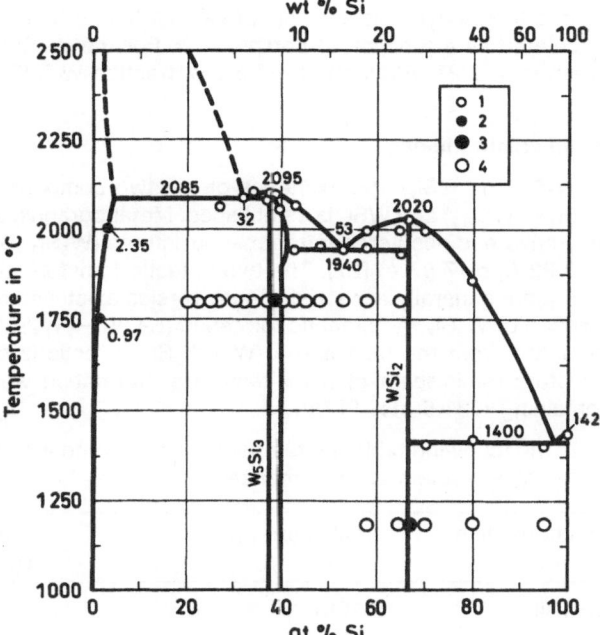

Fig. 126. Phase diagram for the W–Si system, obtained from results of thermal analysis (1), of microhardness measurements (2), and of microstructural analysis (3, one phase and 4, two phases), from [1].

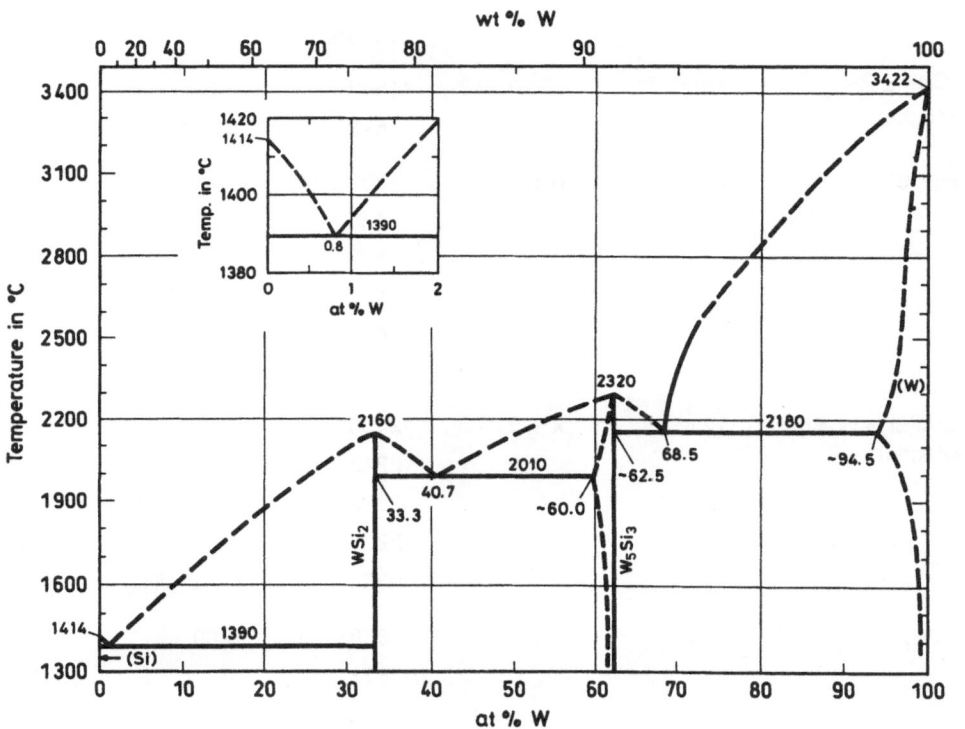

Fig. 127. Assessed Si–W phase diagram [10].

In a thermodynamic analysis of various binary Si-metal systems, the heat of mixing was calculated as a function of composition. For the W–Si system the heat of mixing was at a minimum (−0.27 eV/atom) at ~0.5 mol fraction W [26].

Intermediate Phases

W_5Si_3 and WSi_2. The existence of the two compounds W_5Si_3 (sometimes formulated as W_3Si_2 or $WSi_{0.7}$) and WSi_2 is established. Metallographic and X-ray diffraction studies showed they have body-centered tetragonal structures (W_5Si_3: D_{4h}^{18}, a = 9.64 Å, c = 4.93 Å; WSi_2: D_{4h}^{17}, a = 3.22 Å, c = 7.87 Å) [27]. The two eutectic transformations, $L \rightleftharpoons (Si) + WSi_2$ and $L \rightleftharpoons WSi_2 + W_5Si_3$, are generally accepted. Controversies exist with regard to the nature of the formation reaction of W_5Si_3. A peritectic formation ($L + (W) \rightarrow W_5Si_3$) was assumed [6, 7], while a direct formation from the melt and a $(W) + W_5Si_3$ eutectic is claimed by [1 to 4, 8, 9], though a structure reminiscent of a peritectic transformation was found in small regions of alloys containing <31.5 at% Si [8].

In the following table the temperature and composition data on the phase transformations in the W–Si system are summarized:

transformation	calculated [24]		experimental		Ref.
	T(K)	x_W	T(K)	x_W	
$L \rightleftharpoons (Si)$	1687	0	1683	0	[8]
			1687		[22]
			1693 ± 20		[1]
$L \rightleftharpoons (Si) + WSi_2$ (eutectic)	1661	$x^L = 0.043$	1663	~0.01	[4, 8]
			~1673	0.008	[1, 3, 28]
$L \rightleftharpoons WSi_2$	2275	0.333	2293 ± 20		[1, 28]
			2423		[29]
			2433	0.33	[4, 8, 18]
			2437		[22]
			2438		[3]
			2455		[30]
$L \rightleftharpoons WSi_2 + W_5Si_3$ (eutectic)	2213	$x^L = 0.475$	>2160		[2]
			2213 ± 20	0.47	[1, 28]
			2223 ± 33		[9]
			2283	~0.40	[4, 6, 8]
			2333	(~82 wt%)	[3]
$L \rightleftharpoons W_5Si_3$	2381	0.625	2368 ± 20	~0.60	[1, 28]
			>2573		[6, 7]
			2593		[3, 8, 18]
			2643		[4]

transformation	calculated [24]		experimental		Ref.
	T(K)	x_W	T(K)	x_W	
L⇌W_5Si_3+(W) (bcc)	2362	$x^L = 0.671$	>2290		[2]
(eutectic)		$x^{bcc} = 0.97$	2343±43		[9]
			2358±20	0.68	[1, 28]
			2453	~0.685	[8]
			2483	~0.78	[4]
			2523	(~96 wt%)	[3]
L⇌(W)	3695	1	3613	(98–99 wt%)	[4]
			3683	(99.5 wt%)	[8]

While WSi_2 is considered to exist only at the stoichiometric composition of 33.3 at% W ("line" compound, see Fig. 127) [10], a range of homogeneity seems to exist for W_5Si_3 [2, 46]; this is reflected in results for the lattice parameters [31, 32]. The homogeneity range decreases with a drop in temperature [6]. A range of homogeneity with a variation of ≤2.5 at% Si at 1800°C is estimated [1]; the composition ranges from 60 to 62.5 at% W at the eutectic temperature of 2010°C [10]. Studies of the crystallization of amorphous W–Si films (prepared by d.c. magnetron sputtering) indicated that the homogeneity range of W_5Si_3 extended to a composition of $WSi_{0.9}$ on the Si rich side at 2000°C, but was very narrow at 850°C [33]. Rather broad homogeneity ranges for the W_5Si_3 and WSi_2 phases were obtained in thin (< 1000 Å) condensed W–Si films (10 to 90 at% Si) heated to 770 K, viz. 30 to 40 and 60 to 70 at% Si (from the diagram in the paper [27]), respectively. The broadening is attributed to the manifestation of a dimensional effect and the far-from-equilibrium conditions under which the films were condensed. In films < 100 Å thick and consisting of 40 to 70 at% Si, a phase formed with an fcc lattice and a=11.0 Å; this was considered a new modification of WSi_2 [27].

The crystallization behavior of d.c. sputter-deposited W–Si alloy films with 5 to 38 at% Si was studied by X-ray diffraction and DTA measurements. Films with 22 to 38 at% Si were amorphous on deposition; their crystallization temperature (the films transformed to their equilibrium phases of W_5Si_3+(W)) depended on the alloy composition and had a maximum value of ~915°C for ~28 at% Si. Alloy films containing 5 to 22 at% Si were crystalline on deposition, but also displayed a phase transition (considered as the transition from a metasta- ble silicide phase to the equilibrium phases) with a maximum transition temperature of ~1010°C for an alloy with ~10 at% Si. The samples with 5 and 10 at% Si transformed to solid solutions of Si in W according to the powder diffraction results [34].

For a plot (and its discussion) of the Schottky-barrier heights vs. the eutectic temperatures of transition metal silicide–Si systems (including WSi_2–Si), see [35].

Other Phases. The formation of cubic **W_3Si** (structural type β-W, space group O_h^3, with lattice parameter a=4.91_0±0.005 Å) was observed at the W/W_5Si_3 boundary during the diffusion silicidation of 99% pure W cylinders [5, 36]. The presence of W_3Si was also indicated by X-ray diffraction after a sputter-deposited W film on polycrystalline Si had been sintered at 1100°C [37]. Though the formation of W_3Si was refuted [7, 8], its existence as a metastable compound is assumed [15, 18, 38, 47]. The formation of **W_2Si** [39, 40] and **WSi** [40] during the pyrolytic deposition of W on Si at 600 and 900°C, respectively, was deduced from Rutherford backscattering spectroscopy (RBS) employing 1.3-MeV N^+ and He^+. However, the phases occurring during silicide formation reactions may be quite different from those obtained in phase diagram determinations, as in the latter only equilibrium conditions are involved while transport kinetics may play a role in the silicide formations [40].

Solid Solubilities

There are no exact data about the mutual solubilities of W and Si. The solubility of W in Si is very small [10, 11, 13, 20], but >0.1 at% according to [41] (no temperature is given). A solubility of $\sim10^{15}$ atom/cm^3 at 1200°C is given [42]. When W overlayers on n-type Si were irradiated with 7 MeV electron beams at ~360°C, a maximum W concentration of $\sim1\times10^{20}$ atom/cm^3 in the surface layers of Si was determined by Rutherford backscattering spectroscopy (RBS) and secondary ion mass spectrometry (SIMS) [43]. For a theoretical estimation of the enthalpies of solution for the neutral W atom and the W$^+$ ion on tetrahedral interstices of the Si lattice (14.93 and 9.91 eV/atom, respectively), see [44].

In the following the available data on the solubility of Si in W are summarized:

temperature in °C	625 to 730	1180	~1230	1750	1800	2000
solubility of Si in W in at%	≤2	~1.5	2	0.97	~5.5	2.35
Ref.	[45]	[46]	[18]	[1]	[3]	[1]

temperature in °C	~2070	2180	~2200	>2300	2370
solubility of Si in W in at%	≤6	5.5	6 to 7	10 to 13	≤8.5
Ref.	[9]	[10]	[4, 14]	[7]	[20]

References:

[1] Kocherzhinskii, Yu. A.; Kulik, O. G.; Shishkin, E. A.; Yupko, L. M. (Dokl. Akad. Nauk SSSR, **212** [1973] 642/43; Doklady Chem. [Engl. Transl.] **212** [1973] 782/4).

[2] Brewer, L.; Searcy, A. W.; Templeton, D. H.; Dauben, C. H. (J. Am. Ceram. Soc. **33** [1950] 291/4).

[3] Kieffer, R.; Benesovsky, F.; Gallistl, E. (Z. Metallk. **43** [1952] 284/91).

[4] Blanchard, R.; Cueilleron, J. (C. R. Hebd. Seances Acad. Sci. **244** [1957] 1782/85).

[5] Matyushenko, N. N.; Efimenko, L. N.; Solopikhin, D. P. (Fiz. Metal. Metalloved. **8** [1959] 878/80; Phys. Met. Metallogr. [Engl. Transl.] **8** No. 6 [1959] 67/9).

[6] Obrowski, W. (J. Inst. Metals **89** [1960/61] 79/80).

[7] Efimenko, L. N.; Verkhorobin, L. F.; Matyushenko, N. N. (Izv. Akad. Nauk SSSR, Met. **1965** No. 4, 163/7; Russ. Metall. [Engl. Transl.] **1965** No. 4, 113/8).

[8] Maksimov, V. A.; Shamrai, F. I. (Izv. Akad. Nauk SSSR, Neorg. Mater. **5** [1969] 1136/7; Inorg. Mater. [Engl. Transl.] **5** [1969] 965/6).

[9] Zotov, Yu. P.; Kroshkina, N. N.; Bukharin, V. E. (Izv. Akad, Nauk SSSR, Neorg. Mater. **16** [1980] 842/5; Inorg. Mater. [Engl. Transl.] **16** [1980] 584/7).

[10] Naidu, S. V. N.; Sriramamurthy, A. M.; Rao, P. R. (J. Alloy Phase Diagrams **5** No. 3 [1989] 149/58).

[11] Samsonov, V. G.; Umanskii, Ya. S. (Tverd. Soedin. Tugoplavkikh Metallov, Metallurgizdat, Moscow 1957, pp. 354/6).

[12] Hansen, M.; Anderko, K. (Constitution of Binary Alloys, McGraw-Hill, New York 1958, pp. 1203/4).

[13] Samsonov, G. V. (Silicides and Their Technical Uses, Izdat. Akad. Nauk Ukr. SSR, Kiev 1959, pp. 92/5).

[14] Vol, A. E. (Structure and Properties of Binary Metal Systems, Fizmatgiz, Moscow 1962, pp. 409/14).

[15] Elliott, R. P. (Constitution of Binary Alloys, 1st Suppl., McGraw-Hill, New York 1965, pp. 821/2).

[16] Shunk, F. A. (Constitution of Binary Alloys, 2nd Suppl., McGraw-Hill, New York 1969, pp. 684/5).

[17] Metals Handbook (Metallography, Structures, and Phase Diagrams, 8th Ed., Vol. 8, American Society for Metals, Metals Park, Ohio, 1973, pp. 335, 375).

[18] Chart, T. G. (Met. Sci. **9** [1975] 504/9).

[19] Samsonov, G. I.; Vinitskii, I. M. (Handbook of Refractory Compounds, Plenum, New York 1980, 555 pp.).

[20] Raynor, G. V.; Rivlin, V. G. (Int. Met. Rev. **26** [1981] 213/49; C.A. **95** [1981] No. 193096).

[21] Murarka, S. P. (Silicides for VLSI Applications, Academic, New York 1983, 200 pp.).

[22] Massalski, T. B. (Binary Alloy Phase Diagrams, American Society for Metals, Metals Park, Ohio, 1986, pp. 2062/3).

[23] Kaufman, L. (CALPHAD: Comput. Coupling Phase Diagrams Thermochem. **3** No. 1 [1979] pp. 45/76, 45/6, 73/6).

[24] Vahlas, C.; Chevalier, P. Y.; Blanquet, E. (CALPHAD: Comput. Coupling Phase Diagrams Thermochem. **13** [1989] 273/92).

[25] Drapala, J.; Kuchar, L. (Sb. Ved. Pr. Vys. Sk. Banske Ostrave, Rada Hutn. **29** [1983] 1/9; C.A. **101** [1984] No. 175933).

[26] Gong, S. F.; Hentzell, H. T. G. (J. Appl. Phys. **68** [1990] 4542/9).

[27] Blokha, V. B.; Gladkikh, N. T.; Glushko, P. I.; Zmii, V. I.; Kartmazov, G. N.; Poltavtsev, N. S. (Izv. Akad. Nauk SSSR, Neorg. Mater. **18** [1982] 805/8; Inorg. Mater. [Engl. Transl.] **18** [1982] 677/80).

[28] Kocherzhinskii, Yu. A.; Kulik, O. G.; Shishkin, E. A.; Yupko, I. M. (Therm. Anal. Proc. 4th Int. Conf., Budapest 1974 [1975], Vol. 1, pp. 425/32; C.A. **87** [1977] No. 29813).

[29] Grinthal, R. D. (J. Electrochem. Soc. **107** [1960] 59/61).

[30] Beaver, W.; Stonehouse, A. J.; Paine, R. M. (Plansee Proc. 5th Seminar, Reutte/Tyrol 1964 [1965], pp. 682/700).

[31] Parthé, E.; Schachner, H.; Nowotny, H. (Monatsh. Chem. **86** [1955] 182/5).

[32] Aronsson, B. (Acta Chem. Scand. **9** [1955] 137/40).

[33] Lahav, A.; Wu, C. S.; Baiocchi, F. A.; Sheng, T. T. (Semicond.-Based Heterostruct.: Interfacial Struct. Stab., Proc. Northeast Reg. Meet. Metall. Soc. **1986** 147/54; C.A. **107** [1987] No. 68423).

[34] Thomas, R. E.; Perepezko, J. H.; Wiley, J. D. (Appl. Surf. Sci. **26** [1986] 534/41).

[35] Ottaviani, G.; Tu, K. N.; Mayer, J. W. (Phys. Rev. Lett. **44** [1980] 284/7).

[36] Verkhorobin, L. F.; Ivanov, V. E.; Matyushenko, N. N.; Nechiporenko, E. P.; Pagachev, N. S.; Somov, A. I. (Fiz. Metal. Metalloved. **13** [1962] 77/81; Phys. Met. Metallogr. [Engl. Transl.] **13** [1962] 67/71).

[37] Murarka, S. P.; Read, M. H.; Doherty, C. J.; Fraser, D. B. (J. Electrochem. Soc. **129** [1982] 293/301).

[38] Nowotny, H.; Brukl, C.; Benesovsky, F. (Monatsh. Chem. **92** [1961] 116/27).

[39] Bayerl, P.; Eichinger, P. (RS 03/75, Institut f. Festkörpertechnologie, München 1975).

[40] Eichinger, P.; Sauermann, H.; Wahl, M. (Ion Beam Surf. Layer Anal. [Proc. Int. Conf.] 1975 [1976], Vol. 1, pp. 353/62; C.A. **86** [1977] No. 181578).

[41] Kornilov, I. I.; Matveeva, N. M.; Pryakhina, D. I.; Polyakova, R. S. (Metallokhim. Svoistva Elementov Period. Systemy (Metal-chemical Properties of the Elements of the Periodic System) Izd. Nauk, Moscow 1966; C.A. **68** [1968] No. 16365).

[42] Zibuts, Yu. A.; Paritskii, L. G.; Ryvkin, S. M. (Fiz. Tverd. Tela **5** [1964] 3301/4; Soviet Phys. Solid State **5** [1964] 2416/9).

[43] Wada, T.; Takeda, M. (Nagoya Kogyo Daigaku Gakuho **36** [1985] 113/25; C.A. **104** [1986] No. 160090).

[44] Fistul', V. I.; Shmugurov, V. A. (Fiz. Tekh. Poluprovodn. [Leningrad] **23** [1989] 688/92; C.A. **111** [1989] No. 84826).

[45] Borders, J. A.; Sweet, J. N. (Appl. Ion Beams Met. Int. Conf., Albuquerque 1973 [1974], pp. 179/91; C.A. **82** [1975] No. 103761).

[46] Gelain, C.; Cassuto, A.; LeGoff, P. (Bull Soc. Fr. Ceram. **80** [1968] 23/7).

[47] Efimov, Yu. V.; Frolova, T. M.; Bodak, O. I.; Kharchenko, O. I. (Izv. Akad. Nauk SSSR, Neorg. Mater. **20** [1984] 1593/5; Inorg. Mater. [Engl. Transl.] **20** [1984] 1374/6).

9.2   Reaction Characteristics

General

The starting materials (e.g., purity, size, and dimensions of the samples, their crystallinity, and the presence of defects and oxide films) and the experimental conditions largely determine the reaction products, the mechanism, and the kinetics of the W–Si reaction. Methods such as metallographic, optical, X-ray analyses, Rutherford backscattering (RBS), and hardness measurements were used to determine the reaction products and the course of the reaction.

9.2.1   Silicide Formation on Compact Tungsten

Solid Phase Reactions Between Bulk Samples

Reaction between solid W and solid Si could not be induced by mechanical activation alone, e.g., by applying pressure $(14 \pm 1\,kbar)$ and deformation to a mixture of W and Si turnings (equal amounts by volume) [1]. Reaction between W and Si occurs on additional application of heat, laser- or electron-beams, or ion bombardment. Close contact between W and single crystal Si was effected by a mechanical load [2, 3], by welding W to the (111) surface of an Si single crystal [5], or by clamping platelets of monocrystalline Si between single crystals of W (initial orientation of the W surface (100)) [4]. The samples were heated under vacuum at 1200 to 1350°C for up to 10 h [2], 1000 to 1300°C for 5 to 50 min [5], 1570 K for 20 h [3], or 1000°C [4]. The extent of the reaction was determined by metallographic, X-ray, optical, electron-microscopic analyses and hardness measurements. The thickness of the WSi_2 layer (x in μm) formed between the W and the Si increased with time (t in min) according to $x^2 = 5.98 \times 10^{-3}$ exp(-162000/RT)t. The activation energy of 162 kJ/mol [5] is below an earlier value of 200 kJ/mol [2] (see also pp. 161/2). Linear growth kinetics were observed for the initial reaction stages [6], cf. below. Thus the reaction rate is mainly determined by diffussion. Marker experiments [7 to 9] and the results of other studies [2, 4, 10] indicate that Si is the diffusing species, and that the reaction takes place at the W/WSi_2 interface.

Mechanical breakdown of the samples at the WSi_2/Si boundary [2, 5] showed that the silicide is bonded more strongly to the W than to the Si. Pores appeared in silicide layers with a thickness of >20 μm, causing cracks in the direction normal to the original Si/W interface [5, 10]. Structural analysis of the composition of the silicide layer showed deviations from stoichiometric WSi_2 (e.g., $WSi_{2.24}$ to $WSi_{1.84}$ [6]), especially in the surface layer, i.e. near the

WSi_2/Si boundary, see [3, 6, 11, 12]. Volume changes due to the formation of the silicide phase, concentration gradients of the Si during the diffusion annealing, and plastic deformation due to the compression of the specimens may cause nonuniformly distributed stresses in the specimens [4].

Lumps or wires of W (purity $\geq$ 99.5%) were heated with single-crystal grains of Si at 1000°C for 24 h in the presence of 100 Torr Br_2 (or Cl_2). W_5Si_3 and WSi_2 layers (in that order) formed on the W, with their layer thicknesses in the ratio 0.2:1 ($WSi_2 \approx 30$ µm thick), respectively; no other W silicide was found. The suggested reaction mechanism is: $2 Si(s) + 2 SiBr_4(g) \rightleftharpoons 4 SiBr_2(g)$ and $4 SiBr_2(g) + W(s) \rightleftharpoons WSi_2(s) + 2 SiBr_4(g)$ [22].

Reactions of W With Si Powder

The reaction of W specimens with Si powder was studied in vacuum ($\sim 10^{-5}$ to 10^{-6} Torr, 1150 to 1350°C) [2, 6, 10, 13] and is considered to take place mainly via the vapor phase. At lower temperatures and shorter reaction times the lower silicides W_3Si (1150°C, 5 h) and W_5Si_3 (1150°C, 5 h and 1240°C, 1 h) were formed; only WSi_2 was formed at temperatures above 1240°C. When a certain silicide layer depth was reached, the reaction slowed down and a noticeable growth of the lower silicides was observed [10]. The addition of activators, viz. ammonium and alkali metal halides, to the Si powder, reduced the temperature range for the W–Si reaction to 850 to 1350°C (in an H_2, and/or Ar atmosphere, or at reduced pressures):

added activator	atmosphere	temperature in °C	Ref.
NaF	H_2	980	[14]
NaF	5% H_2 in Ar	855 to 1100	[7]
NaF	H_2/Ar	850 to 1100	[8]
KF, NaF	(reduced pressure)	1094	[15]
$NaHF_2$	Ar	1240 $\pm$ 10	[16]
NaF (or NaCl) + NH_4Cl	H_2	900	[17]
K_2SiF_6	Ar	1100	[18]
NaCl	vacuum	1250	[12]
NaCl	vacuum	1200 to 1350	[11, 21]

A suggested activator mechanism includes [14]: $4 NaF + Si \rightarrow SiF_4 + 4 Na$; $SiF_4 + 2 H_2 \rightarrow Si + 4 HF$. Nonlinear kinetics apply as long as the rate of silicide formation is greater than the rate of Si diffusion into the W and through the WSi_2 layer [7, 8].

At temperatures below 1200°C only WSi_2 is detected as the rate of formation of WSi_2 is much faster than that of W_5Si_3. The rate of formation of WSi_2 is given by $v_{WSi_2} = 6.1 \exp(-105\,000/RT)$ cm/min$^{1/2}$ with an activation energy of 105 kJ/mol [7, 8]; an activation energy reported by [19] appears to be too low.

At higher temperatures (e.g., 1300°C) W_5Si_3 (with initially W_5Si_3 as the inner zone and WSi_2 as the outer zone phase) was formed also [16]; on annealing in vacuum at temperatures of 1300 to 1870°C WSi_2 layers were converted completely to W_5Si_3 at 1600°C [18], 1800°C [16], and 1870°C [20]. Suggested reactions are $5 WSi_2 \rightarrow W_5Si_3 + 7 Si$ and $3 WSi_2 + 7 W \rightarrow 2 W_5Si_3$ with an apparent activation energy of 180 kJ/mol [20]. The growth of the W_5Si_3 phase is largely controlled by diffusion processes [16, 18, 20]; for $x^2 = kt$ (x = thickness of the W_5Si_3 layer in cm, t = time in h), the temperature dependence of the rate constant is given by: $k = 17.5 \exp(-272\,000/RT)$ cm^2/h (activation energy in J/mol) [16].

A W rod was reacted with a mixture of Si powder and B_4C (0 to 100 wt%) with 5% NaF as activator at 1100°C for 10 h: For mixtures containing < 20% Si only a W boride layer formed on the W, for mixtures with ~ 20 to ~ 95% Si an outer WSi_2 layer formed on top of the inner boride layer, and for mixtures with >95% Si only WSi_2 formed on the W; for details, see [23]. Annealing of such boride–WSi_2 layers on W at 1400 to 1600°C [24] and 1300 to 1800°C [16] showed that the boride phase slowed down or even prevented the formation of W_5Si_3. The reaction of solid W with Si powder in the presence of B_4C and fluoride activators, forming WSi_2 (and W borides), is thought to proceed with the help of B subfluorides in the gas phase above the reaction mixture [46]. Similarly, W_5Si_3 and WSi_2 (besides borides) were formed when W disks and a powder mixture of $MoSi_2$ with B_4C or amorphous B were heated at 1400°C in a vacuum furnace [47].

Slices of high-purity W were heated with Si powder at 1000 to 1200°C in iodine vapor (~ 300 Torr) in a closed capsule. Only WSi_2 formed, the layer increased with the square root of time with an activation energy of 179 ± 10 kJ/mol. The WSi_2 crystallites showed a strong preferred orientation; the WSi_2 textures formed at 1000 and 1200°C showed considerable differences and were independent of the texture and grain size of the polycrystalline W substrate [25].

The kinetics of the diffusion saturation of solid W with Si (as powder) were studied by Zmii et al. [2, 3, 6, 11 to 13, 21, 26], cf. also Section 9.3, pp. 186/91. When solid W and Si powder are heated under vacuum at 1200 to 1350°C, linear kinetics are observed for reaction times shorter than 2 h. For reaction times larger than 2 h, however, the layer thickness increased with the square root of time, see **Fig. 128**. The temperature dependencies of the rate constants for the initial (linear) and the nonlinear regimes are given by (activation energies in J/mol) 2.4×10^{-4} exp(−64 400/RT) cm/s and 1.5×10^{-2} exp(−180 000/RT) cm²/s, respectively [6, 21].

The effect of activators on the silicide formation was studied with regard to the activator added, its concentration, and reaction time [11, 14, 15, 17, 21]; kinetic parameters were evaluated and tabulated for NaCl activator concentrations of 0 to 2.5 wt% in the temperature range 1200 to 1350°C [11, 21]. Thus, e.g., the addition of 2.5 wt% NaCl to Si powder increases the 1200°C rate constant in the nonlinear regime from 0.6×10^{-8} cm²/s for the pure Si powder to 2.6×10^{-8} cm²/s; correspondingly, the activation energy decreases from 180 to 64 kJ/mol. It is considered that the activator causes a change in the diffusion mechanism [11, 21].

Reactions in Mixtures of W and Si Powders

W and Si powders (in the atomic ratio of WSi_2, mean particle size of the order of 1 to 10 μm) were ground together for 24 h under N_2 at room temperature in a vibratory ball mill made of WC. The concentrations of O_2 and H_2O were in the ppm range. X-ray diffraction results showed that W and Si had reacted only partially to form tetragonal WSi_2 and β-W_5Si_3 [27, 28]. WSi_2 containing only 90 ppm of oxygen was obtained when W and Si powders were heated in vacuum (10^{-5} Torr) at 1250°C and subsequently annealed at 1500°C for 2 h [29]. WSi_2 and tetragonal $W_3Si_2(W_5Si_3)$ were formed when mixtures of W and Si powders (in the atomic ratios 1:2 or 3:2, respectively) were heated at 800 to 1300°C in H_2, or in H_2 or N_2 mixed with $SiCl_4$. In the H_2 atmosphere, silicide formation started at 1100 to 1200°C, while in the presence of $SiCl_4$ silicides formed at ~ 900°C. For details, see [30]. When a powder mixture of W (~ 99.98% pure) and Si (containing 0.26% Al, 0.23% Fe) (in the atomic ratio 1:2) was heated at 600 to 1200°C in Ar, homogeneous WSi_2 had formed after 24, 4, and 0.5 h at 900, 1000, and 1100°C, respectively. The rather high activation energy of 264 kJ/mol is attributed to the hindered diffusion in powders and/or to the presence of oxide films on the powders [31].

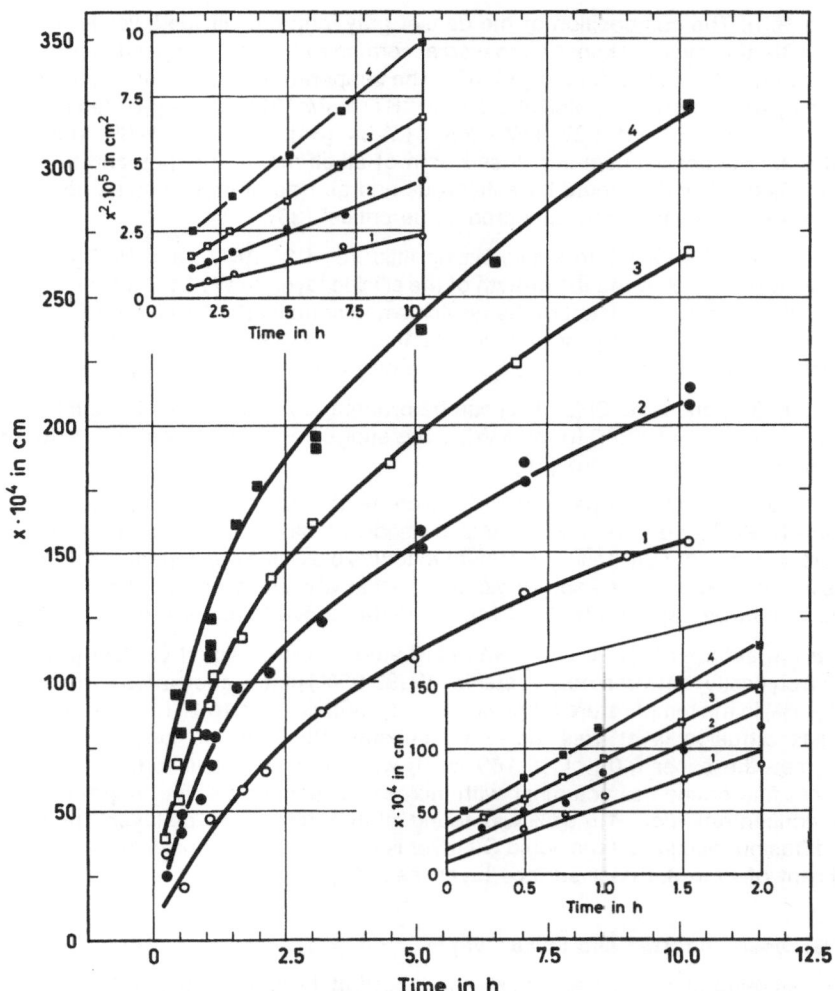

Fig. 128.   Reaction of solid W with Si powder. Time dependence of
the thickness (x) of the silicide layer formed at 1200°C (1), 1250°C (2),
1300°C (3), and 1350°C (4) from [6].

Reactions Involving Melts

A high yield of a W silicide is formed by reacting Si and W granules in molten Si [32]. When liquid Si (99.9999%) was made to flow over plates of rolled W (annealed and etched in an HNO$_3$–HF solution) in a vacuum furnace (1×10^{-4} to 5×10^{-4} Torr) at 1430°C and flow rates of ~0.10 cm/s to ~0.02 cm/s, silicide layers formed: an inner layer of W$_3$Si$_2$ (W$_5$Si$_3$) and an outer layer of WSi$_2$ with an Si concentration gradient in both silicides [33].

The reaction of solid W (rods or wires) with Si powder dissolved in a molten mixture of NaF, Na$_2$SiF$_6$, and equimolar NaCl–KCl (14, 14, and 72 wt%, respectively) under Ar was studied by X-ray diffraction and metallography in the temperature range of 800 to 950°C and for reaction times of 3 to 10 h. The composition of the dense, gray, smooth silicide layer formed corresponded to WSi_2; the layer thickness increased according to a $\sqrt{\tau}$-time dependence (maximum thickness ~30 μm after 10 h at 950°C) [34, 39]. The temperature dependence of the rate constant is given by $k = 1.2 \times 10^2 \exp(-268\,000\ J \cdot mol^{-1}/RT)\ cm^2/s$ [35]. The suggested mechanism includes: $Si + Si^{4+} \rightleftharpoons 2\,Si^{2+}$ and $4\,Si^{2+} + W \rightarrow 2\,Si^{4+} + WSi_2$ [34]. Similarly, WSi_2 formed on a W panel at 800°C under vacuum in a fused bath of alkali metal fluorides and fluosilicates containing Si. The salt bath should be anhydrous and as free of impurities as possible; the chemical composition of the bath appeared to be critical [36].

Addition of 1 to 30 wt% Cu to a siliconizing mixture of Si, Al_2O_3, and NaF (40, 57, and 3 wt%, respectively) accelerated the growth of the silicide layer on W at 1100°C during 2, 4, 8, or 12 h. The silicide layer consisted of the disilicide and contained 0.1 to 0.5 wt% Cu. For Cu additions >10 wt% the quality of the silicide layer deteriorated. The growth rate was approximately linear with time and was governed by the rate of the chemical reaction [37].

Solid W (99.93%) and solid Si (99.9%) can be brought to react in liquid Zn under Ar in the temperature range 900 to 1200°C forming WSi_2. It is suggested that the W silicide is formed by a series of topochemical reactions [38].

Si plates and W samples, separated by Sn were heated under vacuum (~1×10^{-4} Torr) to 900, 950, and 1000°C for up to 4 h. X-ray analysis showed that WSi_2 had formed at the boundary between the W and the Sn melt. The growth followed a $\sqrt{\tau}$-time dependence. The activation energy was 210 kJ/g atom. The results indicate also that W is well wetted by liquid Sn containing Si; thus W can be soldered by Sn–Si melts in the above temperature range [45].

The diffusion saturation of W by Si was also studied with a liquid Cu film (2 μm thick, applied electrolytically) between polycrystalline W (99.98%) plates and plates of monocrystalline Si (99.999%) in the temperature range of 1150 to 1300°C for up to 10 h. The rate of growth of the silicide diffusion layer was increased substantially. For the kinetic and diffusion parameters, see the paper [40], cf. p. 189. WSi_2 was the only phase formed, its structure changing from fine grained to columnar with increasing temperature and depth of diffusion layer. A Cu concentration of ~1.5 wt% was found all through the silicide layer. The acceleration of the diffusion reaction by the liquid Cu layer is attributed to a considerably higher diffusion coefficient of Si in the near surface layer, see [40].

Reactions Between Tungsten and Silicon Vapor

Solid W was reacted with Si vapor in a vacuum cell at 1200 to 1250°C. A thin W_5Si_3 layer formed between the W and the outer WSi_2 layer. The silicide layer growth on the W followed a $\sqrt{\tau}$-time dependence [42, 43]. Similarly, monocrystalline W disks (with faces (100), (110), and (111)) were vacuum siliconized by Si vapor at 1300°C, and a two-layer structure (inner layer W_5Si_3, outer layer WSi_2) formed on the W. The silicide crystals showed a preferred direction of growth ((001)) at all stages. For details of the texture characteristics of the silicide coating, see [41]. In a glow discharge at 1000°C the reaction rate of W siliconization is much higher due to the ionized atmosphere and an activated condition of the W surface. The outer silicide layer consisted of WSi_2; intermediate layers (between the WSi_2 and the W) had a lower Si content [44].

Reactions of Tungsten With Silicon Films

The reactions of W with thin Si films deposited on the W by, e.g., e-beam evaporation, sputtering, pyrolytic decomposition of an Si compound, or by a chemical reaction, were studied by a variety of methods, e.g., photoemission, Auger electron spectroscopy (AES), field ion microscopy (FIM), field electron microscopy (FEM), X-ray analysis, etc. In general, a certain temperature ("threshold" temperature), viz. ~500 to 800°C, was found necessary for the reaction to start. However, photoemission, Auger, and work function measurements showed that very thin Si films (0.3 to 50 Å) evaporated onto a clean W(110) surface under ultrahigh vacuum interacted strongly with the W at room temperature to form WSi_2 with a strong metallic character. For Si films with 10 monolayers and more, the reaction gradually stopped as an Si oxide species (resulting from Si reaction with O impurities) appeared on the surface and prevented the diffusion of Si atoms to the W/silicide interface [48].

Si was vapor-deposited (by reacting $SiCl_4$ with H_2) onto cleaned and degassed W wire (99.95% pure) at ~980 to 1400°C. Silicides formed on subsequent heating in Ar at 1200°C, but WSi_2 was the only silicide identified [49]. W foil and wire (after heating in H_2 at 800°C for 1 h) were covered with thin (e.g., ~0.4 μm) Si layers by decomposition of SiH_4 in H_2 (0.3 and 0.12% SiH_4) at 700 and 800°C. At 800°C the Si diffused into the W forming WSi_2; the silicide coating reduced the oxidation rates in air of W foil (at 600°C) and W wire (at 900°C) considerably [50].

In ultrahigh vacuum (UHV) and with perfectly clean surfaces the W–Si reaction proceeds faster and at lower temperatures, cf. [51]. Si was electron-beam evaporated onto cleaned, fine-grained W (grain size 5 to 90 μm) in the temperature range 400 to 850°C in UHV. Silicide formation started in the temperature range 625 to 670°C; W_5Si_3 is formed at the lower temperatures and during the early reaction stages; at 780 to 850°C tetragonal WSi_2 is the only product [52, 53]. Initially linear kinetics are observed (nucleation phenomena compete with the interface reaction), later the silicide growth is diffusion-controlled, cf. [6]. P coevaporated with the Si influenced the interface reaction, viz. the disilicide growth was retarded [53]. Similarly, formation of highly oriented WSi_2 was observed when Si was evaporated onto W (grain size 35 to 50 μm, orientation (100)) in UHV at temperatures >600°C [54]. Low energy electron diffraction (LEED) studies of thin Si films on the (110) face of W (for structures see the paper [55]) showed that WSi_2 formed on annealing at temperatures ≥500°C [55]. When small amounts of Si were evaporated in UHV onto a clean W emitter tip, FEM studies showed that annealing at 1150 to 1400 K caused bulk diffusion of Si into the W and formation of silicide phases, probably W_5Si_3 at first and then WSi_2 [56]. Work function, FEM, and FIM studies of Si films on single-crystal W emitter tips in UHV showed that W silicides form on W(100) at temperatures above 600 K for a minimum coverage of more than 4 monolayers; for details, see [57, 58].

Early Stages of Silicide Formation. Detailed studies of the initial stages of silicide formation on W were made possible by the method of the time-of-flight atom-probe field ion microscope [59 to 63]. Single atoms and atomic layers on a selected sample surface can be identified; thus W images can be distinguished easily from WSi_2 images as they field evaporate at 5.3 and 4.8 V/Å, respectively. Si was vapor-deposited onto a W tip at temperatures of 20 to 900 K, and the tip was then heated for 1 to 120 min for silicide formation. WSi_2 (of tetragonal C 11b structure) formation took place in the optimum temperature range of 900 to 1000 K preferably on the (001) plane of the tip surface. The sharp W/Si and silicide/W interfaces indicated the absence of transitional atomic layers with graded composition [59, 60].

Four stages of silicide formation were observed for Si evaporated onto carefully prepared W emitter tips at the "threshold" temperature of ~700°C [61 to 63]: 1) For a low Si coverage and for short annealing times (10 s to several min) silicide (possibly WSi_2) starts to grow epitaxially on W(001) planes, while on the W(112) plane a one-atomic silicide layer forms, probably WSi with a p(2×1) structure. The W(110) planes remain bare. 2) As the Si dose is increased, three atomic layers of silicide are formed, but the W(110) planes still show no reaction. This stage is very stable with regard to heating. 3) When tens to hundreds of layers of Si atoms have been deposited (annealing time 1 to 3 min), a nearly uniform silicide layer covers the W emitter surface, the WSi_2 growth proceeds from the W(001) planes (their lattice size is comparable to the basal plane of the C11b structure of tetragonal WSi_2). This state is metastable and reverts to stage 2) on prolonged heating or proceeds to the next stage if more Si is deposited. 4) Finally many well-ordered polycrystallites of WSi_2 composition are formed. The W/WSi_2 interface is very sharp, there are no layers with graded composition, the orientation of the WSi_2 microcrystals is not closely related to the orientation of the W, and the dominant silicide faces are (112) and (101). For more details, see [61 to 63].

WSi clusters are considered to constitute one stage in the process of silicide formation [64]; for their properties and their role in the silicidation process, see the paper. Similarly, silicide formation in UHV was studied on the W(100) face of a textured W ribbon by high resolution AES, thermodesorption spectrometry, and contact potential difference methods [65 to 67]. The initial Si coverages, the W temperature during Si deposition, and the duration of the Si deposition were varied. The results show that Si dissolution in the W takes place neither for Si coverages less than 1 monolayer (estimated $N_{Si} \sim 1 \times 10^{15}$ at/cm²) in the temperature range 1000 to 1400 K nor at temperatures lower than 1100 K, even at high Si coverages. In the temperature range 1100 to 1400 K the fast dissolution of the Si in excess of the monolayer concentration is indicated, though bulk diffusion of Si into the W (with formation of WSi_2) does not start before the surface silicide formation has been completed [65]. At temperatures above 1500 K Si is desorbed from the W ribbon and the silicide layer decomposes; for more details, see [66, 67].

Further studies were carried out with W single crystals (W(100) [68], W(110) [69]) by a combination of AES, LEEDS, and EELS methods (EELS = electron energy loss spectroscopy). The Si films (1 to 20 monolayers) were e-beam evaporated onto the carefully pretreated W at room temperature and then annealed for 3 min (followed by cooling and analysis) at increasing temperatures. Up to ~ 900 K (for W(110) [69]) and ~ 950 K (for W(100) [68]) the growth of the disordered Si films and the Si coverage are not changed by annealing; the W/Si boundary is sharp. With the onset of silicide formation at the above temperatures a marked change takes place in composition and structure of the near-surface region within a temperature interval of ~ 100 degrees. The results indicate the direct formation of the disilicide WSi_2. In the case of the W(100) face a monocrystalline film of WSi_2(100) was formed, oriented parallel to that face. In the case of the W(110), the diffraction patterns indicated the absence of long-distance order in the silicide along the ⟨110⟩ direction of the W. Changes in the Si content in the surface layers and structural rearrangements appeared after annealing at higher temperatures (up to 1500 K). The composition, structure, and formation temperature of the surface phases were found to be specific for each W crystal face. For more details, see the papers [68, 69].

References:

[1] Neverov, V. V.; Burov, V. N. (Izv. Sib. Otd. Akad. Nauk SSSR Ser. Khim. Nauk **1979** No. 4, pp. 3/8; C.A. **91** [1979] No. 203612).

[2] Zmii, V. I.; Seryugina, A. S. (Zashch. Pokrytiya Met. No. 2 [1968] 195/201; Prot. Coat. Met. [Engl. Transl.] No. 2 [1970] 158/63).

[3] Zmii, V. I.; Kartmazov, G. N.; Poltavtsev, N. S.; Semenov, N. A. (Izv. Akad. Nauk SSSR Neorg. Mater. **17** [1981] 916/7; Inorg. Mater. [Engl. Transl.] **17** [1981] 654/5).

[4] Bogdanov, E. I.; Larikov, L. N.; Maksimenko, E. A. (Metallofizika Akad. Nauk Ukr. SSR Otd. Fiz. Astron. **10** No. 2 [1988] 116/8; C.A. **109** [1988] No. 41996).

[5] Popov, V. S.; Khoklov, V. P.; Alimov, S. A.; Titukh, Yu. I.; Noga, N. A.; Romanovskii, V. F. (Zharostoikie Pokrytiya Zashch. Konstr. Mater. Tr. 7th Soveshch., Kalinin 1975 [1977], pp. 184/8; C.A. **88** [1978] No. 175345).

[6] Zmii, V. I. (Zashch. Pokrytiya Met. No. 11 [1977] 14/8; C.A. **88** [1978] No. 10746).

[7] Bartlett, R. W.; Gage, P. R. (ASD-TDR-63-753 – Pt. II [1964] 136 pp., 5/48, 124/7; N.S.A. **19** [1965] No. 7848).

[8] Gage, P. R.; Bartlett, R. W. (Trans. Met. Soc. AIME **233** [1965] 832/4).

[9] Baglin, J.; Dempsey, J.; Hammer, W.; D'Heurle, F.; Petersson, S.; Serrano, C. (J. Electron. Mater. **8** [1979] 641/61).

[10] Verkhorobin, L. F.; Ivanov, V. E.; Matyushenko, N. N.; Nechiporenko, E. P.; Pagachev, N. S.; Somov, A. I. (Fiz. Met. Metalloved. **13** [1962] 77/81; Phys. Met. Metallogr. [Engl. Transl.] **13** [1962] 67/71).

[11] Zmii, V. I.; Glushko, P. I.; Trofimov, V. F. (Izv. Akad. Nauk SSSR Neorg. Mater. **17** [1981] 644/6; Inorg. Mater. [Engl. Transl.] **17** [1981] 427/9).

[12] Zmii, V. I.; Efimenko, L. N.; Poltavtsev, N. S.; Snezhko, I. A. (Zashch. Met. **20** No. 1 [1984] 141/3; Prot. Met. [Engl. Transl.] **20** [1984] 123/4).

[13] Poltavtsev, N. S.; Zmii, V. I.; Snezhko, I. A. (Izv. Akad. Nauk SSSR Neorg. Mater. **16** [1980] 674/7; Inorg. Mater. [Engl. Transl.] **16** [1980] 464/6).

[14] Goetzel, C. G.; Landler, P. (WADD-TR-60-825; AD-258574 [1960] 1/42; N.S.A. **15** [1961] No. 23986).

[15] Nolting, H. I.; Jeffreys, R. A. (N 63-21802; ASD-TDR-63-459 [1963] 1/140, 23, 121; C.A. **60** [1964] 6560).

[16] Zotov, Yu. P.; Davydova, A. D.; Kroshkina, N. N.; Gorlov, V. I.; Ablogin, M. I. (Izv. Akad. Nauk SSSR Neorg. Mater. **14** [1978] 879/83; Inorg. Mater. [Engl. Transl.] **14** [1978] 688/92).

[17] Goetzel, C. G.; Landler, P. (Planseeber. Pulvermet. **9** [1961] 36/8).

[18] Burykina, A. L.; Dyadykevich, Yu. V.; Sosnovskii, L. A.; Epik, A. P.; Gorskii, V. V. (Izv. Akad. Nauk SSSR Met. **1975** No. 1, pp. 153/7; Russ. Metall. [Engl. Transl.] **1975** No. 1, pp. 126/9).

[19] Samsonov, G. V.; Solonnikova, L. A. (Fiz. Met. Metalloved. **5** [1957] 565/6; Phys. Met. Metallogr. [Engl. Transl.] **5** No. 3 [1957] 177; C.A. **1958** 19804).

[20] Bartlett, R. W.; Gage, P. R.; Larssen, P. A. (Trans. Metall. Soc. AIME **230** [1964] 1528/34).

[21] Zmii, V. I.; Glushko, P. I.; Trofimov, V. F. (Izv. Akad. Nauk SSSR Neorg. Mater. **13** [1977] 1896/7; Inorg. Mater. [Engl. Transl.] **13** [1977] 1525/6).

[22] Nickl, J. J.; Koukoussas, J. D. (J. Less-Common Met. **23** [1971] 73/81).

[23] Sosnovskii, L. A.; Saplina, G. S. (Zashch. Pokrytiya Met. No. 19 [1985] 3/5; C.A. **104** [1986] No. 134172).

[24] Dzyadykevich, Yu. V. (Izv. Akad. Nauk SSSR Met. **1977** No. 3, pp. 182/7; Russ. Metall. [Engl. Transl.] **1977** No. 3, pp. 156/60).

[25] Maas, J. H.; Rieck, G. D. (High Temp. High Pressures **10** No. 3 [1978] 297/304).

[26] Zmii, V. I. (Fiz. Khim. Obrab. Mater. **1986** No. 3, pp. 96/101; C.A. **105** [1986] No. 83544).

[27] Matteazzi, P. (Ceramurgia **20** No. 6 [1990] 227/33; C.A. **115** [1991] No. 55173).

[28] Matteazzi, P.; LeCaer, G.; Bauer-Grosse, E. (Mater. Sci. Monogr. B **66** [1991] 793/802; C.A. **115** [1991] No. 261793).

[29] Kyono, I.; Sugano, M.; Nagase, R.; Hosaka, K.; Kato, A. (Jpn. Kokai Tokkyo Koho 62-171911 [87-171911] [1986/1987] 5 pp.; C.A. **107** [1987] No. 157691).

[30] Sasahara, T.; Someno, M.; Nagasaki, H. (Nippon Kinzoku Gakkaishi **23** [1959] 30/4; C.A. **60** [1964] 3711).

[31] Samsonov, G. V.; Koval'chenko, M. S.; Verkhoglyadova, T. S. (Zh. Neorg. Khim. **4** [1959] 2759/65; Russ. J. Inorg. Chem. [Engl. Transl.] **4** [1959] 1276/9).

[32] Sawada, S.; Kuroki, M.; Kanano, O. (Jpn. Kokai Tokkyo Koho 63-179028 [88-179028] [1987/88] 5 pp.; C.A. **109** [1988] No. 214764).

[33] Kostikov, V. I.; Levin, N. P.; Levin, V. Ya. (Izv. Akad. Nauk SSSR Neorg. Mater. **5** [1969] 152/4; Inorg. Mater. [Engl. Transl.] **5** [1969] 123/6).

[34] Ilyushchenko, N. G.; Anfinogenov, A. I.; Belyaeva, G. I.; Plotnikova, A. F.; Kornilov, N. I. (Zharostoikie Teplostoikie Pokrytiya Tr. 4th Vses. Soveshch., Leningrad 1968 [1969], pp. 105/20; C. A. **72** [1970] No. 103156).

[35] Belyaeva, G. I.; Anfinogenov, A. I.; Plotnikova, A. F.; Ilyushchenko, N. G. (Tr. Inst. Elektrokhim. Ural. Nauchn. Tsentr Akad. Nauk SSSR No. 26 [1978] 40/2; C. A. **92** [1980] No. 133460).

[36] Cook, N. C. (U.S. 3024177 [1959/62]; C. A. **56** [1962] 15292).

[37] Dzyadykevich, Yu. V.; Medyukh, R. M.; Gorskii, V. V.; Iofan, A. A.; Starodubtsev, Yu. Ya.; Reznikov, V. S. (Izv. Akad. Nauk SSSR Met. **1980** No. 4, pp. 203/6; C. A. **93** [1980] No. 136291).

[38] Obukhov, A. P.; Gurin, V. N.; Kozlova, I. R.; Terent'eva, Z. P.; Mazina, T. I. (Izv. Akad. Nauk SSSR Neorg. Mater. **4** [1968] 527/31; Inorg. Mater. [Engl. Transl.] **4** [1968] 452/5).

[39] Anfinogenov, A. I.; Ilyushchenko, N. G.; Belyaeva, G. I.; Finkel'shtein, S. D. (Tr. Inst. Elektrokhim. Ural. Fil. Akad. Nauk SSSR **1968** No. 11, pp. 67/71; Electrochem. Molten Solid Electrolytes **8** [1970] 51/5).

[40] Zmii, V. I.; Kovtun, N. V.; Matyukhina, L. G. (Poverkhnost **1989** No. 8, pp. 148/53; C. A. **111** [1989] No. 199747).

[41] Tsirlin, M. S.; Bakhtina, I. P. (Izv. Akad. Nauk SSSR Met. **1989** No. 2, pp. 141/5; C. A. **110** [1989] No. 217367).

[42] Ivanov, V. E.; Nechiporenko, E. P.; Krivoruchko, V. M.; Mitrofanov, A. S. (Fiz. Met. Metalloved. **17** [1964] 862/5; Phys. Met. Metallogr. [Engl. Transl.] **17** No. 6 [1964] 62/5).

[43] Ivanov, V. E.; Nechiporenko, E. P.; Zmii, V. I.; Krivoruchko, V. M. (in: Samsonov, G. V.; Diffusion Cladding of Metals, Naukova Dumka, Kiev 1965, pp. 29/35).

[44] Prokoshkin, D. A.; Arzamasov, B. N.; Ryabchenko, E. V. (in: Samsonov, G. V.; Diffusion Cladding of Metals, Naukova Dumka, Kiev 1965, pp. 25/8).

[45] Kuznetsov, G. M.; Latysheva, Z. R. (Izv. Vyssh. Uchebn. Zaved. Tsvetn. Metall. **1968** No. 5, pp. 92/5; C. A. **70** [1969] No. 31269).

[46] Fenochka, B. V.; Gordienko, S. P.; Sosnovskii, L. A. (Zashch. Pokrytiya Met. No. 24 [1990] 11/4; C. A. **114** [1991] No. 211923).

[47] Tsirlin, M. S.; Zakharova, L. I. (Zashch. Pokrytiya Met. No. 15 [1981] 3/6; C. A. **96** [1982] No. 221757).

[48] Weng, S. L. (Phys. Rev. B Condens. Matter **29** [1984] 2363/5).

[49] Goetzel, C. G.; Venkatesan, P. S.; Bunshah, R. F. (WADC-TR-59-405 [1960] 57 pp.; N.S.A. **14** [1960] No. 15024).

[50] Cabrera, A. L.; Kirner, J. F.; Miller, R. A.; Pierantozzi, R.; Armor, J. N. (U.S. 4822642 [1985/89] 15 pp.; C. A. **111** [1989] No. 158757).

[51] Bomchil, G.; Goeltz, G.; Torres, J. (Thin Solid Films **140** [1986] 59/70).

[52] Bevolo, A. J.; Schmidt, F. A.; Shanks, H. R.; Campisi, G. J. (J. Vac. Sci. Technol. **16** [1979] 13/9).

[53] Campisi, G. J.; Bevolo, A. J.; Shanks, H. R.; Schmidt, F. A. (J. Appl. Phys. **53** [1982] 1714/9).

[54] Racette, G. W.; Frost, R. T. (J. Cryst. Growth **47** [1979] 384/8).

[55] Boiko, B. A.; Gorodetskii, D. A.; Yas'ko, A. A. (Fiz. Tverd. Tela [Leningrad] **15** [1973] 3145/53; Sov. Phys. Solid State [Engl. Transl.] **15** [1974] 2101/6).

[56] Swenson, O. F.; Sinha, M. K. (J. Vac. Sci. Technol. **9** [1972] 942/6).

[57] Janssen, A. P.; Jones, J. P. (Surf. Sci. **41** [1974] 257/76).

[58] Janssen, A. P.; Jones, J. P. (Thin Solid Films **28** [1975] L25/L28).

[59] Nishikawa, O.; Tsunashima, Y.; Nomura, E.; Horie, S.; Wada, M.; Shibata, M.; Yoshimura, T.; Uemori, R. (J. Vac. Sci. Technol. B **1** [1983] 6/9).

[60] Nishikawa, O.; Tsunashima, Y.; Nomura, E.; Wada, M.; Horie, S.; Shibata, M.; Yoshimura, T.; Uemori, R. (Surf. Sci. **126** [1983] 529/33).

[61] Tsong, T. T.; Wang, S. C.; Liu, F. H.; Cheng, H.; Ahmad, M. (J. Vac. Sci. Technol. B **1** [1983] 915/22).

[62] Tsong, T. T. (Nucl. Instrum. Methods Phys. Res. **218** [1983] 383/90).

[63] Tsong, T. T. (Mater. Res. Soc. Symp. Proc. **25** [1984] 363/74; C. A. **101** [1984] No. 101 342).

[64] Wrigley, J. D.; Ehrlich, G. (Mater. Res. Soc. Symp. Proc. **48** [1985] 47/53; C. A. **104** [1986] No. 24 710).

[65] Ageev, V. N.; Afanas'eva, E. Yu.; Gall, N. R.; Mikhailov, S. N.; Rut'kov, E. V.; Tontegode, A. (Pis'ma Zh. Tekh. Fiz. **12** No. 9 [1986] 565/70; C. A. **105** [1986] No. 30 609).

[66] Ageev, V. N.; Afanas'eva, E. Yu.; Gall, N. R.; Mikhailov, S. N.; Rut'kov, E. V.; Tontegode, A. Ya. (Poverkhnost **1987** No. 5, pp. 7/14; C. A. **107** [1987] No. 13 357).

[67] Ageev, V. N.; Afanas'eva, E. Yu. (Poverkhnost **1987** No. 7, pp. 30/7; C. A. **107** [1987] No. 103 418).

[68] Ageev, V. N.; Gomoyunova, M. V.; Pronin, I. I.; Khoruzhii, S. V. (Poverkhnost **1988** No. 5, pp. 57/63; C. A. **109** [1988] No. 80 304).

[69] Ageev, V. N.; Gomoyunova, M. V.; Grigor'ev, A. K.; Pronin, I. I.; Rodnyanskii, A. E. (Poverkhnost **1990** No. 8, pp. 88/93; C. A. **113** [1990] No. 139 404).

9.2.2 Reactions of Tungsten Films With Silicon

General

The divergence of the results published for the reaction of W films with Si substrates reflects the large number of factors that influence this interaction: the properties of the Si substrate (e.g., purity, doping, presence of oxide films, orientation of crystal faces), the mode of preparation of the W film on the Si surface (e.g., preparation by chemical reactions, sputtering, electron-beam evaporation, ion-beam irradiation), the nature of the W film (e.g., thickness, crystallinity, grain size, impurities), the type of annealing, etc., cf. [1 to 3].

The methods employed to study the Si substrate–W film systems include X-ray diffraction (XRD), Rutherford backscattering (RBS), Auger electron spectroscopy (AES), transmission electron microscopy (TEM), secondary ion mass spectroscopy (SIMS), scanning electron microscopy (SEM), sheet resistivity measurements, field-ion microscopy (FIM), high resolution electron energy loss spectroscopy (HREELS), reflection high energy electron diffraction (RHEED), and others.

Silicide Formation

In general, the W–Si reaction starts at a threshold temperature. This temperature depends considerably on the mode of preparation of the W film. W films can be produced by chemical vapor deposition (CVD) or low pressure chemical vapor deposition (LPCVD), such as the reduction of W halides by Si (e.g. [4 to 6]) or by H_2 (e.g. [5, 7 to 12]), and the pyrolysis of $W(CO)_6$ [3]. W film deposition by electron-beam evaporation (e.g. [13 to 20]) and sputtering (e.g. [21 to 30]) may involve a local rise in temperature and thus the W–Si reaction may require a lower temperature than in the case of CVD or LPCVD deposited W films. For the preparation of a practically inert W/Si interface, see [30]. Annealings are carried out in conventional furnaces, or by rapid thermal annealings (RTA), or rapid thermal processes (RTP) in vacuum, inert, or reducing atmospheres. The latter processes use irradiation from, e.g., a quartz halogen W lamp for short times (of the order of seconds). In addition, ion- and electron-beams or lasers are applied. The advantage of RTA and RTP is that silicide formation is completed before dopant atoms in the Si have diffused out [31]. For a comparison of CVD generated, of evaporated, and of sputtered W films, see [26]. For the influence of annealing parameters, see [9, 185, 186].

Silicide formation was observed even at the very low temperature of 300°C [32]. Usually, the reaction of **CVD or LPCVD generated W films** with Si starts at temperatures between 550 and 700°C [6, 8, 10, 11, 33 to 36]; for higher reaction temperatures in the 700 to 800°C range, see [12, 37 to 40]. In all cases, WSi_2 was the product. At the lower temperatures the WSi_2 product was found to have an amorphous structure [33, 34]. Rapid silicide formation on polycrystalline Si was observed at temperatures above 800°C [41, 42]. The formation of tetragonal W_5Si_3 was observed at 1000°C (for 600 and 1000 Å W on Si(100)) [33] and at 800°C (metastable β-W film) [43]. No WSi_2 was formed by CVD generated W films on p-type Si(111) at temperatures up to 1000°C [44]. For the formation of W_2Si and WSi during the CVD preparation of W films on Si at temperatures of 300 to 900°C, see [4, 5, 187]. W films prepared by the reduction of WF_6 by GeH_4 contained 12 at% Ge and did not react with Si in a vacuum furnace below 700°C. At 800°C badly adhering WSi_2 and α-W formed [45].

Sputter-deposited W films reacted with monocrystalline Si substrates at 600°C [27], 650°C [21], 690°C [46], 1100°C [24, 25] with the formation of tetragonal WSi_2. The formation of tetragonal W_5Si_3 was observed at temperatures above 950°C; at 1100°C, however, only tetragonal WSi_2 existed [23]. The composition of the silicide layer obtained in the temperature range from 845 to 1100°C (by RTA) was studied in detail. WSi_2 and W_5Si_3 formation starts at 900°C, the W_5Si_3 content of the silicide layer reaches a maximum at 1000°C, and completely disappears at 1100°C. All the W has reacted at ca. 1050°C [23, 47]. W layers deposited by ion-beam sputtering at temperatures above 400°C intermixed at the W/Si interface. Hexagonal and tetragonal WSi_2 formed at temperatures between 400 and 500°C. At deposition temperatures above 650°C only tetragonal WSi_2 formed [29].

Sputter-deposited W films on P-doped polysilicon structures (polycrystalline Si film on a monocrystalline Si substrate with or without a silicon oxide interlayer) reacted at 1100°C (30 min under vacuum) to form a layer consisting of W, some W_3Si, and/or W_5Si_3 [22].

W films generated **by electron-beam evaporation** reacted with single crystalline Si at 550°C [14], 450 to 550°C [13, 15] to form tetragonal WSi_2. No intermixing at temperatures up to ~500°C, but the growth of epitaxial hexagonal (0001) WSi_2 at temperatures above 750°C was observed on Si(111) [20]. At 400°C Si atoms were found to segregate in the W layer. At 700°C tetragonal WSi_2 formed [18, 19], cf. [188].

W films deposited on single crystal Si(100) surfaces by electron-beam evaporation and annealed by various techniques were studied in detail. The W film was capped by a 200 Å Si film to prevent its oxidation by the annealing atmosphere. When the W film was deposited onto cold substrates, no WSi_2 formation occurred at annealing temperatures below 800°C; annealing at 1000°C for 1 h resulted in a partial reaction to WSi_2 and W_5Si_3 with blister formation. Deposition onto heated substrates initiated silicide formation above 500°C; the low-temperature hexagonal phase of WSi_2 appears along with tetragonal WSi_2 at temperatures above 575°C. When W was deposited at 500°C and annealing was carried out in a furnace at 675 to 760°C, the reaction was controlled by nucleation at the beginning followed by an Si diffusion-controlled step with an activation energy of 328 ± 19 kJ/mol. For RTP annealing no true activation energy could be determined. For laser annealing (temperature 1140 to 1350°C) an activation energy of 210 ± 19 kJ/mol was obtained [17].

Effects of Laser Annealing. Irradiation of a 480 Å W layer on Si with 30 ns laser pulses yielded polycrystalline WSi_2 needles projecting into spherical voids in the Si matrix, indicating that the crystal needles grew from the W surface into molten Si [48]. Continuous wave (CW) laser beams were used to irradiate a 440 Å W film (e-beam evaporated) on single crystal Si at 350°C (the W was covered by ~200 Å Si as an antireflection coating). Silicidation was complete after 10 laser scans [49 to 51], cf. [52]. If an Ar-sputtered W film was used, the reaction was limited, see [50]. The impurity distribution was studied in experiments with Si substrates doped with As and B [51]. A 700 Å (sputter-deposited) W layer on n-type Si(100) (with a 400 Å amorphous Si antireflection film on the W) was IR laser heated at 10^{-2} Torr. The native Si oxide had been removed by presputtering. Tetragonal WSi_2 with a smooth surface was formed; the reaction was shown to be diffusion-controlled, as the growth rate depended on the square root of the number of scans. The activation energy was 260 ± 20 kJ/mol [53], cf. [54].

Effects of Electron-Beam Irradiation. When a 2900 Å thick W film on Si(100) was irradiated with electron-beam pulses (60 ns), a mixture of W_5Si_3 and WSi_2 was obtained, with W_5Si_3 and WSi_2 dominating at lower (2.8 J/cm²) and higher (3.5 J/cm²) fluences, respectively. It is considered that the electron-beam irradiation produces a quasi-liquid phase, from which the silicide phases form on rapid cooling by nucleation and growth from a considerably super-cooled liquid [55]. Electron beams (fluence $\sim 1.0 \times 10^{18}$ electrons/cm²) were used to introduce W into n-type Si(100) at 360°C up to a depth of ~0.3 μm, see [56].

Effects of Pulsed Proton-Beam Irradiation. The application of a 280 keV pulsed proton beam at 3 J/cm² produced very little interaction at the W/Si interface [57].

Influence of Oxygen. For a comprehensive review on the influence of oxygen on the reaction of W (and other refractory metal) films with Si substrates, see [58]. Si is normally covered by a native oxide film; oxygen is introduced into the W layer during film preparation and the annealing process; e.g., the oxygen content of electron-beam evaporated W layers was 6 and 1 at% for deposition at room temperature and 500°C, respectively [17]. Sputtered W usually contains less oxygen than evaporated W [46]. The oxygen tends to accumulate at the W/Si interface during the annealing process, and the resulting Si oxide layer acts as a barrier to the W–Si reaction up to temperatures of 1000 to 1100°C (e.g. [23 to 26, 31, 39, 40, 47, 59, 60]). At ~1100°C the oxide barrier breaks up, see [23 to 25, 31, 47, 60]. The delaying effect of the interfacial oxygen on the WSi_2 formation is ascribed to the fact that the O atoms prevent the Si diffusion in the W layer [10].

A 20 Å SiO_2 film on mono- and polycrystalline Si was found to inhibit W silicide formation up to temperatures of ~1050°C when annealing was carried out in H_2 atmospheres containing up to 15 to 20 ppm H_2O [61]; similarly, no silicide was formed when a 100 nm W layer separated from the Si substrate by a 10 nm SiO_2 film was annealed at 1000°C for 30 min [26].

The oxygen content of sputter-deposited W was reduced by increasing the substrate temperature and the rate of deposition, viz. from ~1 at% (room temperature, rate of deposition ~100 Å/min) to less than 0.2 and 0.03 at% at 200°C and deposition rates of ~15 and ~100 Å/min, respectively [62], cf. [63]. W films, LPCVD prepared with Si as the reducing agent, showed no silicide formation when annealed in He at temperatures up to 900°C, while W films prepared by H_2 reduction (and hence of lower oxygen content) started to react after 30 min annealing at 750°C [40].

The influence of the oxygen content of the Si substrate was studied with W films (LPCVD, 160 to 300 nm) on p-type Si(100), viz. Czochralski crystals (CZ), float zone crystals (FZ), and 5 μm epitaxial Si on CZ (epi) with respective oxygen contents of $\sim 7 \times 10^{17}$, less than 1×10^{16}, and $\sim 2 \times 10^{17}$ atom/cm³. Isochronous and isothermal annealing studies (in 99.9992% pure Ar at 700 to 760°C for 8 to 105 min) showed that the silicide growth rate decreased in the order FZ > epi > CZ, i.e. with increasing oxygen content. Oxygen was found to pile up at the W/WSi$_2$ interface. Oxygen additionally implanted into the W or the Si was found to retard the reaction [39].

Ultrahigh vacuum (UHV) conditions also reduced the oxygen content, viz. by a factor of 4 compared to high vacuum conditions. The reaction temperature was lowered and the silicide growth rate was increased [23]. Additionally, WSi$_2$ may grow with a well-defined epitaxial relationship under UHV conditions at temperatures above the threshold temperature [58].

Capping the W film with thin layers of Si [23] or Al [40] also reduced the oxygen content: the Si prevented gettering of oxygen by the W, while the Al presumably acted as O getter. A 150 nm Al film on a 110 nm W layer (CVD) on As$^+$ implanted Si(100) decreased the necessary annealing temperature from 850 to 500°C [64]; for detailed studies with Al capping, see [10, 11].

The barrier action of Si oxide at the W/Si interface can be inhibited by interface mixing induced by ion implantation [58]. Silicide formation was not prevented in RTA or FRP annealings, as the Si diffuses faster than O in these processes [25].

The doping level (but not the type of doping) of the Si substrate was found to influence the barrier action of oxygen near the W/Si interface; for P- or B-doped Si the temperature of beginning silicide formation was above 850 and 1100°C for 20 and 50 at% interfacial O, respectively. P and B dopants tend to decrease the influence of the oxygen [65].

For the influence of oxygen in sandwich layers, see p. 174.

Studies with a 20 Å thermally generated Si oxide layer interposed between crystalline Si and a 300 Å W layer showed that the stability of the SiO$_2$ diffusion barrier between W and Si was enhanced by O incorporated in the W during deposition and by the presence of H$_2$O in the annealing gas [66].

Influence of Other Impurities and Dopants. For a discussion of the influence of dopants and impurities in Si substrates on the W silicide formation, see [67]. The influence of dopants and doping levels in the Si was studied [6, 11, 65, 68]; for the influence of F and implanted N, see [39] and [17], respectively. The redistribution of dopant and impurity concentrations during annealing processes was investigated [31, 60, 69], for the redistribution of F, see [35, 42].

Influence of Ion Implantation and Ion-Beam Irradiation. As$^+$ implanted Si substrates needed higher temperatures and longer annealing times for silicide formation, while an As$^+$ implantation in the W layer had no noticeable effect on the silicide growth kinetics [27, 70]. Similar results were observed for B$^+$ implantations in the Si substrate or in the W film [71].

Ion-beam mixing (prior to annealing procedures) is used to initiate silicidation by dispersing the interfacial oxide barrier and thus rendering it less effective [72]. Even if high doses are applied, e.g. of As$^+$ ions [73], no silicide formation takes place unless annealing is applied, though intermixing is observed at the W/Si interface. The silicide surfaces formed are smoother and more uniform if ion-beam mixing precedes the annealing. A theoretical treatment of ion-beam mixing for silicide formation and a comparison of the results with experimental data are given [74].

The silicide formation temperature was decreased by ion-beam treatment. The low-temperature, metastable hexagonal WSi$_2$ was found to form at ca. 350°C after the bombardment with As$^+$ ions. At 675°C all the hexagonal WSi$_2$ had changed to the tetragonal WSi$_2$ [75, 76]. It is suggested that As$^+$ irradiation increases the diffusion coefficient of Si in silicides by a factor of $\sim 10^6$ [77]. Similarly, As$^+$ ion irradiation was studied [78, 79]. For BF$_2^+$ ion irradiation, see [80], for studies with As$^-$, P$^+$, B$^+$, Si$^+$, and BF$_2^+$ ions, see [81]. No intermixing was observed at the W/Si interface when a ~300 Å W film on Si was bombarded with Ar+ ions at room temperature [82]. Hexagonal WSi_2 was also formed, when a W film (~55 nm thick) on Si was treated with P+ ions. Some WSi_2 formation was observed prior to furnace annealing. The reaction rate depended on the square root of time. An activation energy of 340 ± 30 kJ/mol was determined [46, 83].

When low energy W ion beams were used to deposit W at 500°C onto cleaned, single crystalline Si(100), a uniform, polycrystalline WSi_2 layer formed [84].

Effect of the Annealing Atmosphere. When 200 Å thick W films (sputter-deposited at 200°C in 20% H_2 – 80% Ar) were annealed in an NH_3 atmosphere for 2 h in the temperature range 500 to 1100°C, no silicide formation could be detected by RBS, XRD, and other methods (only W nitrides were formed) [85, 86]. Negligible W–Si interdiffusion occurred at 1100°C in the case of a 1000 Å W film [86], while tetragonal WSi_2 (covered by W nitride) formed with 2000 Å W films at temperatures above 800°C, the thickness of the WSi_2 layer increased with increasing annealing temperature [85, 86]. Similarly, no W–Si interaction took place when 50 nm W films (sputter-deposited onto unoxidized Si wafers) were annealed in NH_3 at 700°C for 1 h, but WSi_2 was formed at 850°C. Nitrogen was found to accumulate at the W/Si interface. In an Ar atmosphere silicide formation was completed within 60 and 15 min at 700 and 750°C, respectively. WSi_2 was formed also when the annealing was carried out at 850°C in a dilute O_2 atmosphere, but annealing in O_2 at atmospheric pressure completely oxidized the W at temperatures above 500°C, impeding WSi_2 formation [87].

The reaction of W films with p-type Si(100) single crystal substrates during RTA (in the temperature range 500 to 1100°C for 10 to 20 s) in NH_3, N_2, and Ar atmospheres was studied by RBS, AES, XRD, TEM, and resistivity measurements [88, 89]. Annealing for 20 s at 650 to 750°C in Ar and N_2 atmospheres produced intermixing of Si and W, but complete conversion of the W film (85 nm) to tetragonal WSi_2 (215 nm) was achieved at 1000 to 1100°C. Pre-annealing (for 10 s) in NH_3 in the temperature range 600 to 900°C inhibited the silicide formation; the effect increased with increasing pre-annealing temperature. Annealing for 10 s in NH_3 at 600°C followed by annealing for 20 s in Ar at 1000°C produced a layer consisting of W, W_5Si_3, and WSi_2. The inhibiting action of annealing in NH_3 is attributed to the inclusion of nitrogen in the W film and at the W/Si interface. Nitrogen was not incorporated into the W during RTA in N_2. The nitrogen incorporation was accompanied by tunnel defects (>1 μm) at the W/Si interface; see [88, 89]. Mixtures of NH_3 and N_2 were used as atmosphere during RTA (at 870 to 1000°C) of 100 nm W layers on monocrystalline Si(100) and polycrystalline Si wafers. Pure NH_3 prevented WSi_2 formation; ~1.5% NH_3 in N_2 at 1000°C yielded a very thin, inhomogeneous WSi_2 layer, and with less than 1000 ppm NH_3 in N_2 more than 90% of the W was transformed to WSi_2. For the effects of interface contaminations and of doping of the Si, see [90].

An N_2 atmosphere was found to influence silicide formation for prolonged RTA treatments. While W films (75 and 110 nm) on clean Si(100) surfaces rapidly formed WSi_2 when annealed at 1100°C for 5 min in flowing Ar, WSi_2 formation was prevented in N_2; only some W_5Si_3 was formed in electron-beam evaporated W films in flowing N_2 (1100°C, 5 min) [16, 91]. The results are attributed to the reaction barrier of an interfacial nitride or subnitride. RBS results indicate that less than 2×10^{16} N atoms/cm² are required to prevent the W–Si reaction [16]. The concentration of nitrogen at the W/Si interface increased with the annealing time. Pre-annealing in N_2 for 5 to 40 min also retarded silicide formation (W_5Si_3 and WSi_2) during subsequent annealing in Ar [91]. For the influence of plasma nitridation, see [139].

For the influence of an NH_3 atmosphere during the annealing of sandwich structures (Si substrate/Si layer/W layer/Si layer), see [92].

Reactions in Sandwich Structures. (Layer systems of the type Si substrate/Si/W/Si.) The temperature of silicide formation is lowered a) by reacting amorphous W and Si layers free of incorporated oxygen by their preparation in an H_2 atmosphere, and b) by subjecting the W and Si layers to ion implantation prior to the annealing process. The Si top layer prevents the formation of an oxide layer on the W, diffusion is enhanced by a higher concentration of defects in the amorphous layers, and the hydrogen incorporated in the Si layers (Si:H) during preparation and annealing ensures a reducing atmosphere [93].

Detailed studies were carried out with the sandwich structures: crystalline Si substrate/ amorphous Si:H (520 to 540 Å)/sputter-deposited W (100 to 300 Å)/amorphous Si:H (100 to 450 Å), using X-ray diffraction, IR transmission, secondary ion mass spectrometry, and Rutherford backscattering methods. Sputter deposition was carried out in H_2 (20 to 40%)–Ar atmospheres. 180 keV W ions were used for the ion-beam mixing at room temperature, and the sandwich structures were annealed in purified H_2 for 1 h in the temperature range of 350 to 1000°C [92 to 100].

The ion implantation produced amorphous WSi_2 (without annealing), which was completely crystallized at 650°C [94]. No WSi_2 was formed by annealing for 10 h at 650°C, unless ion implantation was applied [93].

Hexagonal WSi_2 was formed after annealing the layers (W ion implanted) for 1 h at 350 or 450°C, while at ~550°C the formation of tetragonal WSi_2 begins [95, 96, 98, 100]. For a higher O content in the W layer (~6 at% instead of ~1 at%), 650°C were necessary to form crystalline WSi_2, though this was still 200 degrees lower than the temperature necessary for unimplanted sandwich structures [96]. For a detailed SIMS study of the concentrations of Si and O across the sandwich during deposition and annealing treatment and a discussion of the mechanism of the reaction, see [99]. When NH_3 was used as the annealing atmosphere, tetragonal WSi_2 formed at 700 to 800°C; at temperatures above 700°C N was incorporated into the silicide film and WN_x and SiN_x compounds were formed. At temperatures above 850°C the WSi_2 decomposed completely [92].

The influence of oxygen was studied with a sandwich structure Si substrate/~540 Å Si:H (containing ~13 at% H, ~1 at% O)/~100 Å W (containing ~1 at% H, ~6 at% O)/~150 Å Si:H (containing ~20 at% H, ~1 at% O). Ion beam mixing was not applied. Oxygen segregation began at annealing (in H_2) temperatures above 750°C, increasing significantly at 850°C when amorphous WSi_2 appeared. Crystalline WSi_2 was formed at 950°C [97].

When "sandwich"-like W/Si layer structures were deposited onto Si substrates at room temperature, some W_5Si_3 and WSi_2 were formed on annealing (30 min) at 1000°C; however, if the layers were deposited at 500°C, hexagonal and tetragonal WSi_2 formed at 575°C; only tetragonal WSi_2 was formed at temperatures above 600°C [63].

Reactions in Multilayers. The interaction at W/Si interfaces in Si/W multilayers (15 to 40 periods of (15 Å W/24 Å Si)pairs) was studied by X-ray reflection and kinetic ellipsometry [101]. Interleaved 5 nm W and 10 nm Si layers (sputter-deposited, nominal thickness 250 nm) yielded tetragonal WSi_2 when annealed in the temperature range 700 to 1100°C for 30 min in ultrapure Ar [102]. For the results of vacuum furnace annealing of W/Si multilayers (288 periods of 3.08 nm W/Si pairs) at 200 to 900°C, see [103]; for furnace and laser annealing of W/Si multilayers (thickness of W/Si bilayer 1.44 to 8.9 nm), see [104].

Low energy ion beam mixing was carried out with W/Si multilayers on polished Si substrates [105]. Electron-beam evaporated layers (50 Å W/3×(250 Å Si/100 Å W)) on p-type Si(100) were irradiated with 180 keV As^+ ions at room temperature or at 350°C, yielding amorphous WSi_2 and (probably metastable) hexagonal WSi_2, respectively. At temperatures of 600°C and above only the stable tetragonal WSi_2 was formed [106, 107].

For a layer system consisting of three different pairs of W and Si layers ion-beam mixing (150 keV As^+) at room temperature reduced the necessary annealing temperature for WSi_2

formation from 1050°C (for the untreated sample) to 950°C. Hexagonal WSi_2 was formed first if the irradiation was carried out at 350°C [108, 109].

Increasing the annealing temperature increased the oxygen diffusion to the surface of a multilayered thin W/Si film. Hexagonal WSi_2 was formed under As^+ ion irradiation at 350°C in samples with a low oxygen content [111].

Layers of WSi_2 were formed by annealing a system of Si substrate/5000 Å thermal Si oxide/4×(200 Å W/525 Å Si) at 950°C (in H_2 for 30 min), but at temperatures above 1000°C uniform WSi_2 was formed [66].

W/Si multilayers (individual thickness less than 0.5 nm) were sputter-deposited onto GaAs and treated by RTA at 900°C. For the nominal compositions $WSi_{0.48}$, $WSi_{0.52}$, $WSi_{0.63}$, and WSi_2, the phases identified were $W_3Si + \alpha$-W, $W_3Si + \alpha$-W, α-W + W_5Si_3, and $WSi_2 + W_5Si_3$, respectively [110].

For a method of producing cosputtered W/Si multilayer films with considerably reduced impurity levels, see [112].

Reactions in Codeposited W–Si Films. Codeposited W–Si films (by CVD, LPCVD, hot pressing (HP), plasma chemical vapor deposition (PD), cosputtering, and coevaporation) react at lower temperatures [22, 113], because of the intimate contact of the elements on an atomic scale.

Codeposition of W and Si (CVD) from SiH_4 and WF_6 onto graphite yielded an amorphous silicide layer if deposited below 300°C. For substrates at 600 and 800°C (and annealing at 900°C for 10 min) W, WSi_2, W_5Si_3, and Si phases can be obtained singly or together, depending on the experimental conditions. WSi_2 coating rates by this method are 20 to 37 times faster than by solid state diffusion at 855°C [114, 115].

W–Si layers were codeposited by LPCVD onto single crystalline Si. On annealing in Ar hexagonal WSi_2 formed at ~500°C, while at 800°C the silicide consisted completely of tetragonal WSi_2; the change of hexagonal $WSi_2 \rightarrow$ tetragonal WSi_2 occurs between 500 and 600°C [116, 117]. Similar results were obtained (by CVD) on oxide covered p-type Si(100) [118]. The formation of hexagonal and/or tetragonal WSi_2 depended on the composition of the film, its thickness, and the annealing temperature, see [119]. Hexagonal WSi_2 formed at an annealing temperature of ~450°C in LPCVD codeposited W–Si films [120].

The kinetics of the two stages, $W + 2Si \rightarrow WSi_2$ (1) and WSi_2 (hexagonal) $\rightarrow WSi_2$ (tetragonal) (2), were studied by isothermal heat treatment and sheet resistance measurements yielding activation energies of ca. 210 kJ/mol and 340 kJ/mol for stages (1) and (2), respectively [121].

For HP and PD formation of W_xSi_{1-x} (x = 0.04 to 0.99) layers on single crystal Si(100) covered by 2000 Å SiO_2, see [122, 123].

When amorphous W–Si films (cosputtered onto oxidized Si) were furnace annealed, hexagonal WSi_2 appeared to crystallize at 300 to 400°C; at ~550°C the hexagonal WSi_2 was transformed into tetragonal WSi_2. At 600°C the rate of this transition showed a square root time dependence; for a suggested mechanism, see [124]. Similar results were obtained by [125] and for coevaporated W–Si films by [126]. A cosputtered $W_{0.85}Si_{0.15}$ amorphous film did not crystallize when annealed below 700°C [127 to 129]. The crystallization temperature of cosputtered amorphous W–Si films was found to depend on the composition of the film. Al and Au overlayers reduced the crystallization temperature by at least 100°C [130]. An Al overlayer on a cosputtered $W_{0.72}Si_{0.28}$ film on Si(111) reduced the crystallization temperature of the film by at least 150°C [131].

WSi$_x$ (x = 2.46 to 2.61) films, cosputtered in Ar onto Si(100) with and without 1000 Å SiO$_2$, contained 7.8×10^{20} and 5.1×10^{20} Ar atoms/cm^3 for Ar pressures of 4 and 20 mTorr, respectively [132].

Furnace annealing and rapid thermal processing (RTP) of cosputtered W–Si films were compared. The RTP produced more readily a stoichiometry of 1:2.1 or 1:2.2 (±0.1) for the annealed film (only WSi$_2$ was formed) [133]. Electron-beam irradiation of cosputtered WSi$_x$ films (~0.25 μm thick) on oxidized Si produced WSi$_2$ [134]. The time necessary to anneal a cosputtered W–Si film by electron-beam irradiation is shortened by about one order of magnitude in comparison to thermal treatment [135].

W–Si films (atomic ratio W:Si = 1:2.1), cosputtered onto oxidized Si, showed no silicide formation when treated by rapid thermal annealing (RTA) for 15 s in vacuum up to temperatures of 700°C. At 800°C mainly W$_5$Si$_3$ was formed. The fraction of WSi$_2$ increased with increased annealing temperature; at 1100°C the film consisted mainly of WSi$_2$ [136]. Furnace annealing in inert atmospheres (Ar, He, H$_2$) at temperatures between 700 and 1100°C produced silicide layers consisting of W$_5$Si$_3$, WSi$_2$, and WSi$_2$ + Si from (electron beam) coevaporated W–Si films of varying compositions [137, 138].

Nucleation of crystalline WSi$_2$ took place uniformly throughout a W–Si film (on a polycrystalline Si layer on oxidized Si(100)) when furnace annealed at temperatures between 400 and 800°C. At annealing temperatures of ~800 to 1000°C small grooves appeared on the WSi$_2$ surface and the WSi$_2$/Si interface became rougher [139].

Cosputtered WSi$_x$ Films on GaAs. WSi$_x$ films (x = 0.57, 0.64, 0.70) cosputtered onto GaAs were annealed at 700 to 900°C. The WSi$_{0.64}$ and the WSi$_{0.70}$ films yielded W$_5$Si$_3$ solid solutions at 800 to 900°C. WSi$_2$ was not detected. Similar results were obtained for wider ranges x = 0.40 to 1.50 [140] and x = 0.5, 0.7, 0.9 [142]. The stresses which arose in the annealed films were discussed [140]. For the crystallization behavior of cosputtered WSi$_x$ (x > 0.25) films on annealing either in flowing H$_2$ or by RTA in an N$_2$ atmosphere, see [143]. RTA at 800 to 850°C of cosputtered WSi$_x$ films (~20 to 50 at% W) on (Al, Ga)As/GaAs heterostructures yielded a crystalline W$_5$Si$_3$ phase [144]. WSi$_{0.45}$ films (250 nm thick) on GaAs remained amorphous during 10 s RTA treatments up to 900°C; WSi$_{0.45}$ films on InGaAs became crystalline (composition: W and W$_5$Si$_3$) when annealed for 10 s at 900°C [145].

Stresses Induced by Silicide Formation. A volume reduction of 27.7% (of the initial volume of W + Si) results during the formation of tetragonal WSi$_2$, giving rise to compressive stresses, which, however, relax at temperatures higher than that necessary for silicide formation. Tensile stresses arise when the silicide is cooled to room temperature because of the relatively large thermal expansion coefficient of WSi$_2$. A tensile stress of 16×10^9 dyn/cm^2 was measured in the WSi$_2$ layer (formed from W films evaporated onto single crystal Si wafers, and annealed for 1 h in an He atmosphere at 1000°C) after cooling to room temperature [146]. The following stresses were measured in 185 nm thick cosputtered WSi$_{1.7}$ films on oxidized Si as deposited and after 30 min annealing at 910°C [147]:

substrate	stress at room temperature	
	as deposited (compressive stress) in Pa	after annealing (tensile stress) in Pa
SiO$_2$/Si	−1.6×10^8	10.3×10^8
polycrystalline Si/SiO$_2$/Si	−0.6×10^8	19.0×10^8

The stresses in WSi$_x$ films (2000 Å thick), cosputtered onto Si substrates at Ar pressures of 4 and 20 mTorr, were 1.6×10^9 to 2.1×10^9 dyne/cm^2 for x = 2.46 to 2.61, respectively, [132].

The absence of oxygen and H_2O in the annealing atmosphere is important, as the formation of oxides leads to compressive stresses [148]. Stresses in silicides also depend on the thickness of the films; for thin films interfacial stresses may dominate, while in thicker films volume generated stresses may prevail [8]. "Buckling" of silicide films was observed at ~1150°C; this was attributed to differences in thermal expansion coefficients [149, 150].

Influence of Ti, TiN, and $TiSi_2$ Interlayers. Studies with Ti layers (5 to 50 nm thick) interposed between W layers (20 to 75 nm thick) and Si(100) substrates, annealed by RTP in vacuum, air, or N_2–H_2 (5%) in the temperature range of 500 to 1000°C, showed that the presence of the Ti layer lowered the formation temperature of WSi_2 (570°C), increased its rate of formation and its surface smoothness, and improved the adhesion of W and WSi_2 films on Si substrates. Si was the diffusing species, and there was no evidence for the formation of W_5Si_3 or hexagonal WSi_2 [151, 152].

TiN films (0.08 μm), containing oxygen as impurity, when interposed between 0.3 μm W films and Si(100) substrates prevented silicide formation even if the sample was annealed for 6 h at 900°C, though some slight diffusion of Ti and Si into the W was observed [153 to 155]. Similarly, TiN layers (500 to 1000 Å thick) interposed between W layers (500 to 2000 Å) and Si prevented WSi_2 formation during annealing for up to 4.75 h at 950°C in an N_2 atmosphere. Only a very small amount of WSi_2 had formed after 6 h; the rate of its formation was 1/25000 and 1/130 of the rates for W/Si and W/TiN/$TiSi_2$/Si structures (under comparable conditions), respectively, [41, 156 to 159]. For the influence of the annealing atmosphere, see [160].

Silicide formation was studied in some detail with interlayers of 500 to 1000 Å TiN + 500 Å $TiSi_2$ between 500 to 2000 Å W layers and Si substrates (B- or As-doped Si(100), or Si covered by 300 to 1000 Å polycrystalline Si layers) and compared with the results for the corresponding W/Si structures [41, 157 to 159, 161]. The kinetics of the growth of the silicide layer (applying for W layers < 2000 Å) were described by $dT_w/dt = 2.5 \times 10^{10} \exp(-2.3\,\text{eV}/kT)$ Å/min where T_w is the thickness (in Å) of the W layer that has reacted, the annealing time in min. The activation energy amounts to 220 kJ/mol; for the corresponding W/Si structure: $dT_w/dt = 1.5 \times 10^{15} \exp(-2.9\,\text{eV}/kT)$ Å/min. Thus the silicidation rate in the W/TiN/$TiSi_2$/Si structure is 2 to 2.5 orders of magnitude lower than for the W/Si structure [41, 159, 161]. X-ray analysis indicated that the Si substrate is the Si source. For a discussion of the mechanism, see [157, 158].

No reaction between W and Si was observed in W/TiN/$TiSi_2$/Si and W(150 to 300 nm)/WSi_x(100 to 150 nm)/Si (x=1.4 and 2.1) structures when annealed at 900°C. Complete silicidation of the W film occurred for the W/WSi_x/Si structures with x=2.5; oxygen was found to segregate at the WSi_x/Si interfaces. The W was silicided rapidly in W/$TiSi_2$/Si structures when these were annealed at temperatures above 700°C (there was no oxygen segregation at the $TiSi_2$/Si interface) [176]. No silicide was formed in W/WSi_2/Si and W/$TiSi_2$/Si structures after annealing at temperatures below 700°C for 30 min [162].

A formation temperature for tetragonal WSi_2 as low as 485°C was claimed for the system p-type Si(100)/W(250 to 1000 Å) (and for Si(100)/Ti: W/Al/Ti/Al) prepared by CVD [163].

Effects of Alloying. WSi_2 formation destroyed the adhesion of W films (1 to 3 μm thick) on Si/Ge thermopiles in the temperature range 550 to 725°C. An activation energy of 300 to 340 kJ/atom was derived for the W silicide formation. Initially the WSi_2 layer increased with time in proportion to t^n with n > 1/2 [164]. Al "capping" of W films on Si significantly lowers the temperature of silicide formation [11, 40]. Alloying W with Pd or Pt also reduced the temperature of W silicide formation, e.g., for $Pd_{80}W_{20}$ and $Pd_{30}W_{70}$ films on Si, WSi_2 formation started at 500 and 650°C, respectively [165, 166]. In the layer system crystalline n-type Si substrate/PtSi/$Ti_{0.1}W_{0.9}$/Al WSi_2 was formed at 550°C and higher temperatures [167]. Annealing of Cr/W bilayer films on Si(100) yielded tetragonal WSi_2 at 600 to 700°C [168]. WSi_2 formed at 800 to

850°C when W/Ta bilayers on n-type Si(100) were annealed in a vacuum furnace for 30 min. No silicide was formed in the case of W/Ta alloy films [169]. For a detailed study of WSi_2 (and $MoSi_2$) formation with Mo/W bilayer films (electron-beam evaporated) on Si single crystal substrates, see [184].

Epitaxy. When W is deposited on the (111), (110), and (100) faces of single crystal Si, oriented growth of silicide layers is observed by electron- and X-ray diffraction. At 1200°C the following epitaxies form: $WSi_2(110)//Si(110)$ substrate, $WSi_2(101)//Si(111)$ [170]. TEM showed epitaxial growth of tetragonal WSi_2 (strongly textured) when thin W films were evaporated onto Si(100) heated to 600 to 800°C: $Si(100)//WSi_2(100)$ (parallel to interface), $Si(011)//WSi_2(001)$ (perpendicular to interface), $Si(01\bar{1})//WSi_2(010)$ (perpendicular to interface). If the W is deposited at room temperature with subsequent annealing, the polycrystalline silicide shows random orientation [171].

Silicide growth for 300 Å W films (electron-beam evaporated or sputtered onto (001), (111) and Si(011) was studied using one-and two-step annealing at 600 to 1100°C in N_2 or vacuum [172, 173]. For annealing in N_2 the reaction started at 800 to 900°C with no epitaxy. Two-step annealing in vacuum at 600 to 1100°C gives the best epitaxial growth of tetragonal and hexagonal WSi_2, the tetragonal WSi_2 always dominating. Sputtered W films produce more hexagonal WSi_2 than electron-beam evaporated W films. Though hexagonal WSi_2 is stable only at relatively low temperatures some epitaxial hexagonal WSi_2 is found at higher temperatures. For details of the different modes of epitaxy, see [172, 173].

During a careful study of the silicide formation on Si(111) with electron-beam evaporated W, an epitaxial hexagonal $WSi_2(0001)$ overlayer developed at annealing temperatures above 750°C [20]. For epitaxial WSi_2 growth on Si(100) at 700 to 750°C, see [17, 18]. For the details of the epitaxial orientations of hexagonal and tetragonal WSi_2 on Si(111), formed at temperatures of 1000 to 1100°C, see [174, 175].

Kinetics. Silicide formation starts at the W/Si interface. Thus any irregularities there considerably affect the mechanism and kinetics of the reaction. An additional reason for the difficulties in getting reproducible kinetic results for the W silicide formation is the native oxide layer (some 10 Å thick) initially present on the Si surfaces, see, e.g. [177] and pp. 171/2. As the mode of the preparation of the W, the condition of the Si substrate, and the method of annealing affect the W/Si interface, these factors also determine the kinetics of the silicide formation.

Mostly, the silicide layer grows linearly with time during the initial (nucleation) stages of the reaction, i.e. for thin silicide films [21, 41, 58, 178 to 182]. Activation energies for this stage are: 270 ± 20 kJ/mol for n+- and p+-doped Si, and 310 ± 40 kJ/mol for n-type Si [21]; 280 kJ/mol [41]. When a continuous WSi_2 layer has formed, the further growth of the silicide layer depends on the square root of the time [58, 178 to 182] indicating that diffusion (of the Si) is the rate limiting process. For this stage the following activation energies (in kJ/mol) were determined: 280 [41]; 330 ± 20 (furnace annealing), 210 ± 20 (laser annealing) [17]; 290 [83]; 330 ± 10 [182]; 340 ± 30 [46]; 250 ± 20 [12, 38]; (obvious misprint in [21]). **Fig. 129** shows the Arrhenius plots for the reaction in different temperature regions; the high-temperature value is in good agreement with the value obtained for the reaction with an Si powder pack with activator [141]. Lower activation energies for higher temperatures have also been found by [179, 183].

For a model for the barrier-controlled reaction stage, see [180].

The results from studies with Mo–W bilayer films on monocrystalline Si are consistent with a process dominated by grain boundary diffusion of Si through the W film [184].

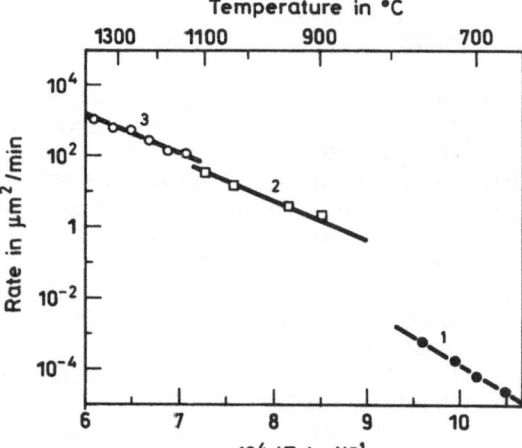

Fig. 129.   Arrhenius plot for WSi$_2$ formation.
1) Low-temperature results (furnace annealing),
2) solid W in Si powder pack [141],
3) high-temperature results (laser annealing); from [182].

References:

[1] Deneuville, A. (Ann. Chim. [Paris] [15] **11** [1986] 603/13).

[2] Murarka, S. P. (Ann. Rev. Mater. Sci. **13** [1983] 117/37; C.A. **99** [1983] No. 114 117).

[3] Pauleau, Y. (Thin Solid Films **122** [1984] 243/58).

[4] Eichinger, P.; Sauermann, H.; Wahl, M. (Ion Beam Surf. Layer Anal. Proc. Int. Conf., Karlsruhe, FRG, 1975 [1976], Vol. 1, pp. 353/62; C.A. **86** [1977] No. 181 578).

[5] Sauermann, H.; Wahl, M. (Valvo Ber. **20** No. 2 [1977] 62/80).

[6] Kamins, T. I.; Laderman, S. S.; Coulman, D. J.; Turner, J. E. (J. Electrochem. Soc. **133** [1986] 1438/42).

[7] Janssen, A. P.; Jones, J. P. (Thin Solid Films **28** [1975] L 25/L 28).

[8] Green, M. L.; Levy, R. A. (J. Electrochem. Soc. **132** [1985] 1243/50).

[9] Pauleau, Y.; Lami, P.; Tissier, A.; Pantel, R.; Oberlin, J. C. (Thin Solid Films **143** [1986] 259/67).

[10] Pauleau, Y.; Dassapa, F. C.; Lami, P.; Oberlin, J. C.; Romagna, F. (J. Vac. Sci. Technol. B **6** [1988] 817/24; C.A. **109** [1988] No. 120 521).

[11] Pauleau, Y.; Dassapa, F. C.; Lami, P.; Oberlin, J. C.; Romagna, F. (J. Mater. Res. **4** [1989] 156/62).

[12] Zhang, S. L.; Buchta, R.; Oestling, M. (Mater. Res. Soc. Symp. Proc. **168** [1990] 173/8; C.A. **113** [1990] No. 101 775).

[13] Oertel, B. (Wiss. Z.-Tech. Hochsch. Ilmenau **25** No. 6 [1979] 151/6; C.A. **91** [1979] No. 150 120).

[14] Oertel, B.; Sperling, R. (Thin Solid Films **37** [1976] 185/94).

[15] Oertel, B.; Kurze, H. J.; Haueisen, H. (Thin Solid Films **52** [1978] 129/35).

[16] Smith, P. M.; Thompson, M. O. (Tungsten Other Adv. Met. VLSI/ULSI Appl. 5, Proc. 6th Workshop, San Mateo, Calif. and Tokyo 1989 [1990], pp. 311/6; C.A. **113** [1990] No. 156 755).

[17] Lajzerowicz, J. (ENST-86E017 [1986], ETN-87-90 290 [1987] 165 pp.; C.A. **108** [1988] No. 196 501).

[18] Cros, A.; Pierrisnard, R.; Pierre, F.; Layet, J. M.; Meyer, F. (Appl. Surf. Sci. **38** [1989] 148/55).

[19] Cros, A.; Pierrisnard, R.; Pierre, F.; Layet, J. M.; Meyer, F. (Appl. Phys. Lett. **55** [1989] 226/8).

[20] Azizan, M.; Nguyen Tan, T. A.; Cinti, R.; Baptist, R.; Chauvet, G. (Surf. Sci. **178** [1986] 17/26).

[21] Locker, L. D.; Capio, C. D. (J. Appl. Phys. **44** [1973] 4366/9).

[22] Murarka, S. P.; Read, M. H.; Doherty, C. J.; Fraser, D. B. (J. Electrochem. Soc. **129** [1982] 293/301).

[23] Santiago-Aviles, J. J. (ARO-22681.8-MS-H [1988] 10 pp.; C.A. **111** [1989] No. 200426).

[24] Siegal, M. P.; Santiago, J. J. (J. Appl. Phys. **63** [1988] 525/9).

[25] Siegal, M.; Santiago, J. J.; Van der Spiegel, J.; Graham, W. R.; Setton, M. (Mater. Res. Soc. Symp. Proc. **74** [1987] 673/8; C.A. **107** [1987] No. 166535).

[26] Ting, C. Y.; Davari, B. (Appl. Surf. Sci. **38** [1989] 416/28).

[27] Torres, J.; Pelissier, A.; Perio, A.; Oberlin, J. C.; Bomchil, G. (Vide Couches Minces **42** No. 236 [1987] 91/3).

[28] Martin, T. L.; Malhotra, V.; Mahan, J. E. (J. Electron. Mater. **13** [1984] 309/25; C.A. **100** [1984] No. 149074).

[29] Meyer, F.; Schwebel, C.; Pellet, C.; Gautherin, G. (Appl. Surf. Sci. **36** [1989] 231/9; C.A. **110** [1989] No. 125839).

[30] Khaidar, M.; Essafti, A.; Bennoura, A.; Ameziane, E. L.; Brunel, M. (J. Appl. Phys. **65** [1989] 3248/52).

[31] Siegal, M. P.; Santiago, J. J. (J. Appl. Phys. **65** [1989] 760/6).

[32] Paine, D. C.; Bravman, J. C.; Saraswat, K. (Tungsten Other Refract. Met. VLSI Appl. Proc. Workshop, Albuquerque 1984/1985 [1986], pp. 117/23; C.A. **106** [1987] No. 144664).

[33] Shioya, Y.; Ikegami, K.; Maeda, M.; Yanagida, K. (J. Appl. Phys. **61** [1987] 561/6).

[34] Delfino, M.; Lehrer, W. I. (J. Electrochem. Soc. **128** [1981] 1071/4).

[35] Whitlow, H. J.; Eriksson, T.; Östling, M.; Petersson, C. S.; Keinonen, J.; Anttila, A. (Appl. Phys. Lett. **50** [1987] 1497/9).

[36] Shenai, K.; Lewis, N.; Smith, G. A.; McConnell, M. D.; Burrell, M. (Tungsten Other Adv. Met. VLSI/ULSI Appl. 5 Proc. 6th Workshop, San Mateo, Calif. and Tokyo 1989 [1990], pp. 317/24; C.A. **113** [1990] No. 156756).

[37] Carlisle, J. A.; Chopra, D. R.; Dillingham, T. R.; Gnade, B.; Smith, G. (J. Appl. Phys. **65** [1989] 2313/20).

[38] Zhang, S. L.; Buchta, R.; Oestling, M. (J. Mater. Res. **6** [1991] 1886/91).

[39] Zhang, S. L.; Smith, U.; Buchta, R.; Oestling, M. (J. Appl. Phys. **69** [1991] 213/9).

[40] Thomas, O.; Charai, A.; D'Heurle, F. M.; Finstad, T. G.; Joshi, R. V. (Thin Solid Films **171** [1989] 343/57).

[41] Suguro, K.; Nakasaki, Y.; Shima, S.; Yoshi, T.; Moriya, T.; Tango, H. (J. Appl. Phys. **62** [1987] 1265/73).

[42] Eriksson, T.; Carlsson, J. O.; Keinonen, J.; Petersson, C. S. (J. Appl. Phys. **64** [1988] 3229/32).

[43] Schmitz, J. E. J.; Buiting, M. J.; Ellwanger, R. C. (Tungsten Other Refract. Met. VLSI Appl. 4 Proc. Workshop, Albuquerque 1988 [1989], pp. 27/33; C.A. **111** [1989] No. 199720).

[44] Morosanu, C.-E.; Soltuz, V. (Thin Solid Films **52** [1978] 181/94).

[45] Van der Jeugd, C. A.; Janssen, G. C. A. M.; Radelaar, S.; Buiting, M. J.; Wolters, R. A. M. (Tungsten Other Adv. Met. ULSI Appl. 1990 Proc. 7th Workshop, Dallas 1990 [1991], pp. 137/41; C.A. **115** [1991] No. 140465).

[46] Ma, E.; Alvi, N. S.; Hamdi, A. H.; Nicolet, M. A. (Mater. Res. Soc. Symp. Proc. **122** [1988] 579/84; C.A. **110** [1989] No. 126555).

[47] Siegal, M. P.; Graham, W. R.; Santiago, J. J. (J. Appl. Phys. **66** [1989] 6073/6).

[48] Von Allmen, M.; Lau, S. S.; Sheng, T. T.; Wittmers, M. (Laser Electron Beam Process. Mater. Proc. Symp., Cambridge, Mass., 1979 [1980], pp. 524/9; C.A. **93** [1980] No. 159996).

[49] Shibata, T.; Sigmon, T. W.; Gibbons, J. F. (Proc. Electrochem. Soc. **80**-2 [1980] 533 pp., 458/68).

[50] Shibata, T.; Sigmon, T. W.; Regolini, J. L.; Gibbons, J. F. (J. Electrochem. Soc. **128** [1981] 637/44).

[51] Wakita, A. S.; Sigmon, T. W.; Gibbons, J. F. (Mater. Res. Soc. Symp. Proc. **13** [1983] 721/6).

[52] Shibata, T.; Wakita, A.; Sigmon, T. W.; Gibbons, J. F. (Semicond. Semimet. **17** [1984] 341/95).

[53] Lee, H. S.; Wolga, G. J. (J. Electrochem. Soc. **137** [1990] 684/90).

[54] Lee, H. S.; Wolga, G. J. (J. Electrochem. Soc. **137** [1990] 2618/23).

[55] Majni, G.; Nava, F.; Ottaviani, G.; Luches, A.; Nassisi, V.; Celotti, G. (Vacuum **32** [1982] 11/8).

[56] Wada, T.; Takeda, M. (Nagoya Kogyo Daigaku Gakuho **36** [1985] 113/25; C.A. **104** [1986] No. 160090).

[57] Baglin, J. E. E.; Hodgson, R. T.; Chu, W. K.; Neri, J. M.; Hammer, D. A.; Chen, L. J. (Nucl. Instrum. Methods Phys. Res. **191** [1981] 169/76; C.A. **96** [1982] No. 95843).

[58] Bomchil, G.; Goeltz, G.; Torres, J. (Thin Solid Films **140** [1986] 59/70).

[59] Haarsta, A.; Carlsson, J. O. (Thin Solid Films **176** [1989] 263/76).

[60] Siegal, M. P.; Santiago, J. J. (Mater. Res. Soc. Symp. Proc. **100** [1988] 701/6; C.A. **109** [1988] No. 140180).

[61] Silversmith, D. J.; Rathman, D. D.; Mountain, R. W. (Thin Solid Films **93** [1982] 413/9).

[62] Deneuville, A.; Kadri, M.; Burnel, M. (Vide Couches Minces **42** No. 236 [1987] 33/5; C.A. **107** [1987] No. 87802).

[63] Goltz, G.; Torres, J.; Lajzerowicz, J., Jr.; Bomchil, G. (Thin Solid Films **124** [1985] 19/26).

[64] Pauleau, Y.; Lami, P.; Dassapa, F.; Romagna, F.; Oberlin, J. C. (Vide Couches Minces **42** No. 236 [1987] 163/5; C.A. **107** [1987] No. 166563).

[65] Blewer, R. S.; Tracy, M. E. (Tungsten Other Refract. Met. VLSI Appl. Proc. Workshop, Albuquerque 1984/1985 [1986], pp. 53/62; C.A. **106** [1987] No. 166969).

[66] Silversmith, D. J.; Rathman, D. D.; Mountain, R. W. (Mater. Res. Soc. Symp. Proc. **10** [1982] 425/31; C.A. **98** [1983] No. 208290).

[67] Vasil'ev, S. V.; Gerasimenko, N. N. (Poverkhnost **1986** No. 7, pp. 57/62; C.A. **105** [1986] No. 144865).

[68] Arena, C.; Papapietro, M. (Tungsten Other Refract. Met. VLSI Appl. Proc. Worhshop, Albuquerque 1984/1985 [1986], pp. 483/95; C.A. **106** [1987] No. 147647).

[69] Gerasimenko, N. N.; Vasil'ev, S. V.; Kalinin, V. V. (Nucl. Instrum. Methods Phys. Res. B **39** [1989] 259/67; C.A. **111** [1989] No. 32197).

[70] Torres, J.; Oberlin, J. C.; Bomchil, G.; Perio, A.; Levy, D.; Saulnier, A.; Ponpon, J. P.; Stuck, R. (in: Soncini, G.; Calzolari, P. U.; Solid State Devices, Elsevier, North Holland, 1988, pp. 357/61).

[71] Torres, J.; Oberlin, J. C.; Stuck, R.; Bourhila, N.; Palleau, J.; Goltz, G.; Bomchil, G. (Appl. Surf. Sci. **38** [1989] 186/94).

[72] Maex, K.; Van den Hove, L.; DeKeersmaecker, R. F. (Thin Solid Films **140** [1986] 149/61).

[73] Hara, T.; Chen, S. C.; Ando, H. (J. Electrochem. Soc. **134** [1987] 3139/42).

[74] Agamy, S. A.; Khalil, M. Y.; Badawi, A. A. (Isotopenpraxis **26** No. 6 [1990] 265/8; C.A. **113** [1990] No. 143127).

[75] D'Heurle, F. M.; Petersson, C. S.; Tsai, M. Y. (J. Appl. Phys. **51** [1980] 5976/80).

[76] Tsai, M. Y.; Petersson, C. S.; D'Heurle, F. M.; Maniscalco, V. (Appl. Phys. Lett. **37** [1980] 295/8).

[77] D'Heurle, F. M.; Tsai, M. Y.; Petersson, C. S.; Stritzler, B. (J. Appl. Phys. **53** [1982] 3067/9).

[78] Kwong, D. L.; Meyers, D. C.; Alvi, N. S.; Li, L. W.; Norbeck, E. (Appl. Phys. Lett. **47** [1985] 688/91).

[79] Saraswat, K. C.; Swirhun, S.; McVittie, J. P. (Proc. Electrochem. Soc. **84**-7 [1984] 409/19; C.A. **101** [1984] No. 121173).

[80] Mimura, H.; Inada, T. (Rep. Res. Cent. Ion Beam Technol. Hosei Univ. Suppl. No. 8 [1990] 97/102; C.A. **113** [1990] No. 32573).

[81] Beale, M. I. J.; Deshmukh, W. G. I; Chew, N. G.; Cullis, A. G. (Physica B **129** [1985] 210/4).

[82] Kurup, M. B.; Bhagawat, A.; Prasad, K. G. (Nucl. Instrum. Methods Phys. Res. B **13** [1986] 473/8; C.A. **105** [1986] No. 47222).

[83] Ma, E.; Lim, B. S.; Nicolet, M. A.; Alvi, N. S.; Hamdi, A. H. (J. Electron. Mater. **17** No. 3 [1988] 207/11; C.A. **109** [1988] No. 102597).

[84] Zuhr, R. A.; Pennycook, S. J.; Haynes, T. E.; Holland, O. W. (Mater. Res. Soc. Symp. Proc. **128** [1989] 47/53; C.A. **111** [1989] No. 107063).

[85] Deneuville, A.; Benyahya, M.; Brunel, M.; Canut, B. (J. Phys. Colloq. [Paris] **49** [1988] C4-499/C4-502; C.A. **110** [1989] No. 16787).

[86] Deneuville, A.; Benyahya, M.; Brunel, M.; Oberlin, J. C.; Torres, J.; Bourhila, N.; Paleau, J.; Canut, B. (Appl. Surf. Sci. **38** [1989] 139/47).

[87] Torres, J.; Palleau, J.; Bourhila, N.; Oberlin, J. C.; Deneuville, A.; Benyahya, M. (J. Phys. Colloq. [Paris] **49** [1988] C4-183/C4-186; C.A. **109** [1988] No. 241887).

[88] Broadbent, E. K.; Morgan, A. E.; Flanner, J. M.; Coulman, B.; Sadana, D. K.; Burrow, B. J.; Ellwanger, R. C. (J. Appl. Phys. **64** [1988] 6721/6).

[89] Broadbent, E. K.; Morgan, A. E.; Ellwanger, R. C.; Flanner, J. M.; Coulman, B.; Sadana, D. K.; Burrow, B. J.; Gutai, L. (Tungsten Other Refract. Met. VLSI Appl. 4 Proc. Workshop, Albuquerque 1988 [1989], pp. 259/68; C.A. **111** [1989] No. 144779).

[90] Bakli, M.; Goltz, G.; Vernet, M.; Torres, J.; Palleau, J.; Bourhila, N.; Oberlin, J. C. (Appl. Surf. Sci. **38** [1989] 441/6).

[91] Smith, P. M.; Thompson, M. O.; Blewer, R. S. (Tungsten Other Adv. Met. ULSI Appl. 1990 Proc. 7th Workshop, Dallas 1990 [1991], pp. 297/303; C.A. **115** [1991] No. 54658).

[92] Deneuville, A.; Benyahya, M.; Brunel, M.; Canut, B. (J. Less-Common Met. **145** [1988] 197/207).

[93] Deneuville, A.; Mandeville, P. (Fr. Demande 2578272 [1985/1986] 14 pp.; C.A. **106** [1987] No. 77081).

[94] Kadri, M.; Deneuville, A.; Brunel, M. (Adv. Mater. Telecommun. Pap. Symp. 13 Eur. Mater. Res. Soc. Meet., Strasbourg 1986, pp. 497/501; C.A. **108** [1988] No. 67648).

[95] Deneuville, A.; Kadri, M.; Brunel, M. (Vide Couches Minces **42** No. 236 [1987] 37/9; C.A. **107** [1987] No. 87843).

[96] Kadri, M.; Deneuville, A.; Brunel, M. (Proc. Conf. Ion Plat. Allied Tech. **1987** 366/70).

[97] Deneuville, A.; Kadri, M.; Brunel, M. (Proc. Int. Symp. Trends New Appl. Thin Films, Strasbourg 1987, pp. 126/30).

[98] Canut, J. B.; Kadri, M.; Pivot, J.; Deneuville, A. (Proc. Eur. Mater. Res. Symp. **1987** 491/6).

[99] Dubois, C.; Deneuville, A.; Kadri, M. (Second. Ion Mass. Spectrom. Proc. 6th Int. Conf., Versailles 1987 [1988], pp. 777/82; C.A. **111** [1989] No. 15924).

[100] Kadri, M.; Deneuville, A.; Brunel, M. (Mater. Sci. Eng. **98** [1988] 79/84; C.A. **109** [1988] No. 97298).

[101] Houdy, P.; Boher, P.; Schiller, C.; Luzeau, P.; Barchewitz, R.; Alehyane, N.; Ouahabi, M. (Proc. SPIE-Int. Soc. Opt. Eng. **984** [1988] 95/103; C.A. **110** [1989] No. 163221).

[102] Magee, T. J.; Woolhouse, G. R.; Kawayoshi, H. A.; Niemeyer, I. C.; Rodrigues, B.; Ormond, R. D.; Bhandia, A. S. (J. Vac. Sci. Technol. [2] B **2** [1984] 756/61; C.A. **102** [1985] No. 37353).

[103] Jiang, Z.; Jiang, X.; Liu, W.; Wu, Z. (J. Appl. Phys. **65** [1989] 196/200).

[104] Dupuis, V.; Ravet, M. F.; Tete, C.; Piecuch, M.; Vidal, B. (J. Appl. Phys. **68** [1990] 3348/55).

[105] King, B. V.; Puranik, S. G.; MacDonald, R. J. (Nucl. Instrum. Methods Phys. Res. B **33** [1988] 657/60; C.A. **109** [1988] No. 220466).

[106] Tsaur, B.-Y.; Anderson, C. H., Jr. (Appl. Phys. Lett. **41** [1982] 877/9).

[107] Tsaur, B.-Y.; Anderson, C. H., Jr. (Thin Solid Films **104** [1983] 383/9).

[108] Wang, Z.; Ding, X.; Tian, Y. (Nucl. Instrum. Methods Phys. Res. B **33** [1988] 653/6; C.A. **109** [1988] No. 221405).

[109] Ding, X.; Wang, Z.; Qian, Y. (Vacuum **39** [1989] 243/5; C.A. **111** [1989] No. 88152).

[110] Eicher, S.; Bruce, R. A. (Can. J. Phys. **65** [1987] 868/71; C.A. **108** [1988] No. 196504).

[111] Ding, X.; Wang, Z.; Qian, Y. (Hejishu **10** No. 9 [1987] 15/9 from C.A. **108** [1988] No. 114775).

[112] Choi, K. W.; Ahn, K. Y. (J. Vac. Sci. Technol. [2] A **3** [1985] 2272/7).

[113] Blokha, V. B.; Gladkikh, N. T.; Glushko, P. I.; Zmii, V. I.; Kartmazov, G. N.; Poltavtsev, N. S. (Izv. Akad. Nauk SSSR Neorg. Mater. **18** [1982] 805/8; Inorg. Mater. [Engl. Transl.] **18** [1982] 677/80).

[114] Lo, J.-S. (Diss. Univ. Utah, Salt Lake City, Utah, 1973, 98 pp.; C.A. **80** [1974] No. 61575).

[115] Lo, J.-S.; Haskell, R. W.; Byrne, J. G.; Sosin, A. (Chem. Vap. Deposition 4th Int. Conf., Boston 1973, pp. 74/90; C.A. **81** [1974] No. 160230).

[116] Saraswat, K. C.; Brors, D. L.; Fair, J. A.; Monnig, K. A.; Beyers, R. (IEEE Trans. Electron Devices **30** [1983] 1497/505; C.A. **100** [1984] No. 60149).

[117] Fang, Y. K.; Hsu, S. L. (J. Appl. Phys. **57** [1985] 2980/2).

[118] Togei, R. (J. Appl. Phys. **59** [1986] 3582/4).

[119] Shioya, Y.; Maeda, M. (J. Appl. Phys. **60** [1986] 327/33).

[120] Le Goues, F. K.; D'Heurle, F. M.; Joshi, R.; Suni, I. (Mater. Res. Soc. Symp. Proc. **54** [1986] 51/6; C.A. **105** [1986] No. 88869).

[121] Nava, F.; Riantino, G.; Galli, E. (J. Non-Cryst. Solids **104** [1988] 195/202).

[122] Akimoto, K.; Watinabe, K. (Appl. Phys. Lett. **39** [1981] 445/7).

[123] Akimoto, K. (Appl. Phys. Lett. **41** [1982] 49/51).

[124] Murarka, S. P.; Read, M. H.; Chang, C. C. (J. Appl. Phys. **52** [1981] 7450/2).

[125] Jiang, X.; Huang, Y.; Wu, Z.; Li, B. (Chin. Phys. Lett. **3** [1986] 569/72; C.A. **106** [1987] No. 187008).

[126] Tsai, M. Y.; D'Heurle, F. M.; Petersson, C. S.; Johnson, R. W. (J. Appl. Phys. **52** [1981] 5350/5).

[127] Guo, K. J.; Wiley, J. D.; Perepezko, J. H.; Nordman, J. E.; Aaron, D. B.; Dobisz, E. A.; Madisen, D. E.; Thomas, R. E. (Proc. Conf. High Temp. Electron. Instrumentation, Houston, Tex., 1981, pp. 137/46).

[128] Wiley, J. D.; Perepezko, J. H.; Nordman, J. E.; Guo, K.-J. (IEEE Trans. Ind. Electron. **29** No. 2 [1982] 154/7).

[129] Wiley, J. D.; Perepezko, J. H.; Nordman, J. E. (SAND-82-7156 [1983] 37 pp.; C.A. **99** [1983] No. 162550).

[130] Thomas, R. E.; Perepezko, J. H.; Wiley, J. D. (Appl. Surf. Sci. **26** [1986] 534/41).

[131] Thomas, R. E.; Perepezko, J. H.; Wiley, J. D. (Mater. Res. Soc. Symp. Proc. **54** [1986] 127/32; C.A. **105** [1986] No. 163890).

[132] Hara, T.; Takahashi, S. (Nucl. Instrum. Methods Phys. Res. B **39** [1989] 302/5; C.A. **111** [1989] No. 32162).

[133] Sridhar, C. G.; Hodul, D. T.; Chow, R. (Tungsten Other Refract. Met. VLSI Appl. Proc. Workshop, Albuquerque 1984/1985 [1986], pp. 267/80; C.A. **106** [1987] No. 147645).

[134] Du, Y. C.; Wang, H.; Huang, W. N.; Sun, D. C.; Yu, Z. Q.; Li, F. M. (Tungsten Other Refract. Met. VLSI Appl. 4 Proc. Workshop, Albuquerque 1988 [1989], pp. 307/13; C.A. **111** [1989] No. 202812).

[135] Du, Y. C.; Lu, Z.; Yu, Z. Q.; Sun, D. C.; Li, F. M.; Collins, G. J. (Chin. Phys. New York **6** No. 1 [1986] 208/14).

[136] Chen. C.; Zhou, H.; Cao, M.; Jiarg, H.; Xu, W. (Tungsten Other Refract. Met. VLSI Appl. 4 Proc. Workshop, Albuquerque 1988 [1989], pp. 363/7; C.A. **111** [1989] No. 164896).

[137] Crowder, B. L.; Zirinsky, S. (Eur. Appl. 0317 [1979] 31 pp.; C.A. **91** [1979] No. 203172).

[138] Ahn, K. Y.; Herd, S. R.; Baglin, J. E. E.; Han, J. U. (J. Vac. Sci. Technol. [2] A **3** [1985] 2268/71).

[139] Lu, J.; Wu, Z. (Wuli Xuebao **38** [1989] 981/6; C.A. **111** [1989] No. 245432).

[140] Ohnishi, T.; Yokoyama, N.; Onodera, H.; Suzuki, S.; Shibatomi, A. (Appl. Phys. Lett. **43** [1983] 600/2).

[141] Gage, P. R.; Bartlett, R. W. (Trans. Metall. Soc. AIME **233** [1965] 832/4).

[142] Takatani, S.; Matsuoka, N.; Shigeta, J.; Hashimoto, N.; Nakashima, H. (J. Appl. Phys. **61** [1987] 220/4).

[143] Lahav, A. G.; Wu, C. S.; Baiocchi, F. A. (J. Vac. Sci. Technol. [2] B **6** [1988] 1785/95).

[144] Cirillo, N. C., Jr.; Chung, H. K.; Vold, P. J.; Hibbs-Brenner, M. K.; Fraasch, A. M. (J. Vac. Sci. Technol. [2] B **3** [1985] 1680/4).

[145] Lahav, A.; Genut, M. (Mater. Sci. Eng. B **7** [1990] 231/5; C.A. **114** [1991] No. 73055).

[146] Angilello, J.; Baglin, J.; D'Heurle, F.; Petersson, S.; Segmüller, A. (Proc.-Electrochem. Soc. **80**-2 [1980] 369/84; C.A. **94** [1981] No. 200987).

[147] Pan, P. J.; Blech, I. (J. Appl. Phys. **55** [1984] 2874/80).

[148] McLachlan, D. R.; Avins, J. B. (Semicond. Int. **7** [1984] 129/38).

[149] Santiago-Aviles, J. J. (ARO-22681.8-MS-H [AD-A 203428] [1988] 10 pp.; C.A. **111** [1989] No. 200426).

[150] Siegal, M. P.; Santiago, J. J. (J. Appl. Phys. **63** [1988] 525/9).

[151] Wei, C. S.; Van der Spiegel, J.; Setton, M.; Santiago, J.; Tanielian, M.; Blackstone, S. (Mater. Res. Soc. Symp. Proc. **52** [1986] 297/303; C.A. **105** [1986] No. 126002).

[152] Wei, C. S.; Setton, M.; Van der Spiegel, J.; Santiago, J. (J. Appl. Phys. **61** [1987] 1429/34).

[153] Mitsuhashi, K.; Yamazaki, O.; Ohtake, K.; Koba, M. (Jpn. J. Appl. Phys. Pt. 2 **27** [1988] L2401/L2403; C.A. **110** [1989] No. 145718).

[154] Mitsuhashi, K.; Shiozaki, K.; Yamazaki, O.; Ohtake, K.; Koba, M. (VLSI Technol. Symp., San Diego, Calif., 1988, pp. 71/2).

[155] Mitsuhashi, K.; Yamazaki, O.; Ohtake, K.; Koba, M. (Jpn. J. Appl. Phys. Pt. 1 **28** [1989] 593/7; C.A. **111** [1989] No. 164963).

[156] Nakasaki, Y.; Suguro, K.; Shima, S.; Kashiwagi, M. (J. Appl. Phys. **64** [1988] 3263/8).

[157] Suguro, K.; Nakasaki, Y.; Inoue, T.; Shima, S.; Kashiwagi, M. (Thin Solid Films **166** [1988] 1/14).

[158] Suguro, K.; Nakasaki, Y.; Yoshii, T.; Itoh, T. (Appl. Surf. Sci. **41/42** [1989] 277/81).

[159] Suguro, K.; Nakasaki, Y.; Katata, T. (Oyo Butsuri **59** [1990] 1474/8; C.A. **115** [1991] No. 219967).

[160] Suguro, K.; Katata, T.; Nakasaki, Y.; Kunishima, I. (Tungsten Other Adv. Met. VLSI/ULSI Appl. 5 Proc. 6th Workshop, San Mateo, Calif. and Tokyo 1989 [1990], pp. 195/200; C.A. **113** [1990] No. 235606).

[161] Suguro, K.; Nakasaki, Y.; Shima, S.; Yoshii, T.; Moriya, T.; Tango, H. (Ext. Abstr. Conf. Solid State Devices Mater. 18th [1986] 503/6; C.A. **106** [1987] No. 42155).

[162] Yasui, Y.; Ogawa, S.; Yoshida, T.; Terui, Y. (Ext. Abstr. 19th Conf. Solid State Devices Mater., Tokyo 1987, pp. 419/22).

[163] Shen, B. W.; Smith, G. C.; Anthony, J. M.; Matyi, R. J. (J. Vac. Sci. Technol. [2] B **4** [1986] 1369/76).

[164] Sweet, J. N. (SLA-73-1087 [1974] 45 pp.; C.A. **82** [1975] No. 33273).

[165] Olowolafe, J. O.; Tu, K. N.; Angilello, J. (J. Appl. Phys. **50** [1979] 6316/20).

[166] Eizenberg, M.; Tu. K. N. (J. Appl. Phys. **53** [1982] 1577/85).

[167] Canali, C.; Celotti, G.; Fantini, F.; Zanoni, E. (Thin Solid Films **88** [1982] 9/23).

[168] Thomas, O.; Finstad, T. G.; D'Heurle, F. M. (J. Appl. Phys. **67** [1990] 2410/4).

[169] Appelbaum, A.; Eizenberg, M.; Brener, R. (Vacuum **33** No. 4 [1983] 227/30).

[170] Hashimoto, N.; Koga, Y. (J. Electrochem. Soc. **114** [1967] 1189/91).

[171] Torres, J.; Perio, A.; Pantel, R.; Campidelli, Y.; Arnaud D'Avitaya, F. (Thin Solid Films **126** [1986] 233/9).

[172] Lin, W. T.; Chen, L. J. (J. Appl. Phys. **58** [1985] 1515/8).

[173] Lin, W. T.; Chen, L. J. (J. Appl. Phys. **59** [1986] 3481/8).

[174] Cheng, H. C.; Lin, W. T.; Chien, C. J.; Shiau, F. Y.; Chen, L. J. (Appl. Surf. Sci. **22/23** [1985] 512/9).

[175] Chen, L. J.; Cheng, H. C.; Lin, W. T.; Chou, L. J.; Fung, M. S. (Mater. Res. Soc. Symp. Proc. **37** [1985] 375/80).

[176] Ogawa, S.; Yamazaki, K.; Akiyama, S.; Terui, Y. (Tech. Dig. Int. Electron Devices **1986** 62/5).

[177] D'Heurle, F. M.; Gas, P. (J. Mater. Res. **1** [1986] 205/21; C.A. **104** [1986] No. 213864).

[178] Borders, J. A.; Picraux, S. T. (Proc. IEEE **62** [1974] 1224/31; C.A. **82** [1975] No. 118362).

[179] Borders, J. A.; Sweet, J. N. (Appl. Ion Beams Met. Int. Conf., Albuquerque 1973 [1974], pp. 179/91; C.A. **82** [1975] No. 103761).

[180] Chang, C. C.; Quintana, G. (J. Electron Spectrosc. Relat. Phenom. **2** [1973] 363/76).

[181] Ottaviani, G. (Thin Solid Films **140** [1986] 3/21).

[182] Lajzerowicz, J., Jr.; Torres, J.; Goltz, G.; Pantel, R. (Thin Solid Films **140** [1986] 23/8).

[183] Petersson, C. S.; Baglin, J. E. E.; D'Heurle, F. M.; Dempsey, J. J.; Harper, J. M. E.; Serrano, C. M.; Tsai, M. Y. (Proc.-Electrochem. Soc. 80-2 [1980] 290/310; C.A. **94** [1981] No. 200985).

[184] Baglin, J.; Dempsey, J.; Hammer, W.; D'Heurle, F.; Petersson, S.; Serrano, C. (J. Electron. Mater. **8** [1979] 641/61; C.A. **91** [1979] No. 132191).

[185] Hara, T.; Ohtsuka, N.; Ishizawa, Y. (Rep. Res. Cent. Ion Beam Technol. Hosei Univ. Suppl. No. 4 [1985] 125/30; C.A. **102** [1985] No. 191969).

[186] Siegal, M.; Santiago, J. J.; Van der Spiegel, J. (Mater. Res. Soc. Symp. Proc. **52** [1986] 289/95; C.A. **105** [1986] No. 125873).

[187] Bayerl, P.; Eichinger, P. (RS 03/75, Institut für Festkörpertechnologie, München 1975, 14 pp.).

[188] Pantel, R.; Campidelli, Y.; Arnaud D'Avitaya, F. (J. Electrochem. Soc. **131** [1984] 2426/30).

[189] Badoz, P. A.; Bakli, M.; Berenguer, M.; Morin, C.; Oberlin, J. C.; Palleau, J.; Pantel, R.; Perret, P.; Straboni, A.; Torres, J.; Vuillermoz, B. (Proc. 2nd Int. Conf. Electron Mater., Newark, N.J., 1990, pp. 527/32; C.A. **115** [1991] No. 219971).

9.2.3 Gas Phase Reactions

The cluster formation of laser vaporized W and Si atoms (from W carbonyl and Si wafers, respectively, [1]) in a supersonic jet expansion apparatus was studied by vacuum UV photoionization and mass spectrometry. The main clusters found were of the type WSi_n, especially WSi_{15} and WSi_{16} [1, 2]. Various Si clusters and smaller amounts of other W–Si clusters were also indicated. The clusters were found to be stable at 300 K, independently of whether they were neutral or positively charged; cluster cations, WSi_n^+, with n >17, were not stable. It was inferred from the results that one W atom can interact strongly with up to 16 Si atoms, and thus at W/Si interfaces, W atoms (diffusing into the Si lattice) may disrupt Si lattice bonds and thereby enhance the diffusion of Si atoms through the lattice of the W film [2].

References:

[1] Beck, S. M. (J. Chem. Phys. **87** [1987] 4233/4).
[2] Beck, S. M. (J. Chem. Phys. **90** [1989] 6306/12).

9.3 Diffusion

Diffusion of Tungsten in Silicon

There is agreement that the diffusion of W in bulk Si under a neutral or reducing atmosphere is very small at least up to temperatures of ~ 900 to 1000°C [1 to 4]. However, marker experiments with a W/Mo bilayer film on Si(100) wafers allowed to detect considerable diffusion of W into Si at temperatures as low as 800°C [2]. Tungsten was deposited at 77 K up to 300 K onto p-type Si(100) wafers by pyrolysis or photolysis of WF_6 and the status of the metallic layer was studied by AES and thermal desorption spectrometry (TDS) up to 1500 K. Above 400 K the W diffused quite rapidly into the Si with formation of WSi_2 [5]. The diffusion parameters for W diffusion in Si were determined for 0.5 to 2 µm W layers, evaporated under high vacuum onto the etched surfaces of n- and p-type monocrystalline Si. The diffusion annealing was carried out under Ar in quartz ampoules in the temperature range 1223 to 1523 K for 1 to 48 h. For p-type Si, the diffusion coefficient ranged from 5×10^{-13} cm²/s at 1223 K to 2×10^{-11} cm²/s at 1473 K, the temperature dependence being given by $D = 2.3 \times 10^{-6} \exp(-2.12\,eV/kT)$ cm²/s (activation energy 205 kJ/mol). The values for n-type Si practically agreed with those for the p-type Si, for details, see [6].

W atoms were introduced as "impurity" into p-type Czochralski Si wafers coated with WO_x by thermal diffusion at 1000°C in a pure N_2 atmosphere for up to 4 h; this was followed by an additional annealing at 1100°C for 1 h. The results of SIMS studies showed that the major part of the W diffused slowly (thus concentrating just beneath the surface) while a smaller part diffused very rapidly into the crystal. The two types of diffusion are attributed to a site conversion mechanism and an interstitial diffusion mechanism, respectively [7].

A layer system Si(≥100 Å)/W(10 Å)/Si(600 Å) (prepared by e-gun deposition) was irradiated with 300 keV Xe ions at 80 and 300 K and the amount of redistribution ("marker broadening") was determined by backscattering spectrometry. From the observations with W, Au, and Pt in Si a correlation was found between the measured marker broadenings under ion irradiation and the thermally activated diffusivities. Comparative measurements at 300, 600, and 900°C showed that W was virtually immobile in Si up to at least 900°C [8].

Diffusion of Silicon in Tungsten

Very little or no diffusion of Si in W is observed below a certain temperature; the values for this temperature range from ~600 to ~1000°C (the differences probably being due to the varying experimental conditions):

minimum temperature (in °C) for observation of Si diffusion	experimental details	Ref.
~600	B- and P-doped Si; 540 to 3000 Å W films, sputter-deposited	[9]
~625	B- and P-doped Si; ~2200 Å W film, sputter-deposited	[10]
	fine-grained W substrate, thin Si film, e-beam evaporated	[11]
≥650	heavily doped Si–Ge alloy substrate, 1000 to 1300 Å W film, sputter-deposited	[12]
	monocrystalline, p-type Si, ~100 nm W film, CVD	[13]
700	W films deposited by CVD	[14]
700 to 800	monocrystalline Si, Mo/W bilayer film	[3, 15]
870	W wire, Si powder pack	[16]

AES studies showed that Si diffusion in W occurred at 950°C (500 to 8000 Å CVD W films on Si) [4]. FEM measurements with Si films on W emitter tips indicate that bulk diffusion of Si into W is improbable at temperatures ≤1050 K [17]. Work function measurements (Si film on W emitter tip) showed that Si diffusion into W occurs at temperatures >600 K [18]. Part of the Si evaporated onto a single crystal W emitter tip diffused into the W bulk when the tip was heated to temperatures at which Si surface diffusion takes place [19, 20].

According to calculations using the layer growth data of [10], a diffusion coefficient of 10^{-14} cm²/s should characterize silicide formation at 560°C [21].

Mechanism. Experimental evidence indicates that Si diffuses mainly along grain boundaries in polycrystalline W, see [3, 13, 22]. Though grain boundary diffusion dominates the diffusion processes at the lower temperatures, bulk effects may be more important at higher temperatures [36]. Thin SiO_2 films between W and Si prevent Si diffusion into W [23].

Activation Energies. AES, TEM, and XRD methods were used to study the diffusion of Si through W films (thickness 540 to 3000 Å) on doped Si substrates. The "out diffusion" of the Si atoms along the grain boundaries of the W film to the surface of the W film was found to follow nucleation growth kinetics, with activation energies of 250, 270, and 310 kJ/mol for p-, p+-, and n+-Si substrates, respectively. The diffusion rate decreases in that order. Interface oxide and/or doping may influence the out-diffusion kinetics considerably, see [9]. An activation energy of ~100 kJ/mol for Si diffusion in polycrystalline W films was assumed [15]. Studies of the evaporation of Si from silicided W ribbons yielded an activation energy of 322 kJ/mol in the temperature range 1500 to 2000°C for the diffusion of Si in W [24].

Reactive diffusion of Si into W to form a silicide layer is described below.

Tungsten-Silicon Interdiffusion

AES studies of the deposition of Si (by e-beam vaporization) onto atomically clean, fine-grained W showed some interdiffusion at ~ 600 to 625°C, but significant interdiffusion at 750 to 780°C [11]. Thin oxide layers hinder the interdiffusion at the Si/W interface when Si is evaporated onto W(110) surfaces at room temperature under ultrahigh vacuum (according to photo-emission, AES, and work function measurements) [26]. Diffusion coefficients were calculated for the interdiffusion between W layers (sputter-deposited) and Si substrates with regard to the applied thermal annealing, using SIMS measurements [27]. Preliminary results on the interdif-fusion between W layers und Si substrates at 600 to 800°C are indicated in [28]. Interdiffusion of W and Si readily occurs during pyrolytic deposition of W on Si [5]. Ion-beam induced atomic mixing in W/Si multilayer structures was studied by SIMS in the temperature range 80 to 775 K. Distinct changes in the depth profiles at 675 and 775 K (disappearing when the sample was returned to room temperature) are attributed to radiation enhanced diffusion effects [29].

Diffusion in the Silicide Layer

Marker experiments with W rods in an Si powder pack (containing NaF as activator) in the temperature range of 850 to 1100°C showed that there was practically hardly any diffusion of W in the silicide layer; thus the reaction between W and Si is determined mainly by the movement of Si atoms in the system and most of the experimental information pertains to the diffusion of Si in the silicide layer [30, 31].

When very pure (99.9999%) molten Si reacted with W at 1430°C, X-ray microanalysis studies yielded the following values for the diffusion coefficient of Si: D_{Si}(in WSi_2)$=1.9 \times 10^{-8}$ cm²/s, D_{Si}(in W_5Si_3)$=1.8 \times 10^{-8}$ cm²/s [32]. According to [24] (Si films on W ribbons), D_{Si} (in W_5Si_3)$\approx 10^4 D_{Si}$(in W) in the temperature range 1400 to 1800°C. The relatively high diffusion coefficients for layers in comparison to bulk material may be due to a higher density of lattice defects in layers [33]. The growth of Si crystallites on W silicide layers when P is coevaporated with Si onto the W suggests that P retards the diffusion of Si in WSi_2 [34].

Activation energies for Si diffusion in the silicide layer:

temperture range in °C	activation energy in kJ/mol	silicide concerned and experimental details	Ref.
1000 to 1700	340	W_5Si_3; Si films on W ribbons	[24]
1355 to 1870	360	W_5Si_3; W rods in Si powder pack + NaF activator	[30, 35]
1200 to 1350	88	WSi_2 bulk; solid contact between single crystals	[38]
	300	WSi_2 near-surface layer; solid contact	[38]
1200 to 1350	42	WSi_2 bulk; W cylinders or plates, Si powder + 2.5 wt% NaCl activator*)	[41]
	110	WSi_2 near-surface layer; solid W, Si powder + 2.5 wt% NaCl activator*)	[41]
1150 to 1300	76.6	WSi_2 bulk; liquid Cu between solid W and Si	[44]
	42	WSi_2 near-surface layer; liquid Cu between solid W and Si	[44]

*) Activation energies for additions of 0.2 and 1.5 wt% NaCl are given in [41].

The reactive diffusion in the system W–Si was treated in detail [37 to 43]. The theoretical treatment for vacuum siliconization of solid W takes into account surface phenomena, heterogeneous reactions, diffusion of components, and movement of interface boundaries. The kinetic parameters were determined for the reaction of solid W with high-purity Si single crystals or Si powder at ~10^{-6} Torr in the temperature range of 1200 to 1350°C by studying the kinetics of Si evaporation from the diffusion layer. The diffusion coefficient in the subsurface layer of WSi_2 was found to be smaller than that in the remainder of the silicide layer; the composition of the silicide layer varied from $WSi_{2.24}$ to $WSi_{1.84}$ with the depth of the layer [37, 38]. The vacuum reactions of solid W (99.98% pure) with Si powder including additions of 0.2, 1.5, and 2.5 wt% NaCl [39, 41], and the reaction between plates of polycrystalline W (99.98% pure) and plates of monocrystalline Si (99.999% pure) with liquid Cu in between in the temperature range 1150 to 1300°C [44] were treated in a similar manner. The following diffusion coefficients (all in cm²/s) were reported:

a) reaction of solid W with single crystal Si in vacuum [37]:
   diffusion coefficient of Si in WSi_2: 1.04×10^{-6} at 1200°C and 3.82×10^{-6} at 1350°C;
   "effective" diffusion coefficient: 0.79×10^{-6} at 1200°C and 2.93×10^{-6} at 1350°C;

b) reaction of solid W with single crystal Si or high-purity Si powder [38, 41]:
   diffusion coefficient of Si in the near-surface layer D′:
   6.4×10^{-10} at 1200°C, 59.4×10^{-10} at 1350°C, $D' = 25.1 \times 10^{-5} \exp(-36\,000/T)$;
   diffusion coefficient of Si in the bulk silicide D:
   6.5×10^{-8} at 1200°C, 12.3×10^{-8} at 1350°C, $D = 7.9 \times 10^{-5} \exp(-10\,500/T)$;

c) reaction of solid W with high-purity Si powder activated by NaCl [41]:

NaCl wt%	Si diffusion in bulk silicide	Si diffusion in near-surface layer
0.2	$D = 10^{-5} \exp(-7050/T)$	$D' = 34 \times 10^{-5} \exp(-17\,600/T)$
1.5	$D = 1.5 \times 10^{-5} \exp(-6150/T)$	$D' = 14.3 \times 10^{-5} \exp(-15\,500/T)$
2.5	$D = 2.3 \times 10^{-5} \exp(-5000/T)$	$D' = 11.3 \times 10^{-5} \exp(13\,000/T)$

d) reaction between plates of W(99.98% purity) and Si(99.999% purity) with intermediate liquid Cu [44]:
   $D = 4.6 \times 10^{-5} \exp(-9135/T)$, $D' = 9.7 \times 10^{-7} \exp(-5000/T)$

An activation energy as low as 24 kJ/mol was reported for activated reactive diffusion in case that bulk W cylinders were impregnated by a solid phase bath composed of Si powder and the activator (not specified) [25].

The structure and composition of 200 to 250 μm thick WSi_2 diffusion layers (preparation as above, [37, 38]) were studied by metallographic and X-ray structural analyses [40, 42]. Structural imperfections (and microstresses) are concentrated near the outer and inner boundaries of the silicide layer; the bulk of the silicide layer shows a recrystallized texture in the ⟨110⟩ direction. The effect of the texture on the mechanism of the Si diffusion is discussed in the paper [40]. The concentration profiles of Si and W were determined in the silicide layer formed when solid W in contact with an Si single crystal was annealed in vacuum at 1570 K for 20 h. In comparison to the stoichiometric composition of WSi_2 the W concentration was too low mainly in the near-surface layer, while the Si concentration varied more uniformly across the silicide layer (decreasing from the outer towards the inner boundary) [42].

The growth of the silicide layer was described by a four-stage kinetic model. Effective diffusion coefficients $\bar{D}$ (in cm²/s) at a selected temperature of 1200°C and appropiate activation energies E (in kJ/mol) were as follows [43]: 1) initial diffusion-controlled stage:

$\overline{D} = 5.1 \times 10^{-6}$, $E = 13.4$; 2) kinetic stage: $\overline{D} = 6.7 \times 10^{-7}$, $E = 41$; 3) (second) diffusion-controlled stage: $\overline{D} = 5.3 \times 10^{-8}$, $E = 84$; 4) diffusion in bulk WSi_2: $\overline{D} = 6.4 \times 10^{-10}$, $E = 301$ (the latter data correspond to D'_{Si} for diffusion in the near surface layer [38, 41], see above).

References:

[1] Shaw, J. M.; Amick, J. A. (RCA Rev. **31** [1970] 306/16).

[2] Baglin, J.; D'Heurle, F.; Petersson, S. (Appl. Phys. Lett. **33** [1978] 289/90).

[3] Baglin, J.; Dempsey, J.; Hammer, W.; D'Heurle, F.; Petersson, S.; Serrano, C. (J. Electron. Mater. **8** [1979] 641/61).

[4] Diem, M.; Fisk, M.; Goldman, J. (Thin Solid Films **107** [1983] 39/43).

[5] Foord, J. S.; Jackman, R. B. (J. Opt. Soc. Am. B: Opt. Phys. **3** [1986] 806/11).

[6] Makhmudov, K.; Kakharov, S. S.; Mamadalimov, A. T. (Izv. Akad. Nauk UzSSR Ser. Fiz.-Mat. Nauk **1984** No. 2, pp. 71/2; C.A. **101** [1984] No. 82319).

[7] Fujisaki, Y.; Ando, T.; Kozuka, H.; Takano, Y. (J. Appl. Phys. **63** [1988] 2304/6).

[8] Barcz, A. J.; Paine, B. M.; Nicolet, M.-A. (Appl. Phys. Lett. **44** [1984] 45/7).

[9] Chang, C. C.; Quintana, G. (J. Electron. Spectrosc. Relat. Phenom. **2** [1973] 363/76).

[10] Borders, J. A.; Sweet, J. N. (Appl. Ion Beams Met. Int. Conf., Albuquerque 1973 [1974] pp. 179/91; C.A. **82** [1975] No. 103761).

[11] Bevolo, A. J.; Schmidt, F. A.; Shanks, H. R.; Campisi, G. J. (J. Vac. Sci. Technol. **16** [1979] 13/9).

[12] Borders, J. A.; Sweet, J. N. (J. Appl. Phys. **43** [1972] 3803/8).

[13] Pauleau, Y.; Lami, P.; Tissier, A.; Pantel, R.; Oberlin, J. C. (Thin Solid Films **143** [1986] 259/67).

[14] Bayerl, P.; Eichinger, P. (RS-03/75, Institut f. Festkörpertechnologie München 1975).

[15] Baglin, J. E. E.; D'Heurle, F. M.; Hammer, W. N.; Petersson, S. (Nucl. Instrum. Methods **168** [1980] 491/7).

[16] Goetzel, C. G.; Landler, P. (WADD-TR-60-825 [AD-258574] [1960] 1/42; N.S.A. **15** [1961] No. 23986).

[17] Garifullin, N. M.; Zubenko, Yu. V. (Vestn. Leningr. Univ. Ser. 4 Fiz. Khim. **1976** No. 2, pp. 59/66; C.A. **85** [1976] No. 135521).

[18] Janssen, A. P.; Jones, J. P. (Surf. Sci. **41** [1974] 257/76).

[19] Sinha, M. K.; Swenson, O. F. (Appl. Phys. Lett. **19** [1971] 493/4).

[20] Sinha, M. K.; Swenson, O. F.; Venkatachalam, G. (Surf. Sci. **33** [1972] 414/8).

[21] Borders, J. A.; Picraux, S. T. (Proc. IEEE **62** [1974] 1224/31; C.A. **82** [1975] No. 118362).

[22] Petersson, C. S.; Baglin, J. E. E.; D'Heurle, F. M.; Dempsey, J. J.; Harper, J. M. E.; Serrano, C. M.; Tsai, M. Y. (Proc. Electrochem. Soc. **80**-2 [1980] 290/310; C.A. **94** [1981] No. 200985).

[23] Silversmith, D. J.; Rathman, D. D.; Mountain, R. W. (Thin Solid Films **93** [1982] 413/9).

[24] Gelain, C.; Cassuto, A.; Le Goff, P. (Bull. Soc. Fr. Ceram. **80** [1968] 23/7).

[25] Samsonov, G. V.; Solonnikova, L. A. (Fiz. Met. Metalloved. **5** [1957] 565/6; Phys. Met. Metallogr. [Engl. Transl.] **5** No. 3 [1957] 177; C.A. **1958** 19804).

[26] Weng, S. L. (Phys. Rev. B: Condens. Matter [3] **29** [1984] 2363/5).

[27] Mitani, E.; Shichi, H.; Yamamoto, N. (Shitsuryo Bunseki **32** [1984] 315/8 from C.A. **102** [1985] No. 53443).

[28] Zirinsky, S.; Hammer, W.; D'Heurle, F.; Baglin, J. (Appl. Phys. Lett. **33** [1978] 76/8).

[29] Tonn, D. G.; Sankey, O. F.; Tsong, I. S. T. (Nucl. Instrum. Methods Phys. Res. B **15** [1986] 193/7; C.A. **104** [1986] No. 234599).

[30] Bartlett, R. W.; Gage, P. R. (ASD-TDR-63-753-Pt. II [1964] 136 pp., 1/48, 124/7; N.S.A. **19** [1965] No. 7848).

[31] Gage, P. R.; Bartlett, R. W. (Trans. Metall Soc. AIME **233** [1965] 832/4).

[32] Kostikov, V. I.; Levin, N. P.; Levin, V. Ya. (Izv. Akad. Nauk SSSR Neorg. Mater. **5** [1969] 152/4; Inorg. Mater. [Engl. Transl.] **5** [1969] 123/6).

[33] Oertel, B.; Sperling, R. (Thin Solid Films **37** [1976] 185/94).

[34] Campisi, G. J.; Bevolo, A. J.; Shanks, H. R.; Schmidt, F. A. (J. Appl. Phys. **53** [1982] 1714/9).

[35] Bartlett, R. W.; Gage, P. R.; Larssen, P. A. (Trans. Metall. Soc. AIME **230** [1964] 1528/34).

[36] Mayer, J. W.; Tu, K. N. (J. Vac. Sci. Technol. **11** [1974] 86/93).

[37] Zmii, V. I.; Seryugina, A. S. (Zashch. Pokrytiya Met. No. 2 [1968] 195/201; Prot. Coat. Met. [Engl. Transl.] **2** [1970] 158/63).

[38] Zmii, V. I. (Zashch. Pokrytiya Met.) No. 11 [1977] 14/8; C.A. **88** [1978] No. 10746).

[39] Zmii, V. I.; Glushko, P. I.; Trofimov, V. F. (Izv. Akad. Nauk SSSR Neorg. Mater. **13** [1977] 1896/7; Inorg. Mater. [Engl. Transl.] **13** [1977] 1525/6; C.A. **88** [1978] No. 27921).

[40] Poltavtsev, N. S.; Zmii, V. I.; Snezhko, I. A. (Izv. Akad Nauk SSSR Neorg. Mater. **16** [1980] 674/7; Inorg. Mater. [Engl. Transl.] **16** [1980] 464/6).

[41] Zmii, V. I.; Glushko, P. I.; Trofimov, V. F. (Izv. Akad. Nauk SSSR Neorg. Mater. **17** [1981] 644/6; Inorg. Mater. [Engl. Transl.] **17** [1981] 427/9).

[42] Zmii, V. I.; Kartmazov, G. N.; Poltavtsev, N. S.; Semenov, N. A. (Izv. Akad. Nauk SSSR Neorg. Mater. **17** [1981] 916/7; Inorg. Mater. [Engl. Transl.] **17** [1981] 654/5).

[43] Zmii, V. I. (Fiz. Khim. Obrab. Mater. **1986** No. 3, pp. 96/101; C.A. **105** [1986] No. 83544).

[44] Zmii, V. I.; Kovtun, N. V.; Matyukhina, L. G. (Poverkhnost **1989** No. 8, pp. 148/53; C.A. **111** [1989] No. 199747).

10   Reactions with Phosphorus

Phase Diagram

Apparently, there are only two stable intermediate phases in the W–P system at <1100°C and low and medium pressures: WP [1 to 12] and WP$_2$ [1, 5, 8, 9, 13, 14]. The existence of phases W$_2$P [2, 15 to 18] and W$_3$P$_4$ [15] could not be confirmed. Phosphides with P contents lower than that of WP evidently are not obtainable by direct synthesis at temperatures below 1100°C. Alloys in this composition range always consisted of WP and W [3, 8]. W$_2$P only exists in the form of solid solutions with other metal phosphides, e.g., Co$_2$P and Ni$_2$P [19, 20]. A thermally unstable and "amorphous" product of the composition W$_4$P was obtained by electrolysis of phosphate melts [17]. Rapidly arc-melted samples contained a phase W$_3$P, which disappeared on annealing at 800 to 1000°C [9]. It seemed to be identical with a phase of the same composition which was obtained by electrolysis of a fused sodium metaphosphate bath containing WO$_3$ and NaCl at temperatures below 900°C. This phase decomposed on heating at 900°C [21]. The existence of phosphides higher in P content than WP$_2$ seemed highly improbable, for excess P over this composition remained unreacted [1, 8]. Under a pressure of

3 GPa, however, a phase WP_4 formed at 1000°C in 1 h. WP_4 is metastable under normal pressure conditions [22].

WP and WP_2 both have only a small range of homogeneity or are even exactly stoichiometric [8, 9, 23]. WP_2 forms a monoclinic low-temperature [9, 12] and an orthorhombic high-temperature [8, 9] modification, depending on whether it is prepared below or above ~700°C [9].

For reviews of the W–P system, see [24 to 28].

Reaction Characteristics

For first studies of the reaction of W with P, see [15, 29, 30].

Tungsten does not markedly change under phosphorus vapor pressures (~15 to 30 atm) corresponding to a temperature of 550°C [14]. At 400 to 700°C under phosphorus (P_4) vapor pressures of 0.05 to 1 atm, there was no change in electric conductivity of tungsten wires over test periods of 100 h. At 1200°C, the conductivity decreased by 10 to 15% within 2 h, and a surface layer was formed [31]. On heating massive tungsten samples in P vapor for 72 h at 900 $\pm$30°C, a WP layer of 4.0 μm thickness was obtained [32]. The phosphidation of W by P vapor of 1 atm (101.3 kPa) at 800, 900, and 1000°C in a sealed quartz tube was studied by measuring the weight increase. The weight increased with time according to a $\sqrt{\tau}$-rate law. The temperature dependence of the rate constant was given by the relation k_p (in $kg^2 \cdot m^{-4} \cdot s^{-1}$) = 3.06×10^{-5} exp[−103 000/RT] (1073 to 1273 k, 1 atm); the activation energy is in J/mol. X-ray diffraction patterns and electron probe microanalyses showed that the phosphide films formed consisted of two layers, i.e., an outer layer of WP_2 and an inner layer of WP [28, 33].

For the preparation of WP, powder mixtures of W and red P are reacted in sealed tubes for 1 h to 1 month at 550 to 1100°C [1], [3, p. 226]. The products are black powders; single crystals could not be obtained [8]. WP is prepared by a procedure in which the P content of the charges is raised in three stages: (I) W : P = 1 : 1.3, 48 h at 970°C; (II) W : P = 1 : 2.3, 110 h at 825°C; and (III) W : P = 1 : 2.9, 150 h at 700°C [1]; see also [34].

The electronic stopping cross-section of W for (channeled) ^{32}P ions with 370 keV incident energy (velocity 1.5×10^8 cm/s) is $(6.4 \pm 0.4) \times 10^{-14}$ $eV \cdot cm^2 \cdot atom^{-1}$ in the $\langle 100 \rangle$ direction and $(7.8 \pm 0.4) \times 10^{-14}$ $eV \cdot cm^2 \cdot atom^{-1}$ in the $\langle 110 \rangle$ direction [35].

Diffusion

Using ^{32}P as tracer, the diffusion of P in W single crystals was measured at 1880 to 2180°C. The diffusion time was 48 h. The diffusion coefficients D were:

t in °C	1880	1980	2080	2180
D in 10^{-11} cm²/s	0.89 to 2.14	4.34 to 7.03	12.3 to 19	40.5 to 50

The temperature dependence of D was determined by the parameters E_{diff} = 122 $\pm$ 7 kcal/mol and $D_0 = (27^{+107}_{-21})$ cm²/s. Upon dissolution of phosphorus no new phases could be detected. The lattice constant of the W host lattice decreased by 0.0007 Å. For the formation of an interstitial solid solution an increase in the lattice constant would be expected [36].

References:

[1] Faller, F. E.; Biltz, W.; Meisel, K.; Zumbusch, M. (Z. Anorg. Allg. Chem. **248** [1941] 209/28, 209/15).

[2] Chêne, M. (Ann. Chim. [Paris] [11] **15** [1941] 187/285, 259/64).

[3] Schönberg, N. (Acta Chem. Scand. **8** [1954] 226/39).

[4] Defacqz, E. (C. R. Hebd. Seances Acad. Sci. **132** [1901] 32/5).
[5] Defacqz, E. (Ann. Chim. Phys. [7] **22** [1901] 238/88, 269/77).
[6] Bachmayer, K.; Nowotny, H.; Kohl, A. (Monatsh. Chem. **86** [1955] 39/43).
[7] Rundqvist, S. (Acta Chem. Scand. **16** [1962] 287/92).
[8] Rundqvist, S.; Lundström, T. (Acta Chem. Scand. **17** [1963] 37/46).
[9] Rundqvist, S. (Nature **211** [1966] 847/8).
[10] Gingerich, K. A. (NYO-2541-1 [1964] 20 pp. from [26]; N.S.A. **19** [1965] No. 2802).

[11] Ripley, R. L. (J. Less-Common Met. **4** [1962] 496/503).
[12] Hulliger, F. (Nature **204** [1964] 775).
[13] Defacqz, E. (C. R. Hebd. Seances Acad. Sci. **130** [1900] 915/7).
[14] Heinerth, E.; Biltz, W. (Z. Anorg. Allg. Chem. **198** [1931] 168/77).
[15] Wöhler, F.; Wright, H. (Justus Liebigs Ann. Chem. **79** [1851] 244/7).
[16] Hartmann, H.; Ebert, F.; Bretschneider, O. (Z. Anorg. Allg. Chem. **198** [1931] 116/40, 121, 125).
[17] Hartmann, H.; Orban, J. (Z. Anorg. Allg. Chem. **226** [1936] 257/64).
[18] Andrieux, J. L. (Rev. Metall. [Paris] **45** [1948] 49/59).
[19] Orishchin, S. V.; Kuz'ma, Yu. B. (Poroshkov. Metall. [Kiev] **1982** No. 5, pp. 59/61; Sov. Powder Metall. Met. Ceram. [Engl. Transl.] **21** [1982] 395/7).
[20] Guerin, R.; Sergent, M. (Mater. Res. Bull. **12** [1977] 381/8).

[21] Hsu, S. S.; Yocom, P. N.; Cheng, T. T. C. (Interim. Rep. Office of Naval Research, Department of the Navy (Univ. of Illinois, Urbana 1955, Contract N 6-ori-071 (50), from [9, 23]).
[22] Kinomura, N.; Terao, K.; Kikkawa, S.; Koizumi, M. (J. Solid State Chem. **48** [1983] 306/7).
[23] Gingerich, K. A. (J. Phys. Chem. **68** [1964] 768/72).
[24] Hansen, M.; Anderko, K. (Constitution of Binary Alloys, McGraw-Hill, New York 1958, p. 1093).
[25] Elliott, R. P. (Constitution of Binary Alloys, 1st Suppl., McGraw-Hill, New York 1965, p. 719).
[26] Shunk, F. A. (Constitution of Binary Alloys, 2nd Suppl., McGraw-Hill, New York 1969, p. 600).
[27] Kornilov, I. I.; Matveeva, N. M.; Pryakhina, L. I.; Polyakova, R. S. (Metallokhimicheskie Svoistva Elementov Periodicheskoi Systemy [Metal-Chemical Properties of the Elements of the Periodical System], Moscow 1966, pp. 179/82; C.A. **68** [1968] No. 16365).
[28] Sasaki, Y.; Ishimura, K.; Matagawa, S. (Z. Anorg. Allg. Chem. **561** [1988] 185/91).
[29] Pelletier, M. (Ann. Chim. Phys. [1] **13** [1972] 121/43, 137).
[30] Moissan, H. (C. R. Hebd. Seances Acad. Sci. **123** [1896] 13/6).

[31] Boosz, H. J. (Metall [Berlin] **21** [1967] 28/30).
[32] Witte, J. H.; Wilhelm, H. A. (IS-854 [TID-4500] [1964] 1/37; C.A. **62** [1965] 7470).
[33] Sasaki, Y.; Ishimura, K.; Matagawa, S.; et al. (Denki Kagaku oyobi Kogyo Butsuri Kagaku **55** [1987] 857/8 from C.A. **108** [1988] No. 42266).
[34] Samsonov, G. V.; Vereikina, L. P. (Fosfidy [Phophides], Izv. Akad. Nauk SSSR, Kiev 1961, pp. 48/55, 119/27).
[35] Eriksson, L.; Davies, J. A.; Jespersgård, P. (Phys. Rev. [2] **161** [1967] 219/34).
[36] Novikov, V. P.; Panov, A. S.; Ryabenko, A. V. (Izv. Akad. Nauk SSSR Met. **1978** No. 1, pp. 78/9; Russ. Metall. [Engl. Transl.] **1978** No. 1, pp. 68/9).

11 Reactions with Arsenic

The intermediate phases W_2As [1] and WAs_2 [1 to 5] have been definitely identified in the W–As system. A phase analogous to Mo_5As_4 does not exist in the binary system, but appears in the W–Ti–As system at the approximate composition $(W_{0.6}Ti_{0.4})_5As_4$. The occurrence of a phase W_4As_5 is suggested by a rather complex X-ray pattern. W_2As always contained free tungsten [1]. A monoarsenide WAs was reported [6]. The existence of a phase W_2As_3 has been postulated for the W–Ga–As system [7]. WAs_2 produced from the elements at 620°C in a sealed tube (5 h, slight excess of As) had the composition $WAs_{1.954}$ (66.15 at% (44.3 wt%) As) [4]. It decomposes readily at low temperatures [1]. For reviews of the system, see [8 to 10].

Tungsten arsenides were generally obtained from powder mixtures of the elements in sealed tubes at 600 to 1100°C [1, 4]. For the preparation of alloys rich in As, some iodine was added as a "mineralizer" [1].

Concentration profiles of As ions implanted with 40 to 160 keV at intensities of 5×10^{15} cm^{-2} into W layers on Si were determined using Rutherford backscattering and secondary ion mass spectrometry. Significant differences were noted between sputtered and chemical vapor-deposited W layers [11].

References:

[1] Boller, H.; Nowotny, H. (Monatsh. Chem. **95** [1964] 1272/82).
[2] Defacqz, E. (C. R. Hebd. Seances Acad. Sci. **132** [1901] 138/40).
[3] Defacqz, E. (Ann. Chim. Phys. [7] **22** [1901] 238/88, 277/9).
[4] Heinerth, E.; Biltz, W. (Z. Anorg. Allg. Chem. **198** [1931] 168/77).
[5] Hulliger, F. (Nature **204** [1964] 775).
[6] Weiss, G. (Diss. Grenoble 1946 from [12]).
[7] Schmid-Fetzer, R. (J. Electron. Mater. **17** No. 2 [1988] 193/200).
[8] Hansen, M.; Anderko, K. (Constitution of Binary Alloys, McGraw-Hill, New York 1958, p. 185).
[9] Shunk, F. A. (Constitution of Binary Alloys, 2nd Suppl., McGraw-Hill, New York 1969, p. 64).
[10] Kornilov, I. I.; Matveeva, N. M.; Pryakhina, L. I.; Polyakova, R. S. (Metallokhimicheskie Svoistva Elementov Periodicheskoi Systemy [Metal-Chemical Properties of the Elements of the Periodical System], Moscow 1966, pp. 179/82; C.A. **68** [1968] No. 16365).

[11] Hara, T.; Chen S.-C.; Ando, H. (J. Electrochem. Soc. **134** [1987] 3139/42).
[12] Andrieux, J. L. (Rev. Metall. [Paris] **45** [1948] 49/59).

Physical Constants and Conversion Factors

Avogadro constant N_A (or L) = 6.02214×10^{23} mol^{-1}		Planck constant	h = 6.62608×10^{-34} J·s	
Faraday constant	F = 9.64853×10^4 C/mol	elementary charge	e = 1.60218×10^{-19} C	
molar gas constant	R = 8.31451 J·mol^{-1}·K^{-1}	electron mass	m_e = 9.10939×10^{-31} kg	
molar volume (ideal gas)	V_m = 2.24141×10^1 L/mol	proton mass	m_p = 1.67262×10^{-27} kg	

(273.15 K, 101 325 Pa)

1 kg = 2.205 pounds

1 m = 3.937×10^1 inches = 3.281 feet

1 m^3 = 2.642×10^2 gallons (U.S.)

1 m^3 = 2.200×10^2 gallons (Imperial)

Force	N	dyn	kp
1 N	1	10^5	1.019716×10^{-1}
1 dyn	10^{-5}	1	1.019716×10^{-6}
1 kp	9.80665	9.80665×10^5	1

Pressure	Pa	bar	kp/m^2	at	atm	Torr	lb/in^2
1 Pa = 1 N/m^2	1	10^{-5}	1.019716×10^{-1}	1.019716×10^{-5}	9.86923×10^{-6}	7.50062×10^{-3}	1.450378×10^{-4}
1 bar = 10^6 dyn/cm^2	10^5	1	1.019716×10^4	1.019716	9.86923×10^{-1}	7.50062×10^2	1.450378×10^1
1 kp/m^2 = 1 mm H$_2$O	9.80665	9.80665×10^{-5}	1	10^{-4}	9.67841×10^{-5}	7.35559×10^{-2}	1.422335×10^{-3}
1 at (technical)	9.80665×10^4	9.80665×10^{-1}	10^4	1	9.67841×10^{-1}	7.35559×10^2	1.422335×10^1
1 atm = 760 Torr	1.01325×10^5	1.01325	1.033227×10^4	1.033227	1	7.60×10^2	1.469595×10^1
1 Torr = 1 mm Hg	1.333224×10^2	1.333224×10^{-3}	1.359510×10^1	1.359510×10^{-3}	1.315789×10^{-3}	1	1.933678×10^{-2}
1 lb/in^2 = 1 psi	6.89476×10^3	6.89476×10^{-2}	7.03069×10^2	7.03069×10^{-2}	6.80460×10^{-2}	5.17149×10^1	1

Work, Energy, Heat	J	kW·h	kcal	Btu	eV
1 J = 1 W·s = 1 N·m = 10^7 erg	1	2.778×10^{-7}	2.39006×10^{-4}	9.4781×10^{-4}	6.242×10^{18}
1 kW·h	3.6×10^6	1	8.604×10^2	3.41214×10^3	2.247×10^{25}
1 kcal	4.1840×10^3	1.1622×10^{-3}	1	3.96566	2.6117×10^{22}
1 Btu (British thermal unit)	1.05506×10^3	2.93071×10^{-4}	2.5164×10^{-1}	1	6.5858×10^{21}
1 eV	1.602×10^{-19}	4.450×10^{-26}	3.8289×10^{-23}	1.51840×10^{-22}	1

$1\,cm^{-1} \triangleq 1.239842 \times 10^{-4}$ eV $\qquad$ $1\,Hz \triangleq 4.135669 \times 10^{-15}$ eV

2 rydberg = 1 hartree = 27.2114 eV $\qquad$ $1\,eV \triangleq 96.485$ kJ/mol

Power	kW	hp	kp·m·s^{-1}	kcal/s
1 kW = 10^3 J/s	1	1.35962	1.01972×10^2	2.39006×10^{-1}
1 hp (horsepower, metric)	7.3550×10^{-1}	1	7.5×10^1	1.7579×10^{-1}
1 kp·m·s^{-1}	9.80665×10^{-3}	1.333×10^{-2}	1	2.34384×10^{-3}
1 kcal/s	4.1840	5.6886	4.26650×10^2	1

References:

Mills, I. (Ed.), International Union of Pure and Applied Chemistry, Quantities, Units and Symbols in Physical Chemistry, Blackwell Scientific Publications, Oxford 1988.

The International System of Units (SI), National Bureau of Standards Spec. Publ. 330 [1972].

Landolt-Börnstein, 6th Ed., Vol. II, Pt. 1, 1971, pp. 1/14.

ISO Standards Handbook 2, Units of Measurement, 2nd Ed., Geneva 1982.

Cohen, E. R., Taylor, B. N., Codata Bulletin No. 63, Pergamon, Oxford 1986.

Key to the Gmelin System
of Elements and Compounds

System Number	Symbol	Element		System Number	Symbol	Element
1		Noble Gases		37	In	Indium
2	H	Hydrogen		38	Tl	Thallium
3	O	Oxygen		39	Sc, Y	Rare Earth
4	N	Nitrogen			La—Lu	Elements
5	F	Fluorine		40	Ac	Actinium
6	**Cl**	**Chlorine**		41	Ti	Titanium
7	Br	Bromine		42	Zr	Zirconium
8	I	Iodine		43	Hf	Hafnium
8a	At	Astatine		44	Th	Thorium
9	S	Sulfur		45	Ge	Germanium
10	Se	Selenium		46	Sn	Tin
11	Te	Tellurium		47	Pb	Lead
12	Po	Polonium		48	V	Vanadium
13	B	Boron		49	Nb	Niobium
14	C	Carbon		50	Ta	Tantalum
15	Si	Silicon		51	Pa	Protactinium
16	P	Phosphorus		**52**	**Cr**	**Chromium**
17	As	Arsenic		53	Mo	Molybdenum
18	Sb	Antimony		54	W	Tungsten
19	Bi	Bismuth		55	U	Uranium
20	Li	Lithium		56	Mn	Manganese
21	Na	Sodium		57	Ni	Nickel
22	K	Potassium		58	Co	Cobalt
23	NH_4	Ammonium		59	Fe	Iron
24	Rb	Rubidium		60	Cu	Copper
25	Cs	Caesium		61	Ag	Silver
25a	Fr	Francium		62	Au	Gold
26	Be	Beryllium		63	Ru	Ruthenium
27	Mg	Magnesium		64	Rh	Rhodium
28	Ca	Calcium		65	Pd	Palladium
29	Sr	Strontium		66	Os	Osmium
30	Ba	Barium		67	Ir	Iridium
31	Ra	Radium		68	Pt	Platinum
32	**Zn**	**Zinc**		69	Tc	Technetium[1]
33	Cd	Cadmium		70	Re	Rhenium
34	Hg	Mercury		71	Np,Pu...	Transuranium
35	Al	Aluminium				Elements
36	Ga	Gallium				

Boxes/labels: HCl, ZnCl₂ (left), CrCl₂, ZnCrO₄ (middle)

Material presented under each Gmelin System Number includes all information concerning the element(s) listed for that number plus the compounds with elements of lower System Number.

For example, zinc (System Number 32) as well as all zinc compounds with elements numbered from 1 to 31 are classified under number 32.

[1] A Gmelin volume titled "Masurium" was published with this System Number in 1941.

A Periodic Table of the Elements with the Gmelin System Numbers is given on the Inside Front Cover